Edited by
Michael Rychlik

**Fortified Foods
with Vitamins**

Related Titles

Zhang, H. Q., Barbosa-Canovas, G. V., Balasubramaniam, V. M. B., Dunne, C. P., Farkas, D. F., Yuan, J. T. C. (eds.)

Nonthermal Processing Technologies for Food

684 pages
2011
Hardcover
ISBN: 978-0-8138-1668-5

Nollet, L. M. L. (ed.)

Analysis of Endocrine Disrupting Compounds in Food

approx. 504 pages
2011
Hardcover
ISBN: 978-0-8138-1816-0

Sasic, S., Ozaki, Y. (eds.)

Raman, Infrared, and Near-Infrared Chemical Imaging

approx. 328 pages
2011
Hardcover
ISBN: 978-0-470-38204-2

Li-Chan, E., Chalmers, J., Griffiths, P. (eds.)

Applications of Vibrational Spectroscopy in Food Science

752 pages in 2 volumes
2010
Hardcover
ISBN: 978-0-470-74299-0

Mine, Y., Li-Chan, E., Jiang, B. (eds.)

Bioactive Proteins and Peptides as Functional Foods and Nutraceuticals

approx. 436 pages
2010
Hardcover
ISBN: 978-0-8138-1311-0

Taylor, A. J., Linforth, R.

Food Flavour Technology

approx. 376 pages
2010
Hardcover
ISBN: 978-1-4051-8543-1

Stanga, M.

Sanitation
Cleaning and Disinfection in the Food Industry

611 pages with 460 figures and 137 tables
2010
Hardcover
ISBN: 978-3-527-32685-3

Popping, B., Diaz-Amigo, C., Hoenicke, K. (eds.)

Molecular Biological and Immunological Techniques and Applications for Food Chemists

488 pages
2010
Hardcover
ISBN: 978-0-470-06809-0

Kaletunç, G. (ed.)

Calorimetry in Food Processing
Analysis and Design of Food Systems

412 pages
2009
Hardcover
ISBN: 978-0-8138-1483-4

Brennan, J. G. (ed.)

Food Processing Handbook

607 pages with 189 figures and 41 tables
2006
Hardcover
ISBN: 978-3-527-30719-7

Edited by Michael Rychlik

Fortified Foods with Vitamins

Analytical Concepts to Assure Better and Safer Products

WILEY-VCH Verlag GmbH & Co. KGaA

The Editor

Prof. Dr. Michael Rychlik
Bioanalytik Weihenstephan
Technische Universität München
Alte Akademie 10
85354 Freising
Germany

■ All books published by **Wiley-VCH** are carefully produced. Nevertheless, authors, editors, and publisher do not warrant the information contained in these books, including this book, to be free of errors. Readers are advised to keep in mind that statements, data, illustrations, procedural details or other items may inadvertently be inaccurate.

Library of Congress Card No.: applied for

British Library Cataloguing-in-Publication Data
A catalogue record for this book is available from the British Library.

**Bibliographic information published by
the Deutsche Nationalbibliothek**
The Deutsche Nationalbibliothek lists this publication in the Deutsche Nationalbibliografie; detailed bibliographic data are available on the Internet at <http://dnb.d-nb.de>.

© 2011 Wiley-VCH Verlag & Co. KGaA, Boschstr. 12, 69469 Weinheim, Germany

Typesetting Toppan Best-set Premedia Limited, Hong Kong
Printing and Binding Strauss GmbH, Mörlenbach
Cover Adam Design, Weinheim

Printed in the Federal Republic of Germany
Printed on acid-free paper

ISBN: 978-3-527-33078-2
ePDF ISBN: 978-3-527-63417-0
ePub ISBN: 978-3-527-63416-3
Mobi ISBN: 978-3-527-63418-7
oBook ISBN: 978-3-527-63415-6

Dedicated to my caring wife and daughters

Contents

Preface

In the nutrition sciences, evidence is accumulating that vitamins are essential not only for maintaining physiological functions but also to prevent hazards resulting from oxidative stress or disorders of cell division and DNA repair. For example, the vitamers of the folate group are considered to prevent neural tube defects, Alzheimer's disease, and colon cancer. Therefore, many foods are supplemented with vitamins, and for cereal products fortification with folic acid is mandatory in the USA.

When considering the recommended intakes of vitamins, consumers have to rely on the labeled contents to make their diet meet the recommendations. Therefore, manufacturers and official laboratories are called upon to control vitamin contents accurately. However, the analysis of vitamins is still demanding due to their occurrence in minute amounts and their lability. The trace amounts to be quantified are sometimes too small even for some highly sensitive instrumental analytical methods, so that bioassays are still important. For the determination of vitamins, organisms can be used that do not produce the analyte compounds by themselves, but need them for their growth. Well into the twentieth century, vertebrates such as chicks were used for vitamin K determination, but nowadays only bacteria are applied. Furthermore, the occurrence of vitamins as conjugates and as various isomers showing different bioactivities render vitamin analysis extremely demanding. In particular, fortification with synthetic vitamins often lacking isomeric purity has aggravated this problem.

The aim of this multi-authored book is, therefore, to review modern analytical approaches to verify the content of all relevant vitamins in fortified foods. Emphasis is placed on fast, sensitive, and accurate methods along with assays that permit the detection of various isomers and multiple vitamins. The individual contributions include both up-to-date reviews and unprecedented reports on new methods. The authors were asked to omit lengthy historical discourses and also laboratory details, as these are easily accessible from original publications.

Internationally renowned experts in vitamin research have contributed to this volume. Their disciplines range from medicine, toxicology, and nutrition science to chemistry, food science, and food chemistry. The authors are active in universities, official laboratories, and companies in many countries throughout the world, including China, Germany, Italy, Japan, Spain, Sweden, and the USA.

The book is intended for professionals, and also advanced students, concerned with and active in the areas of food science, nutraceuticals, human health, and biochemistry. They will benefit from this book which clearly describes the concepts and analytical and assay methods to study fortification, and applications to create better and safer foods.

Acknowledgments

This volume was inspired by the food science and nutrition team of the publisher, the support and assistance of which during the whole production of the book is gratefully acknowledged.

Many thanks are due to all the co-contributors, who have spent so much of their precious time writing high-quality chapters. Many chapters were included as last-minute contributions and have therefore contributed significantly to ensuring that this volume is up-to-date.

List of Contributors

Sameh Ahmed
Assiut University
Faculty of Pharmacy
Department of Pharmaceutical
Analytical Chemistry
Assiut 71526
Egypt

Irmgard Bitsch
University of Giessen
Institute of Nutritional Science
Wilhelmstrasse 20
35392 Giessen
Germany

Roland Bitsch
Friedrich Schiller University Jena
Institute of Nutrition
D-07743 Jena
Germany

Volker Böhm
Friedrich Schiller University Jena
Institute of Nutrition
07737 Jena
Germany

Fulvia Caretti
Università "La Sapienza" di Roma
Dipartimento di Chimica
Piazzale Aldo Moro 5
00185 Rome
Italy

Bo Chen
Hunan Normal University
College of Chemistry and Chemical
Engineering
Changsha 410081
China

Jonathan W. DeVries
Medallion Laboratories
General Mills
9000 Plymouth Avenue North
Minneapolis, MN 55427
USA

Remedios Fernández Fernández
Almeria University
Department of Analytical Chemistry
Group "Analytical Chemistry of
Contaminants"
04071 Almeria
Spain

Antonia Garrido Frenich
Almeria University
Department of Analytical Chemistry
Group "Analytical Chemistry of
Contaminants"
04071 Almeria
Spain

Alessandra Gentili
Università "La Sapienza" di Roma
Dipartimento di Chimica
Piazzale Aldo Moro 5
00185 Rome
Italy

Jelena Jastrebova
Swedish University of Agricultural
Sciences (Uppsala BioCenter)
Department of Food Science
Undervisningsplan 6C
750 07 Uppsala
Sweden

Afaf Kamal-Eldin
Swedish University of Agricultural
Sciences (Uppsala BioCenter)
Department of Food Science
Undervisningsplan 6C
750 07 Uppsala
Sweden

Current address

United-Arab Emirates University
Department of Food Science
Al Ain
United Arab Emirates

Dorit Kern
Landesamt für Verbraucherschutz des
Landes Sachsen-Anhalt
Freiimfelderstrasse 66–68
061112 Halle/Saale
Germany

Naoya Kishikawa
Nagasaki University
Graduate School of Biomedical
Sciences
Department of Environmental and
Pharmaceutical Sciences
1-14 Bunkyo-machi
Nagasaki 852-8521
Japan

Padmanaban G. Krishnan
South Dakota State University
Department of Health and Nutritional
Sciences
Wagner Hall 415, Rotunda Lane
Brookings, SD 57007
USA

Naotaka Kuroda
Nagasaki University
Graduate School of Biomedical
Sciences
Department of Environmental and
Pharmaceutical Sciences
1-14 Bunkyo-machi
Nagasaki 852-8521
Japan

John L. MacDonald
NP Analytical Labs
Checkerboard Square
Saint Louis, MO 63164
USA

José Luis Martínez Vidal
Almeria University
Department of Analytical Chemistry
Group "Analytical Chemistry of
Contaminants"
04071 Almeria
Spain

Sabine Mönch
Technische Universität München
Chair of Food Chemistry
Lise-Meitner-Str. 34
85354 Freising
Germany

David R. Nelson
Caravan Ingredients
7905 Quivira Road
Lenexa, KS 66215
USA

Anthony O'Kane
Queens University Belfast
Institute of Agri-Food and Land Use
David Keir Building
Stranmillis Road
Belfast BT 95AG
UK

Kaname Ohyama
Nagasaki University
Graduate School of Biomedical
Sciences
Department of Environmental and
Pharmaceutical Sciences
1-14 Bunkyo-machi
Nagasaki 852-8521
Japan

Sudheer R. Musukula
South Dakota State University
Department of Health and Nutritional
Sciences
Wagner Hall 425, Rotunda Lane
Brookings, SD 57007
USA

Roberto Romero-González
Almeria University
Department of Analytical Chemistry
Group "Analytical Chemistry of
Contaminants"
04071 Almeria
Spain

Dora Roth-Meier
Technische Universität München
Animal Nutrition Center of Life and
Food Sciences Weihenstephan
Liesel-Beckmann-Str. 1
Alte Akademie 10
85350 Freising
Germany

Michael Rychlik
Technische Universität München
Chair of Analytical Food Chemistry
Center of Life and Food Sciences
Weihenstephan
Bioanalytik Weihenstephan
Research Center for Nutrition and
Food Sciences
Alte Akademie 10
85350 Freising
Germany

Sylvia Stengl
R-Biopharm AG
An der neuen Bergstrasse 17
64297 Darmstadt
Germany

Lennart Wahlström
GE Healthcare
Björkgatan 30, 751 84 Uppsala
Sweden

Fumio Watanabe
Tottori University
Faculty of Agriculture
School of Agricultural
Biological and Environmental Sciences
4-101 Koyama-Minami
Tottori 680-8550
Japan

Wolfgang Weber
Institut für Produktqualität GmbH
Teltowkanalstrasse 2
12247 Berlin
Germany

Cornelia M. Witthöft
Swedish University of Agricultural
Sciences
Uppsala BioCenter
Department of Food Science
Undervisningsplan 6C
75007 Uppsala
Sweden

Yukinori Yabuta
Tottori University
Faculty of Agriculture
School of Agricultural
Biological and Environmental Sciences
4-101 Koyama-Minami
Tottori 680-8550
Japan

Da-jin Yang
National Institute for Nutrition and
Food Safety
Chinese Center for Disease Control
and Prevention
Beijing 100021
China

Part I
Perspectives and General Methodology in Vitamin Analysis

1

Stable Isotope Dilution Assays in Vitamin Analysis – A Review of Principles and Applications

Michael Rychlik

1.1
Principle of Stable Isotope Dilution Assays

1.1.1
General Remarks

The past decades have seen an increasing use of compounds labeled with stable isotopes in research. For instance, labeled precursors facilitate metabolism studies as the label can be followed on its way into different metabolites by mass spectrometry. Another application is the labeling of high molecular weight compounds such as proteins to elucidate their three-dimensional structures by modern nuclear resonance spectrometric methods. The third important use of stable isotopes is in trace analysis by stable isotope dilution assays (SIDAs), which is the topic of this review on vitamin quantitation.

The origin of SIDAs can be traced back to the beginning of the twentieth century when Soddy [1] discovered the existence of isotopes and George Hevesy used radioactive isotopes to determine the content of lead in rocks and the solubility of lead salts in water [2]. The conviction that elements are composed of atoms containing identical nuclei was refuted by Aston [3], who detected different atomic species of the noble gas neon by mass spectrometry. This resulted in a new definition of elements, which accordingly comprise mixtures of nuclei showing identical charge but different masses. As the nuclei with identical charges have the same (Greek *isos*) place (*topos*) in the periodic system of the elements, Soddy introduced the term "isotopes" [1]. An element has a natural isotopic distribution and there are two types of isotopes, namely the stable and radioactive ones. For example, carbon shows a natural distribution of C-12 (98%), of C-13 (1.1%), and of C-14, the last of which is radioactive and undergoes β-decay with a half-life of 5370 years.

Radioisotopes have the advantages over their stable analogs of their sensitive detectability and the possible use of low degrees of labeling. However, stable isotopes thereafter found their place as analytical tools when Hevesy and Jacobsen used deuterium oxide to quantify the percentage of extracellular liquid [4]. Interestingly, the percentage of deuterated water was determined by measuring the

Fortification of Foods with Vitamins, First Edition. Edited by Michael Rychlik.
© 2011 Wiley-VCH Verlag GmbH & Co. KGaA. Published 2011 by
Wiley-VCH Verlag GmbH & Co. KGaA.

density of the water, as the mass spectrometers at that time showed very low precision.

The term "stable isotope dilution assay" was first introduced in 1940 by Rittenberg and Foster [5], who quantified amino acids in protein hydrolyzates. SIDAs at that time were very tedious, as mass spectrometry required purification of the compounds to be analyzed. Therefore, several chromatographic and recrystallization steps were essential. Finally, coupling of mass spectrometry (MS) to gas chromatography (GC) [6] opened the door to faster and more sensitive methods. In that way, the first modern type of SIDA was performed by Sweeley *et al.* [7], who quantified glucose by GC–MS after trimethylsilylation and used [^{2}H$_7$]glucose as the internal standard.

Although they are very similar in their properties, isotopes can be enriched or depleted due to their different masses. Mixing an element or compound showing a natural isotopic distribution with such an isotopically different material (Figure 1.1) results in a smaller proportion of the naturally abundant isotopes in the resulting material – which led to the term "dilution" in SIDA.

The principle of SIDA is simply explained in Figure 1.2. After addition of the labeled standard and its equilibration with the analyte, the ratio of the isotopologs remains stable throughout all subsequent analytical steps. This is due to their almost identical chemical and physical properties. A final MS step enables the isotopologs to be differentiated. Consequently, the content of the analyte in the sample can be calculated with the known amount of the internal standard (IS)

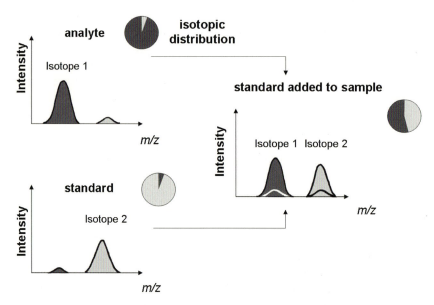

Figure 1.1 Origin of the term "dilution" in stable isotope dilution assays: addition of a standard with a different isotopic distribution – the original isotopic distribution of the analyte has been "diluted."

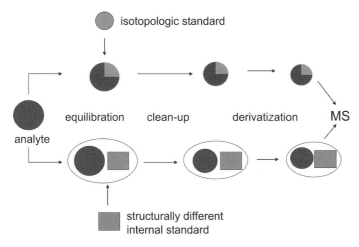

Figure 1.2 The ratio of isotopologic analyte and standard remains stable until final mass spectroscopic analysis. For a structurally different IS, however, the ratio between standard and analyte can alter during sample preparation.

added at the beginning. In contrast, a structurally different internal standard may be discriminated against and, therefore, cause systematic errors and imprecision. Hence losses of the analyte are completely compensated for by identical losses of the isotopolog, whereas the structurally different IS may show different losses.

As SIDAs require more or less elaborate syntheses of labeled compounds, their development was at first restricted to very few applications, in particular to those fields in which highest sensitivity and accuracy were essential. Therefore, toxicology, clinical chemistry, and environmental analysis were the first disciplines to use SIDAs. Subsequently, these methods were transferred to foods and emerged as reference methods for food compounds such as lignans [8] and steroids [9].

However, more recently the direction of research changed and assays developed for foods have opened up new prospects in toxicology and nutrition research. In addition to vitamins such as pantothenic acid [10] and folates [11], further examples have been centered on mycotoxins such as trichothecenes [12, 13] and patulin [14] and on odorants [15].

1.1.2
Benefits and Limitations of Using an Isotopologic Internal Standard

As detailed before, due its ideal compensation for losses, SIDA is a perfect tool for a series of analytical applications, in particular for trace analyses. The latter often demand tedious clean-up procedures due to matrix interferences, which typically evoke losses of the analyte. The use of structurally different ISs requires additional recovery and spiking experiments, which often result in imprecise data. In all these cases, SIDA offers significant benefits.

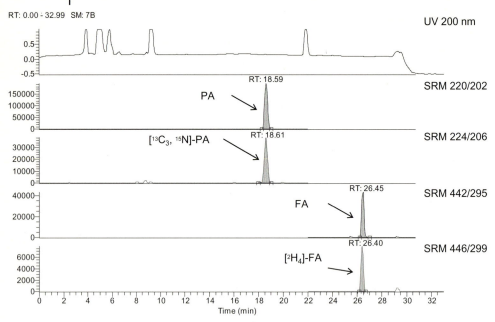

Figure 1.3 MS/MS chromatograms of fortified breakfast cereals containing 7.94 mg per 100 g of pantothenic acid (PA) and 270 µg per 100 g of folic acid (FA). The isotopically labeled internal standards show retention times identical with those their unlabeled isotopologs in the respective traces below.

In addition to compensation for losses, thus resulting in improved accuracy, the use of an isotopologic standard enhances the specificity of the determination. In addition to the specific MS information on the analyte, the IS is eluted at an almost identical retention time and shows a distinct mass shift. Therefore, the analyte can be unequivocally assigned in the chromatogram from a SIDA showing the coeluting peaks in the respective mass traces (Figure 1.3).

A further advantage of adding isotopologic material is to enhance sensitivity by the so-called "carrier effect." Due to adsorption phenomena on glassware or chromatographic columns, a definite amount of the analyte is likely to be lost during sample clean-up. If the total amount of the analyte in an extract is lower than this loss, the compound will no longer be detectable. However, if an isotopologic standard is added in an amount that exceeds this loss, the total sum of standard and analyte is higher than the loss and, therefore, the isotopologs can be detected. Although there are some conflicting opinions on this topic (for a review, see [16]), there are some applications showing a significant enhancement of sensitivity [17]. For vitamins, however, this effect has not yet been demonstrated.

Regarding the major benefits of SIDAs, that is, specificity and ideal compensation for losses, the question arises of whether they have the potential to be "definite methods." According to a definition by Cali and Reed [18], "a definite method is one that, after exhaustive investigation, produces analytical results that are accurate, that is, free of systematic errors, to the extent required for the intended end-use(s)." This definition holds true especially for primary methods, which are "methods having the highest metrological qualities, for which a complete uncertainty statement can be written down in terms of SI units, and whose results are, therefore, accepted without reference to a standard of the quantity being measured" [19]. A SIDA can be traced back to a gravimetric (i.e., primary) measurement and, therefore, is considered a primary method. In the case of being validated intensively for the absence of systematic errors, it has the potential to be accurate, that is, to produce the "true" value.

A SIDA of vitamins is only possible if a combination of chromatography with MS and a labeled IS are available. However, whereas the former has increasingly become basic instrumentation in most laboratories, the latter is a narrow bottleneck for the wider application of SIDAs. Of all the vitamins discovered to date, about 20 have been synthesized as labeled analogs, but only seven of these are commercially available. Hence the compound aimed at often has to be synthesized. There are two aspects that arise as a hindrance before starting with these syntheses. The first is that analysts may hesitate to perform chemical syntheses due to a lack of experience. Although the envisaged syntheses on the microscale require often intense purifications by chromatography, analysts are normally familiar with the necessary methods such as distillation *in vacuo*, purification by high-performance liquid chromatography (HPLC) and checking purity and yield by GC–MS. The procedures are completely detailed in the literature and usually the groups having published these syntheses are willing to give advice in case of practical problems.

The second psychologic hindrance that prevents analysts from synthesizing labeled mycotoxins is the price of labeled educts. However, this is not a convincing argument, as can be explained by the following example: 1 g of a labeled educt may cost around US$1000 and the yield of a multi-step synthesis may be only 1%, both of which are realistic figures. Then, the price for 10 mg of the labeled product is $1000. However, as less than 1 µg of the labeled standard is required for a SIDA, 10 mg of the standard enables at least 10 000 analyses to be performed. Hence the material cost for using a labeled standard is just $0.10 per sample, which is negligible compared with the cost of labor and equipment.

1.1.3
Prerequisites for Isotopologic Standards

As outlined before, SIDA is based on an isotopolog ratio that remains stable during all analytical steps. Therefore, a stable labeling step is essential for an IS. As carbon–carbon and carbon–nitrogen bonds are very unlikely to be cleaved, ^{13}C and ^{15}N labels are considered to be very stable. In contrast, losses of ^{18}O or ^{2}H at labile positions can occur. On the one hand, ^{18}O in carboxyl moieties can be exchanged in acidic or basic solutions. On the other, deuterium is susceptible to so-called protium–deuterium exchange if it is activated by adjacent carbonyl groups or aromatic systems.

Moreover, an isotopolog ratio may be altered by isotope effects (IEs), that is, small differences in physical or chemical properties of the isotopologs. IEs are due to different energy contents that are caused by the mass differences of the isotopes. The lowest energy level of a molecule, namely the zero point energy, is given by the ground-state vibration of the bonds at absolute zero, where the population in excited vibrational levels is negligible. As frequencies and energies of vibrations are proportional to the encountered masses, heavier isotopologs possess a lower energy content, resulting in higher energies of bond dissociation. IEs are mainly observable in the case of hydrogen, as the mass difference between ^{1}H and ^{2}H is proportionally much higher than between ^{13}C and ^{12}C or ^{15}N and ^{14}N. In the case of chemical reactions involving C–H bond rupture, IEs result in monodeuterated isotopologs reacting up to seven times more slowly then their light analogues. For labeling with ^{13}C, these effects are several hundred times smaller. However, the kinetic IE of ^{13}C is a valuable diagnostic tool in enzyme studies, as ^{12}C isotopologs show lower reaction rates than their ^{13}C counterparts. This discrimination is commonly measured as the δ^{13}C value of the products and is dependent on the transitions states of the encountered enzymatic reactions. Therefore, the measurement of δ^{13}C values allows the characterization of the encountered enzymes and the determination of the products' origin. For instance, the authenticity of alcoholic beverages [20] or valuable spices such as vanilla [21] can be verified.

The primary IE, which affects direct bond cleavage between the label and adjoining atoms, has to be distinguished from the secondary IE, which has an influence on the bonds between unlabeled atoms. A primary IE would cause either loss of labeling or, in MS, would result in different intensities of the fragments bearing the label. Therefore, all derivatizations or MS fragmentations including a primary IE have to be avoided. Although much less pronounced, a secondary IE, however, cannot be excluded in these reactions, but has not yet been reported in the literature.

In contrast to chemical IEs, a physical IE is more often observable. The latter particularly affects chromatographic behavior and results in different retention times of the isotopologs. In particular for multiple labelings with ^{2}H, the IE may evoke even baseline separation, as shown in the case of eightfold deuterated β-carotene [22], which was clearly separated from its unlabeled isotopolog upon HPLC with ultraviolet (UV) detection (Figure 1.4) In most cases, the heavier isotopologs are eluted earlier than their light analogs, which is unexpected, as the heavier isotopologs should have higher boiling temperatures and, therefore, should be eluted later in GC. This behavior therefore is referred to as an inverse IE.

In order to prevent chromatographic separations of the isotopologs during clean-up and thus changes in the isotopolog ratio, isotope effects have to be minimized either by choosing labelings with ^{13}C or ^{15}N or by introducing only the necessary number of ^{2}H atoms.

For unequivocal quantification, the standard has to be distinguishable from the analyte by MS. This requires the presence of the mass increment introduced by the labels either in the molecular ion or in its fragments. Therefore, a loss of the label prior to detection has to be avoided. However, quantitation by liquid chro-

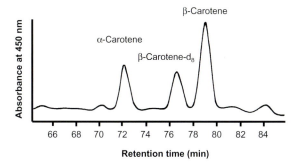

Figure 1.4 Baseline separation of eightfold deuterated β-carotene from its unlabeled isotopolog using HPLC–UV. Reproduced with permission from [22].

matography (LC) coupled with tandem mass spectrometry (MS/MS) is feasible even if the label is lost during the monitored mass transition, as the precursor ions are still differentiated.

A further problem in isotopolog differentiation may arise from spectral overlaps between the standard and analyte. In case of the analyte, the natural abundance of isotopes, in particular of ^{13}C, ^{18}O and ^{34}S, results in isotope clusters of each fragment showing not only the nominal mass m_A, but also to a lesser extent $m_A + 1$, $m_A + 2$, or even higher masses. In particular, ^{13}C in compounds consisting of a higher number of carbons causes a significant abundance of $m + 1$ and $m + 2$ due to the relatively high natural abundance. To avoid overlap of the signals of those natural isotopologs with m_S of the standard, the mass difference therefore has to be sufficient. For vitamins, in which the number of carbon atoms often exceeds 10, generally a mass increment of at least three units is required. However, the number of labels, especially of deuterium, should not be too high, in order to minimize the mentioned chromatographic isotope effect.

In contrast, signals of the labeled material contributing to those of the analyte may also result in spectral overlaps. This may be due to the low isotopic purity of labeled educts or to inadequate labeling during synthetic steps, thus leading to incomplete labels in the standard.

If a spectral overlap cannot be avoided, calculation procedures have been developed that still permit quantification to be achieved. However, these procedures are more complicated the greater the overlap is.

Another important requirement for accurate quantification is complete equilibration between the analyte and standard in the sample to be extracted. As the labeled standard usually is not contained in the sample matrix after its addition, it will likely be recovered to a high extent during the extraction procedure. For the analyte, this might not be true, as it can be trapped in compartments of the matrix and is less extractable by the solvent. Therefore, sufficient time for equilibration of the standard and analyte in all parts of the sample has to be allowed to assure that the analyte and standard show the same concentration ratio in all compartments as far as possible.

1.1.4
Calibration Procedures

From the intensity ratio of suitable ions measured by MS and a calibration function, the isotopolog ratio can be assessed, which directly allows the amount of analyte present in the sample to be calculated. The relation between isotopolog ratio and intensity ratio has to be elucidated by analyzing the intensity ratios of a series of defined standard–analyte mixtures. If there is no spectral overlap, the calibration function is assumed to be linear. However, typically there are still residues of unlabeled analytes in the labeled material and low intensities of natural isotopologs of the analyte contributing to the signals of the standard. Therefore, the calibration function can be expected to be linear only in a restricted region, as illustrated in Figure 1.5. The outlined function appears under the supposition that the standard may contain 2% of unlabeled material und natural isotopologs may contribute 5% intensity to the signal of the standard. In this case, the calibration function can only be assumed to be a straight line for molar ratios of analyte to standard ranging between 0.2 and 5. With excess of the standard or of the analyte, the function is dominated by the unlabeled residues in the standard or by the natural isotopologs of the analyte, respectively.

However, for complicated structures or elaborate syntheses, a spectral overlap often cannot be avoided and, therefore, suitable procedures for calculations are required. In general, there are several ways to cope with this problem: First, hyperbolic or polynomial models have been elaborated [23, 24], which approximate the real calibration relation by a mathematical function. As these procedures are rather complicated, several authors have proposed linearization methods, which convert the nonlinear function into a linear function. In a study on odorants, Fay *et al.* [25] compared four of these methods using [^{2}H$_1$]benzaldehyde as the IS for quantification of benzaldehyde. This assay shows a spectral overlap of about 12% between

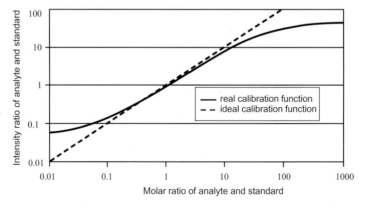

Figure 1.5 "Real" calibration function (dashed line) for a SIDA under the supposition that the standard contains 2% of unlabeled material with natural isotopologs contributing 5% intensity to the standard's signal.

the analyte and the standard. The authors demonstrated that the method of Colby and McCaman [26] did not yield a linear function. By contrast, both the average mass approach [27] and linearization using isotopic enrichment factors [28] gave straight lines, whose calibration points, however, were spread very inhomogeneously. The only procedure giving a calibration line with homogeneously distributed calibration points was the method of Bush and Trager [29], which notably includes only rather simple calculations.

The third way to cope with a nonlinear calibration function is the "bracketing" approach [30], which requires the measurement of further calibration points lying in the proximity of the intensity ratio measured in the sample. Although this method is widely accepted as the most accurate one, it is very elaborate and, therefore, is only seldom applied.

1.2
Application of Stable Isotope Dilution Assays to Vitamins

1.2.1
Fat-Soluble Vitamins

This group of vitamins is naturally embedded in food lipids. For their analysis, these compounds have to be separated from fat prior to detection. Due to their low volatility, GC applications are scarce. Moreover, the hydrophobicity of these vitamins is responsible for a high retention on common reversed phases in HPLC, which limits their separation on these phases. As a consequence thereof and as LC–MS combinations need aqueous mobile phases, reports on LC–MS detection are also rare. However, with the development of new reversed-phase materials, LC–MS applications have become increasingly popular.

For fat-soluble vitamins, the use of stable isotopologs is restricted to β-carotene, vitamin A, α-tocopherol, vitamin K, and vitamin D. To date, nearly all applications have been directed towards quantitation of these vitamins in blood or studies of their bioavailability. SIDAs in foods have not been reported so far.

1.2.1.1 Vitamin A
For vitamin A, most applications of labeled compounds have been dedicated to the administration of labeled β-carotene and quantification of the generated retinol in blood serum. For this purpose, β-carotene and retinol have been labeled with deuterium [22] or carbon-13 [31]. These compounds have been applied in several studies, and the bioefficacy of β-carotene was found to range between 30 and 40% [32].

1.2.1.2 Vitamin E
Because it functions as an antioxidant in tissues, an important part of tocopherol research is dedicated to studying its reaction with oxygen and to elucidating the respective reaction products. An extensive SIDA for the quantification of

α-tocopherol and its oxidation products such as α-tocopherolquinone, α-tocopherolhydroquinone, and several epoxy-α-tocopherolquinones by using their deuterated analogs was presented by Liebler *et al.* [33]. They oxidized endogenous α-tocopherol in rat liver microsomes with the radical generator azobis(amidino-propane) and quantified the isotopologs by GC–MS after trimethylsilylation.

1.2.1.3 Vitamin D

Cholecalciferol (vitamin D_3) and its major metabolites calcidiol and calcitriol were quantified in human blood by two research groups in the 1980s. Whereas Zagalak and Borschberg [34] used [$^2H_8$]cholecalciferol along with [$^2H_3$]calcitriol as IS, Coldwell *et al.* [35] applied $^2H_6$-labeled isotopologs of cholecalciferol, calcidiol, calcitriol, and some other metabolites for SIDAs. As a further isotopolog of vitamin D_3, [$^2H_7$] cholecalciferol was synthesized by Kamao *et al.* [36] and used in a SIDA for several fat-soluble vitamins in human breast milk.

An actual LC–MS/MS method for vitamin D in mineral tablets and baby food is presented in Chapter 17.

1.2.1.4 Vitamin K

One of the seldom used SIDAs for fat-soluble vitamins has been reported for vitamin $K_{1(20)}$ in plasma using GC–MS [37]. The IS was [$^2H_3$]phylloquinone, which was prepared by deuteromethylation of 1,4-naphthoquinone and subsequent coupling with phytol. The suitability of this IS, however, has been called into question for LC–MS, as the deuterium label is at the acidic 2-methyl position and, therefore, is highly susceptible to H–D exchange in aqueous eluents. In contrast, Suhara *et al.* synthesized $^{18}O_2$-labeled phylloquinone homologs starting from 2-methyl-1,4-naphthoquinone diacetate in a four-step procedure including an oxidation in the presence of $H_2^{18}O$ [38]. These compounds were used as ISs in a multiple SIDA for the quantitation of vitamin K along with vitamin K_1, K_4, and K_7 compounds in human breast milk [36]. Further applications of these compounds to foods have not yet been reported.

1.2.2
Water-Soluble Vitamins

Because of their low hydrophobicity and low retention on reversed phases, this group of vitamins is very suited to LC–MS in aqueous solvents. Numerous applications of LC–MS have been published, so it is not surprising that some SIDAs have also been developed. In the following sections, such assays for pyridoxin, niacin, folic acid, and pantothenic acid are described.

1.2.2.1 Vitamin B_6

Due to the wide variety of pyridoxine vitamers, SIDA applications for this group of vitamins are rare. The only assay reported in the literature dates back to 1985, when Hachey *et al.* [39] used deuterated analogs of pyridoxine, pyridoxal, pyridoxamine, and pyridoxic acid to quantify these compounds and their phosphates in guinea pig liver, human urine, feces, and goat milk. In the last case, pyridoxal phosphate was found to be the main vitamer contributing almost 70% to total pyridoxine.

1.2.2.2 Niacin

As [^{2}H$_4$]nicotinic acid is commercially available, it has been used as the basis of SIDAs for the niacin group. Following the first application to determine nicotinic acid and six of its metabolites in urine by Li *et al.* [40], the first SIDA for foods was presented by Goldschmidt and Wolf [41]. However, the latter assay did not measure nicotinamide, which might also contribute to niacin activity in food samples. Nevertheless, the assay showed excellent accuracy for the analysis of certified reference materials such as wheat flour, milk powder, and multivitamin tablets.

1.2.2.3 Ascorbic Acid

Although its importance for the diet is well documented, LC–MS quantitations of ascorbic acid are surprisingly scarce. Examples of the rare reports are a study of degradation products of dehydroascorbic acid and ascorbic acid [42] and the most recent quantitation of vitamin C along with nine other vitamins in multivitamin products by Chen *et al.* [43].

Moreover, labeled ascorbic acid for LC–MS has not yet been used. The only SIDA for vitamin C was reported in 1988, when Ellerbe *et al.* applied ^{13}C-labeled ascorbic acid to milk powder analysis by GC–MS of the *tert*-butyldimethyl derivatives [44].

1.2.2.4 Folic Acid

The first SIDA for folic acid in fortified foods was published by Pawlosky and Flanagan [45], who used commercially available [^{13}C$_5$]folic acid ([^{13}C$_5$]PteGlu) as the IS. One year later, we synthesized fourfold deuterated folic acid along with the most abundant folate monoglutamates for quantifying endogenous food folates and folic acid in fortified foods [46]. Further developments included the use of chicken pancreas in addition to rat plasma for improved deconjugation [47] and the use of 4-morpholinoethanesulfonic acid (MES) for enhancing folate stability [48]. In particular, the need for labeled folate tracers in bioavailability research spurred the generation of new stable isotope-labeled folates. Starting with the first dual label study using [^{2}H$_2$]PteGlu and [^{2}H$_4$]PteGlu by Gregory and Quinlivan [49], early investigations were restricted to the sole measurement of total urinary folate isotopologs by GC–MS [50, 51] and often were hampered by spectral overlap due to insufficient mass increments interfering with naturally occurring isotopologs. For this reason, syntheses of differently labeled folates have been developed, such as [^{2}H$_4$]PteGlu and [^{13}C$_6$]PteGlu labeled in the glutamate moiety and the benzene moiety, respectively [52]. For tracer studies, these folates were suitable for so-called extrinsic labeling, that is, by simple addition to foods. However, as added substances may not behave like the endogenously occurring folates, so-called intrinsically labeled foods were produced by growing spinach in a ^{15}N-labeled environment

[53], which generated $^{15}N_{1-7}$-labeled 5-methyltetrahydrofolate (5-CH$_3$-H$_4$folate) in the latter vegetable.

A further methodological improvement arose from the use of LC–MS in folate isotopolog analysis, when Wright *et al.* [54] measured $^{13}C_6$-labeled, $^{15}N_{1-7}$-labeled, and unlabeled 5-CH$_3$-H$_4$folate using [^{2}H$_2$]folic acid as the IS in the single ion monitoring mode. However, the use of a structurally different IS such as [^{2}H$_2$] PteGlu may decrease the accuracy by ion suppression and, moreover, quantitation was hampered by spectral overlap of [$^{15}N_{1-7}$]-5-CH$_3$-H$_4$folate with, on the one hand, [$^{13}C_6$]-5-CH$_3$-H$_4$folate, and, on the other, unlabeled 5-CH$_3$-H$_4$folate in single-stage LC–MS. This handicap was overcome by Melse-Boonstra *et al.* [55], who measured $^{13}C_6$-labeled along with $^{13}C_{11}$-labeled 5-CH$_3$-H$_4$folate as tracer isotopologs and simultaneously quantified unlabeled 5-CH$_3$-H$_4$folate by using [$^{13}C_5$]-5-CH$_3$-H$_4$folate as the IS. In the latter study, spectral overlaps of 5-CH$_3$-H$_4$folate isotopologs were avoided by labeling different moieties of the target molecule and their differentiation by LC–MS/MS. However, this investigation was restricted to plasma 5-CH$_3$-H$_4$folate without an application to food samples.

With more ^{13}C-labeled folates now being offered commercially, SIDA for folates in foods is currently being used by various groups throughout the world. Of all folate vitamers, H$_4$folate, 5-CH$_3$-H$_4$folate, 5-formyl-H$_4$folate, 5,10-methenyl-H$_4$folate and folic acid are currently available as $^{13}C_5$-labeled isotopologs. Whereas SIDA for monoglutamates can be considered a fairly well-established technology, the analysis of endogenous folates is still challenging. In particular, the polyglutamates and their complete conversion to the monoglutamates are still not fully understood and are the target of further research. The most recent approach is the use of [$^{13}C_5$]pteroylheptaglutamate for confirmation and quantitation of the degree of deconjugation. The latter compound has been produced by a Merrifield-like solid phase synthesis [48].

1.2.2.5 Pantothenic Acid
Along with folic acid, pantothenic acid was one of the first vitamins to be quantified by SIDA based on LC–MS. An extensive description of [$^{13}C_3$,^{15}N]pantothenic acid used as internal standard and recent results are presented in Chapter 9.

1.3
Outlook

Although numerous labeled vitamins are used in clinical chemistry and bioavailability research, they have been only marginally applied in the SIDA of foods (Table 1.1). However, with the advances in LC–MS/MS in vitamin analysis, labeled ISs will gain more importance. In particular for the analysis of multivitamin products, multiparametric SIDAs will be developed, when the labeled standards become available, as can be seen from the past for folic acid and pantothenic acid [56]. In particular for the analysis of reference materials, SIDAs are essential to establish the certified reference contents. In this context, a SIDA for the analysis

Table 1.1 Applications of SIDAs to vitamins in foods.

Vitamin	Internal standard	Food sample	Reference
Retinyl acetate	[^{2}H$_6$]Retinyl acetate	Human milk	Kamao *et al.* [36]
β-Carotene	[^{2}H$_6$]-β-Carotene	Human milk	Kamao *et al.* [36]
Cholecalciferol	[^{2}H$_7$]Cholecalciferol	Human milk	Kamao *et al.* [36]
α-Tocopherol	[^{2}H$_6$]-α-Tocopherol	Human milk	Kamao *et al.* [36]
Phylloquinone, menaquinone-4, menaquinone-7	^{18}O$_2$-labeled analogs	Human milk	Kamao *et al.* [36]
Pyridoxine, pyridoxamine, pyridoxal, pyridoxic acid	^{2}H$_2$-labeledanalogs	Liver, goat milk	Hachey *et al.* [39]
Ascorbic acid	[^{13}C$_6$]Ascorbic acid	Bovine milk	Ellerbe *et al.* [44]
Pantothenic acid	[^{13}C$_3$,^{15}N] Pantothenic acid	Various foods	Rychlik [10, 56]
Folic acid	[^{13}C$_5$]Folic acid	Various foods	Pawlosky and Flanagan [45]
5-Methyltetrahydrofolate, folic acid	^{13}C$_5$-labeled analogs	Various foods	Pawlosky *et al.* [58]
Tetrahydrofolate, 5-methyltetrahydrofolate, 5-formyltetrahydrofolate, 10-formylfolate, folic acid	^{2}H$_4$-labeled analogs	Various foods	Freisleben *et al.* [11]
Niacin	[^{2}H$_4$]Nicotinic acid	Various foods	Goldschmidt and Wolf [41]

of seven water-soluble vitamins in an infant/adult nutritional formula has been reported very recently [57].

References

1 Soddy, F. (1913) Intra-atomic charge. *Nature*, **92**, 399–400.

2 Hevesy, G. and Paneth, F. (1913) The solubility of lead sulfide and lead chromate. *Zeitschrift für anorganische und allgemeine Chemie*, **82**, 323–328.

3 Aston, F.W. (1919) Neon. *Nature*, **104**, 334.

4 Hevesy, G. and Jacobsen, C.F. (1940) Rate of passage of water through capillary and cell walls. *Acta Physiologica Scandinavica*, **1**, 11–18.

5 Rittenberg, D. and Foster, G.L. (1940) A new procedure for quantitative analysis by isotope dilution with application to the determination of amino acids and fat acids.

Journal of Biological Chemistry, **133**, 737–744.

6 Holmes, J.C. and Morell, F.A. (1957) Oscillographic mass spectrometry monitoring of gas chromatography. *Applied Spectroscopy*, **11**, 86–87.

7 Sweeley, C.C., Elliott, W.H., Fries, I., and Ryhage, R. (1966) Mass spectrometric determination of unresolved components in gas chromatographic effluents. *Analytical Chemistry*, **38**, 1549–1553.

8 Mazur, W., Fotsis, T., Wahala, K., Ojala, S., Salakka, A., and Adlercreutz, H. (1996) Isotope dilution gas chromatographic–mass spectrometric method for the determination of isoflavonoids, coumestrol, and lignans in food samples. *Analytical Biochemistry*, **233**, 169–180.

9 Covey, T., Maylin, G., and Henion, J. (1985) Quantitative secondary ion monitoring gas chromatography/mass spectrometry of diethylstilboestrol in bovine liver. *Biomedical Mass Spectrometry*, **12**, 274–287.

10 Rychlik, M. (2003) Pantothenic acid quantification by a stable isotope dilution assay based on liquid chromatography–tandem mass spectrometry. *Analyst*, **128**, 832–837.

11 Freisleben, A., Schieberle, P., and Rychlik, M. (2003) Comparison of folate quantification in foods by high-performance liquid chromatography–fluorescence detection to that by stable isotope dilution assays using high-performance liquid chromatography–tandem mass spectrometry. *Analytical Biochemistry*, **315**, 247–255.

12 Asam, S. and Rychlik, M. (2006) Synthesis of four carbon-13-labeled type A trichothecene mycotoxins and their application as internal standards in stable isotope dilution assays. *Journal of Agricultural and Food Chemistry*, **54**, 6535–6546.

13 Asam, S. and Rychlik, M. (2006) Quantitation of type B-trichothecene mycotoxins in foods and feeds by a multiple stable isotope dilution assay. *European Food Research and Technology*, **224**, 769–783.

14 Rychlik, M. and Schieberle, P. (2001) Model studies on the diffusion behavior of the mycotoxin patulin in apples, tomatoes, and wheat bread. *European Food Research and Technology*, **212**, 274–278.

15 Zeller, A. and Rychlik, M. (2006) Character impact odorants of fennel fruits and fennel tea. *Journal of Agricultural and Food Chemistry*, **54**, 3686–3692.

16 De Leenheer, A.P., Lefevere, M.F., Lambert, W.E., and Colinet, E.S. (1985) Isotope-dilution mass spectrometry in clinical chemistry. *Advances in Clinical Chemistry*, **24**, 111–161.

17 Haskins, N.J., Ford, G.C., Grigson, S.J.W., and Waddell, K.A. (1978) A carrier effect observed in assays for antidiarrheal drug compounds. *Biomedical Mass Spectrometry*, **5**, 423–424.

18 Cali, J.P. and Reed, W.P. (1976) The role of the National Bureau of Standards, standard reference materials in accurate trace analysis. In Accuracy in Trace Analysis: Sampling, Sample Handling, Analysis. National Bureau of Standards Special Publication 422, National Bureau of Standards, Gaithersburg, MD, pp. 41–63.

19 Quinn, T.J. (1997) Primary methods. *Metrologia*, **34**, 61.

20 Bricout, J., Fontes, J.C., Merlivat, L., and Pusset, M. (1975) Stable isotope composition of ethanol. *Industries Alimentaires et Agricoles*, **92**, 375–378.

21 Kaunzinger, A., Juchelka, D., and Mosandl, A. (1997) Progress in the authenticity assessment of vanilla. 1. Initiation of authenticity profiles. *Journal of Agricultural and Food Chemistry*, **45**, 1752–1757.

22 Dueker, S.R., Jones, A.D., Smith, G.M., and Clifford, A.J. (1994) Stable isotope methods for the study of β-carotene-d$_8$ metabolism in humans utilizing tandem mass spectrometry and high-performance liquid chromatography. *Analytical Chemistry*, **66**, 4177–4185.

23 Jonckheere, J.A. and Leenheer, A.P.D. (1983) Statistical evaluation of calibration nonlinearity in isotope dilution gas chromatography/mass spectrometry. *Analytical Chemistry*, **55**, 153–155.

24 Sabot, J.-F. and Pinatel, H. (1993) Calculation of the confidence range in oder to obtain a linear calibration graph in stable isotope dilution mass spectrometry: application to reference methods and pharmaceutical studies. *Analyst*, **118**, 831–834.

25 Fay, L.B., Metairon, S., and Baumgartner, M. (2001) Linearization of second-order calibration curves in stable isotope dilution-mass spectrometry. *Flavour and Fragrance Journal*, **16**, 164–168.

26 Colby, B.N. and McCaman, M.W. (1979) A comparison of calculation procedures for isotope dilution determinations using gas chromatography–mass spectrometry. *Biomedical Mass Spectrometry*, **6**, 225–230.

27 Blom, K.F. (1987) Average mass approach to stable isotope dilution mass spectrometry. *Organic Mass Spectrometry*, **22**, 530–533.

28 Staempfli, A.A., Blank, I., Fumeaux, R., and Fay, L.B. (1994) Study on the decomposition of the Amadori compound *N*-(1-deoxy-D-fructos-1-yl) glycine in model systems: quantification by fast atom bombardment tandem mass spectrometry. *Biological Mass Spectrometry*, **23**, 642–646.

29 Bush, E.D. and Trager, W.F. (1981) Analysis of linear approaches to quantitative stable isotope methodology in mass spectrometry. *Biomedical Mass Spectrometry*, **8**, 211–218.

30 Cohen, A., Hertz, H.S., Mandel, J., Paule, R.C., Schaffer, R., Sniegoski, L.T., Sun, T., Welch, M.J., and White, E.V. (1980) Total serum cholesterol by isotope dilution/mass spectrometry: a candidate definitive method. *Clinical Chemistry*, **26**, 854–860.

31 Wang, Y., Xu, X., van Lieshout, M., West, C.E., Lugtenburg, J., Verhoeven, M.A., Creemers, A.F.L., Muhilal, and van Breemen, R.B. (2000) A liquid chromatography–mass spectrometry method for the quantification of bioavailability and bioconversion of β-carotene to retinol in humans. *Analytical Chemistry*, **72**, 4999–5003.

32 van Lieshout, M., West, C.E., and van Breemen, R.B. (2003) Isotopic tracer techniques for studying the bioavailability and bioefficacy of dietary carotenoids, particularly β-carotene, in humans: a review. *American Journal of Clinical Nutrition*, **77**, 12–28.

33 Liebler, D.C., Burr, J.A., Philips, L., and Ham, A.J.L. (1996) Gas chromatography–mass spectrometry analysis of vitamin E and its oxidation products. *Analytical Biochemistry*, **236**, 27–34.

34 Zagalak, B. and Borschberg, H.J. (1988) Preparation of deuterated cholesterol, calciol, and calcitriol as internal standards for GC–MS-assays. *Spectroscopy*, **6**, 203–211.

35 Coldwell, R.D., Trafford, D.J.H., Varley, M.J., Makin, H.L.J., and Kirk, D.N. (1988) The measurement of vitamins D_2 and D_3 and seven major metabolites in a single sample of human plasma using gas chromatography/mass spectrometry. *Biomedical and Environmental Mass Spectrometry*, **16**, 81–85.

36 Kamao, M., Tsugawa, N., Suhara, Y., Wada, A., Mori, T., Murata, K., Nishino, R., Ukita, T., Uenishi, K., Tanaka, K., and Okano, T. (2007) Quantification of fat-soluble vitamins in human breast milk by liquid chromatography–tandem mass spectrometry. *Journal of Chromatography B*, **859**, 192–200.

37 Fauler, G., Leis, H.J., Schalamon, J., Muntean, W., and Gleispach, H. (1996) Method for the determination of vitamin $K_{1(20)}$ in human plasma by stable isotope dilution/gas chromatography/mass spectrometry. *Journal of Mass Spectrometry*, **31**, 655–660.

38 Suhara, Y., Kamao, M., Tsugawa, N., and Okano, T. (2005) Method for the determination of vitamin K homologues in human plasma using high-performance liquid chromatography–tandem mass spectrometry. *Analytical Chemistry*, **77**, 757–763.

39 Hachey, D.L., Coburn, S.P., Brown, L.T., Erbelding, W.F., Demark, B., and Klein, P.D. (1985) Quantitation of vitamin B_6 in biological samples by isotope dilution mass spectrometry. *Analytical Biochemistry*, **151**, 159–168.

40 Li, A.C., Chen, Y.-L., Junga, H., Shou, W.Z., Jiang, X., and Naidong, W. (2003) Separation of nicotinic acid and six metabolites within 60 seconds using

high-flow gradient chromatography on silica column with tandem mass spectrometric detection. *Chromatographia*, **58**, 723–731.

41 Goldschmidt, R.J. and Wolf, W.R. (2007) Determination of niacin in food materials by liquid chromatography using isotope dilution mass spectrometry. *Journal of AOAC International*, **90**, 1084–1089.

42 Pastore, P., Rizzetto, T., Curcuruto, O., Dal Cin, M., Zaramella, A., and Marton, D. (2001) Characterization of dehydroascorbic acid solutions by liquid chromatography/mass spectrometry. *Rapid Communications in Mass Spectrometry*, **15**, 2051–2057.

43 Chen, Z., Chen, B., and Yao, S. (2006) High-performance liquid chromatography/electrospray ionization-mass spectrometry for simultaneous determination of taurine and 10 water-soluble vitamins in multivitamin tablets. *Analytica Chimica Acta*, **569**, 169–175.

44 Ellerbe, P., Sniegosky, L.T., Miller, J.M., and White, E.V. (1988) An isotope dilution mass spectrometric (IDMS) method for the determination of vitamin C in milk. *Journal of Research of the National Bureau of Standards*, **93**, 367.

45 Pawlosky, R.J. and Flanagan, V.P. (2001) A quantitative stable-isotope LC–MS method for the determination of folic acid in fortified foods. *Journal of Agricultural and Food Chemistry*, **49**, 1282–1286.

46 Freisleben, A., Schieberle, P., and Rychlik, M. (2002) Syntheses of labelled vitamers of folic acid to be used as internal standards in stable isotope dilution assays. *Journal of Agricultural and Food Chemistry*, **50**, 4760–4768.

47 Rychlik, M., Englert, K., Kapfer, S., and Kirchhoff, E. (2007) Folate contents of legumes determined by optimized enzyme treatment and stable isotope dilution assays. *Journal of Food Composition and Analysis*, **20**, 411–419.

48 Mönch, S., and Rychlik, M. (2011) Method developments for improving the guantitation of food locates by stable

isotope dilution assays. Analytical Biochemistry, in preparation.

49 Gregory, J.F. and Quinlivan, E.P. (2002) In vivo kinetics of folate metabolism. *Annual Review of Nutrition*, **22**, 199–220.

50 Rogers, L.M., Pfeiffer, C.M., Bailey, L.B., and Gregory, J.F. (1997) A dual-label stable-isotopic protocol is suitable for determination of folate bioavailability in humans: evaluation of urinary excretion and plasma folate kinetics of intravenous and oral doses of [C-13(5)] and [H- 2(2)] folic acid. *Journal of Nutrition*, **127**, 2321–2327.

51 Pfeiffer, C.M., Rogers, L.M., Bailey, L.B., and Gregory, J.F. (1997) Absorption of folate from fortified cereal-grain products and of supplemental folate consumed with or without food determined by using a dual-label stable-isotope protocol. *American Journal of Clinical Nutrition*, **66**, 1388–1397.

52 Maunder, P., Finglas, P.M., Mallet, A.I., Mellon, F.A., Razzaque, M.A., Ridge, B., Vahteristo, L., and Witthoft, C. (1999) The synthesis of folic acid, multiply labelled with stable isotopes, for bio-availability studies in human nutrition. *Journal of the Chemical Society, Perkin Transactions 1*, 1311–1323.

53 Wolfe, C.A., Finglas, P.M., Hart, D., Wright, A.J.A., and Southon, S. (2000) Isotopic methods to detect food folates. *Innovative Food Science Emerging Technologies*, **1**, 297–302.

54 Wright, A.J.A., Finglas, P.M., Dainty, J.R., Hart, D.J., Wolfe, C.A., Southon, S., and Gregory, J.F. (2003) Single oral doses of ^{13}C forms of pteroylmonoglutamic acid and 5-formyltetrahydrofolic acid elicit differences in short-term kinetics of labelled and unlabelled folates in plasma: potential problems in interpretation of folate bioavailability studies. *British Journal of Nutrition*, **90**, 363–671.

55 Melse-Boonstra, A., Verhoef, P., West, C.E., van Rhijn, J.A., van Breemen, R.B., Lasaroms, J.J.P., Garbis, S.D., Katan, M.B., and Kok, F.J. (2006) A dual-isotope-labeling method of studying the bioavailability of hexaglutamyl folic acid relative to that of monoglutamyl folic acid in humans by using multiple orally administered low doses. *American Journal of Clinical Nutrition*, **84**, 1128–1133.

56 Rychlik, M. (2003) Simultaneous analysis of folic acid and pantothenic acid in foods enriched with vitamins by stable isotope dilution assays. *Analytica Chimica Acta*, **495**, 133–141.

57 Goldschmidt, R.J. and Wolf, W.R. (2010) Simultaneous determination of water-soluble vitamins in SRM 1849 in/

ant/adult nutritional formula powder by liquid chromatography-isotope dilution mass spectrometry. *Analytical and Bioanalytical Chemistry*, **397**, 471–481.

58 Pawlosky, R.J., Flanagan, V.P., and Doherty, R.F. (2003) A mass spectrometric validated high-performance liquid chromatography proceduce for the determination of locates in foods. *Journal of Agricultural and Food Chemistry*, **51**, 3726–3730.

2
Analytical Methods to Assess the Bioavailability of Water-Soluble Vitamins in Food – Exemplified by Folate

Cornelia M. Witthöft

2.1
Introduction

Folate is the generic term for a large group of vitamers (Figure 2.1) participating in one-carbon transfer reactions required for thymidylate and purine biosynthesis and amino acid interconversions. Folate forms differ in the oxidation status of the pteridine moiety being oxidized or reduced as 7,8-dihydro- (H_2folate) and 5,6,7,8-tetrahydrofolate (H_4folate) and their one-carbon substituent at N-5 or N-10. Native food folates are mainly folyl polyglutamates with up to seven glutamic acid residues [1, 2]. Food composition tables indicate that the folate content in food is usually low, in the region of several micrograms per 100 g of food [3]. Foods are sometimes classified into good and moderate folate sources with folate concentrations ranging from 50 to 100 µg and from 15 to 50 µg per serving, respectively [4].

Individual native folates are susceptible – to different extents depending on their chemical nature – to oxidation leading to cleavage of the molecule at the C-9–N-10 bond and subsequent loss of their biological activity (summarized in [5, 6]). Some storage and food processing, under both household and industrial conditions, can result in substantial losses of the vitamin activity of dietary folates, as summarized in [2]. For fortification purposes, the synthetic and fully oxidized vitamer folic acid is used for economic reasons (low cost of synthesis) and due to its high stability. In retention trials using food model systems and different food processing methods, folic acid exhibited higher retention and a half-time ($t\frac{1}{2}$) up to more than 100-fold compared with native folates [2]. Regarding the analysis of folic acid in fortified foods, for further information see Chapter 10.

Several health benefits are associated with a good folate status, due to which this vitamin has attracted both scientific and public interest in recent years. The role of folate in the prevention of neural tube defects (NTDs), such as spina bifida, is now well established [7–10]. With respect to the prevention of coronary heart disease, folate is assumed to play a key role in risk reduction by lowering serum homocysteine concentration, which has been suggested as an independent risk factor [11, 12]. However, evidence about the health protective role of folate – or the mechanisms behind it – is less consistent with respect to certain cancers, for

Fortification of Foods with Vitamins, First Edition. Edited by Michael Rychlik.
© 2011 Wiley-VCH Verlag GmbH & Co. KGaA. Published 2011 by
Wiley-VCH Verlag GmbH & Co. KGaA.

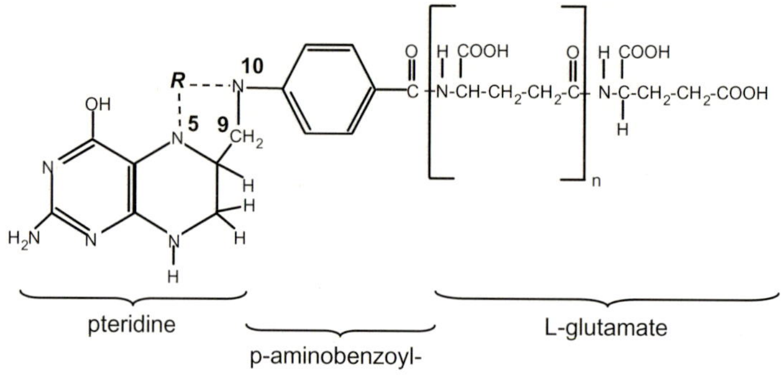

Name	Abbreviation	Substituent	Position
5-methyltetrahydrofolate	5-CH_3-H_4folate	-CH_3	N-5
5-formyltetrahydrofolate	5-HCO-H_4folate	-HCO	N-5
10-formyltetrahydrofolate	10-HCO-H_4folate	-HCO	N-10
5-formiminotetrahydrofolate	5-CHNH-H_4folate	-CH=NH	N-5
5,10-methylenetetrahydrofolate	5,10-CH_2-H_4folate	-CH_2-	N-5 & N-10
5,10-methenyltetrahydrofolate	5,10-CH^+-H_4folate	-CH=	N-5 & N-10

Figure 2.1 Structure of native food folates and their substituent groups and positions. n, number of glutamates.

example, the role of folate status regarding cancer initiation, progression and growth of subclinical cancers, efficacy of antifolate treatment drugs [13–17], and neuro-psychiatric disorders such as dementia and Alzheimer's disease [18, 19]. As summarized by de Bree *et al.* [20] in 1997, adults in several European countries had an average daily folate intake from below 200 to around 300 μg, which was in line with recommendations at that time aimed at preventing folate deficiency. However, the authors stated that – with respect to health benefits from a good folate status – the desired dietary intake should be higher than 350 μg per day, and that only a small part of the studied European populations reach that goal. They also pointed out methodological difficulties in the estimation and comparison of micro-nutrient intake. The observation of suboptimal folate intakes by European populations has also been pointed out in recent studies [21], and this is commonly attributed to the low bioavailability of natural food folates. Data from the US National Health and Nutrition Examination Survey (NHANES) show that before the introduction of mandatory fortification, 20–30 and 35–60% of the population were at risk for low serum and erythrocyte folate concentrations, respectively [22]. In the USA and Canada, the concept of dietary reference intakes was developed for folates and other micronutrients [23] with the aim of including in current concepts the role of nutrients in long-term health, going beyond deficiency diseases. According to new recommendations, based in large part on information about bioavailability, the recommended intake levels were increased to 400–600 μg per day for females of childbearing age and during pregnancy and lactation [23]. A similar development of increased intake recommendations, but to a somewhat lower extent, has been observed in several European countries since 1998. The development of harmonized best practice guidance for a science base for setting micronutrient recommendations is also an aim of the European network of excellence EURRECA [24, 25].

The introduction of nationwide mandatory fortification of cereal-grain products and ready-to-eat cereals in the USA and Canada in 1998 had the particular aim of reducing the incidence of NTD-affected pregnancies. The US folic acid fortification level of 140 μg per 100 g, chosen to increase the intake of women of childbearing age while preventing excessive intake by other groups in the population, was aimed to provide an additional 100 μg of folic acid to the average diet. In several other countries, such as Chile and Costa Rica, which introduced folic acid fortification in staple foods to reduce the number of births complicated by NTD, this public health goal was reached [26, 27]. In the USA, NTD prevalence was reduced by 26% [28]. Furthermore, in most population groups, folate status improved in the post-fortification period so that less than 1 and 5% are at risk for too low serum and erythrocyte folate concentrations, respectively [22]. However, the same report stated that with the current fortification practice, the part of the US population having serum folate concentrations above the normal range of 20 ng ml^{-1} increased from 5–9% to 20–35%. 5-Methyltetrahydrofolate is the dominant folate form in the circulating human plasma, and under normal circumstances dietary folate is deconjugated to its monoglutamate form during an active absorption process, as reviewed in [29]. Folic acid has to undergo in the mucosa cell a saturable process of reduction to H$_4$folate and methylation or formylation. After the introduction of mandatory folic acid fortification in the USA, it was reported that 78% of a certain population group had unmetabolized folic acid in fasted blood samples [30] and that folic acid concentrations in some (non-fasted) plasma samples ranged from 2 to 16% [31]. Kelly *et al.* [32] and Sweeney *et al.* [33, 34] demonstrated the post-prandial appearance of synthetic folic acid from single and multiple doses of folic acid given either as a bolus dose or in the form of fortified food after doses of 400 μg of folic acid. Currently, there is lively discussed as to whether high blood folate concentrations and the presence of unmetabolized folic acid in peripheral plasma lead to adverse health effects. Several hypotheses regarding potential risks from long-term exposure to high dietary folic acid intake from fortification, with respect to progression and growth of pre-neoplastic cells and subclinical cancers, reduced response to antifolate drugs, decreased killer cell cytotoxicity, increased risk of undiagnosed vitamin B$_{12}$ deficiency in the elderly, and the risk of twin pregnancies have been reviewed and scrutinized [15, 17, 35]. In this respect, the issue of folate bioavailability has real importance, and it was stated that "a consideration of the absorption and metabolism of folic acid in human subjects is timely" ([36], p. 1). This review chapter aims to summarize briefly analytical methods and models for the determination of *in vivo* folate bioavailability. Advantages and limitations of the different models are exemplified with respect to current data on the bioavailability of native food folates and folic acid fortificant.

2.2
Folate Bioavailability

2.2.1
Definition

Bioavailability is commonly defined as the absorption and metabolic utilization of a nutrient, referring to the proportion of an ingested nutrient being absorbed, metabolized, and stored [29, 37, 38]. Using different methodological approaches, folate bioavailability is widely studied. Whereas in some cases postprandial intestinal absorption is referred to, others have a wider concept including metabolic processes beyond absorption [37, 38]. The bioavailability of a nutrient is dependent on both host- and food-related factors. Related to the host organism are folate status, health, age, gender, gastrointestinal function, genetic factors [e.g., folate-related genetic polymorphism of 5,10-methylenetetrahydrofolate reductase (MTHFR677C→T) increasing the risk for NTD as reviewed in [10, 39]], and the use of medication and alcohol. Extrinsic factors are related to the chemical nature of the ingested folate (oxidation status of the pteridine ring, substituent, and length of the glutamate chain) or to the food matrix affecting the stability of the vitamin by, for example, the presence of pro- or antioxidants, and by other matrix compounds affecting release of the nutrient during absorption, such as fiber or binding proteins. Food processing can also influence bioavailability by affecting the food matrix and folate stability [2]. The many factors affecting bioavailability are difficult to control and account for when designing bioavailability studies.

2.2.2
Analytical Methods for Folate Quantification and Characterization

Analytical methods play a crucial role when estimating folate bioavailability, both when quantifying and characterizing the folate content in the dose or test food and when determining folate status parameters or vitamin levels in body fluids. Methods and principles for folate quantification have been reviewed in detail by others [1, 6], but the most common methods are briefly summarized in Table 2.1.

The microbiological assay, with a high sensitivity and throughput when automated or performed on microtiter plates, is still the widely accepted standard method [40] used for food composition table data. It provides a value of "total" folate, as the microorganism used in the assay responds to all folate forms with up to three glutamic acid residues, but not to folate degradation products [1, 40]. For clinical purposes, automated competitive binding assays with a high throughput are suitable for quantitation of the dominant plasma folate form 5-methyltetrahydrofolate monoglutamate. High-performance liquid chromatography (HPLC) procedures provide data on individual folate forms and the sum of

Table 2.1 Common procedures for folate analysis – overview.

Total folate		Individual folate forms	
Method	**Application**	**Method**	**Application**
Microbiological assay	Food analysis, e.g., food tables	HPLC (FLD, MWD, UV, EC)	Food analysis
(*Lactobacillus rhamnosus* ATCC 7469) [40]	Clinical analysis; plasma/ serum or red cell folate	LC–MS, SIDA	Clinical analysis/body fluids
Competitive binding assays:			
– radioprotein-binding assay			
– radioimmunoassay			

FLD, fluorescence detection; MWD, multiwavelength detection; EC, electrochemical detection; SIDA, stable isotope dilution assay [41].

folates. They have the limitation of requiring commercially available folate calibration standards. Therefore, the sum of folates determined by HPLC is often lower than the total folate content determined by microbiological assay [42, 43], but the use of recent liquid chromatography (LC)–tandem mass spectrometry (MS/MS) techniques improve specificity and sensitivity [31, 41, 44, 45]. A biosensor-based inhibition immunoassay with monoclonal antibodies was introduced for folate quantification in folic acid fortified foods [46, 47]. Both HPLC and microbiological assay (MA) procedures require sample preparation before analysis (Figure 2.2). A crucial step is the deconjugation of folate polyglutamates to monoglutamates (HPLC) and di- or triglutamates (MA). Also, procedures for heat treatment for deproteination, further enzyme treatment (e.g., trienzyme extraction) for release of folates from the matrix, and procedures for folate stabilization and sample extract purification are required for several food matrices, as summarized in [1]. Sample purification is important for HPLC and LC–mass spectrometry (MS) quantification, and in addition to centrifugation and filtration, solid-phase extraction and affinity chromatography procedures are used. Different approaches to sample preparation omitting or replacing individual steps have been developed, for example, sonication, microwave treatment, and incubation of samples at lower temperatures were introduced instead of heat extraction [1]. However, today there is still no "golden" standard method available for folate quantification and determination. Thorough quality control and documentation are required during all steps of sample preparation and analysis [48, 49].

2.2.3
Models/Methods to Determine Folate Bioavailability

The issue of folate bioavailability has been reviewed in detail by others, and different approaches to experimental nutrition studies have been critically assessed [3, 6, 29]. It is commonly agreed that bioassays have delivered valuable information

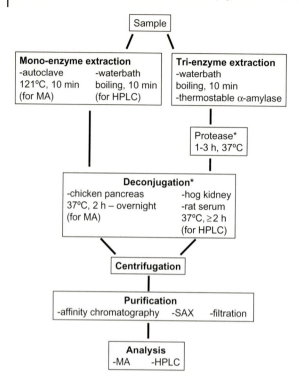

Figure 2.2 The principals of mono- and trienzyme sample preparation for folate analysis by MA or HPLC. For folate quantitation by HPLC and MA, sample pretreatment procedures are required; however, a different order of enzymes during trienzyme extraction is possible. *Enzymes are usually deactivated by an additional heating step (e.g., 5 min in a boiling water-bath), which is not indicated in the flow chart after protease and deconjugase treatment. SAX, solid-phase extraction on strong anion-exchange material.

since the early 1930s regarding the biological role of folates and antifolates with respect to growth and reproduction. Nevertheless, the prediction of human bioavailability from these data is limited due to interspecies differences. *In vitro* models, which provide information cost-efficiently on folate bioaccessibility [50, 51], can only to a limited extent reflect the complexity of *in vivo* folate absorption and metabolism. From observational epidemiological studies, for example, measuring dietary intake and disease in groups or individuals, hypotheses can be generated but no causal relationship can be established due to confounding factors and biases. Therefore, experimental studies involving human volunteers are still required.

The numerous approaches, having specific advantages and limitations, can provide information on folate bioavailability to different extents (Table 2.2). In food-related research, it is difficult to carry out double-blind randomized controlled intervention trials [58]. It is important to control compliance, choice of the

Table 2.2 Advantages and limitations of bioavailability models.

Model type	Principle/studied effect	Advantages	Limitations
In vitro			
TIM [50, 51]	Simulating GIT; bioaccessible folate	Cheap/simple	Do not reflect complexity of *in vivo* absorption and metabolism
Everted sac/ dissected loop/ cell culture models (Caco-2)	Intestinal uptake and transport	No ethic constrictions	
Animal model/bioassay			
Chicken, rat (depletion/ repletion), pig, monkey [52]	Biological role of folate on growth and reproduction	Information on folate metabolism and deficiency, effects from drugs and dietary antifolate compounds can be studied	Limited (quantitative) information due to physiological differences between species
	Effects on tissue folate concentrations and distribution		Relevance to human bioavailability questioned
	Deficiency symptoms		
Human short-term (*in vivo*)			
Human (investigator-controlled) feeding trial	Random application of single doses (foods/ supplements), subjects serve as own controls, carried out in research unit	Determination of (relative) folate absorption/ bioavailability from (fortified) foods and supplements	Expensive
			Problems with compliance
	Quantifying effects of a meal/test food/ nutrient on a variable, e.g., plasma AUC, folate excretion with urine and stomal effluent [53]		Merely absorption studied
			Investigations including pharmacological doses or saturation of body stores (to increase sensitivity) do not reflect physiological conditions
			Strictly standardized trial conditions due to hepatic first-pass effect required

(Continued)

Table 2.2 (*Continued*)

Model type	Principle/studied effect	Advantages	Limitations
Human long-term (*in vivo*)			
Randomized controlled (clinical) trials (RCTs)	Ideally blind intervention	Evidence of effect from diet/intervention on status	Expensive, long duration
	Effect of treatment (supplement/food/diet) on end-point/folate status (erythrocyte or serum folate, plasma homocysteine)	Allows control of confounding factors	Problems with compliance
			Choice of suitable intervention period and status parameter crucial
Short-term or long-term (*in vivo*)			
Stable isotope protocols	Often performed as dual-label stable isotope protocol with simultaneous intravenous/oral (single) doses [54, 55]	Enables discrimination of endogenous folate from body pools and labeled folate from the dose and thereby allows one to evaluate metabolic/ physiological fate of administered compound, absorption kinetics, and turnover	High costs of isotopes and analytical equipment
	Determining dual-label folate (or metabolite) concentrations/ratios in urine and/or plasma		Intrinsic labeling of test food difficult, mainly extrinsic label used, possibly problems with low isotopic purity of labeled compounds
	Assessment of absorption kinetics of various labeled folate compounds and different routes of administration		Requires careful choice of reference doses due to different metabolic handling of differently labeled folates (secondary isotopic effects) or intravenously/orally administrated doses
	When performed as stable isotope tracer method with multiple dosing, determination of isotopic enrichment in body fluids [56, 57]		

TIM, TNO gastrointestinal model [50]; GIT, gastrointestinal tract; AUC, area under the curve.

intervention period (preferably three or more months, especially when slow-responding erythrocyte folate concentrations are used for status assessment), and choice of suitable endpoint/status parameters when designing an intervention study. High bioavailabilities of endogenous folates from vegetables and fruits [59, 60], rye cereal products and orange juice [61], folic acid-fortified milk [58, 62], and folic acid-fortified breakfast rolls [63] have been reported using different intervention models. Also, the effects of dietary polyglutamyl folates [64] and different folate supplements [65] on folate status have been studied. Cross-over designs [66] are unusual in folate intervention studies with healthy volunteers, but improved plasma homocysteine levels after intervention with a folate-rich diet and supplemental folic acid in volunteers with hyperhomocysteinemia could be demonstrated.

In human short-term feeding trials, often (relative) postprandial folate absorption is determined using biokinetic methods and the area under the plasma (concentration) curve (area under the curve, AUC) [3, 6]. For each individual, the size of the AUC is equivalent to the ingested nutrient dose. A drawback of the AUC model is that volunteers' liver folate stores, the hepatic first-pass effect, and enterohepatic circulation affect systemic blood folate concentrations. As shown in Figure 2.3, for the oral application of fortified milk products, dose-corrected AUCs (depicted as staples) show high inter-individual variation. Therefore, strictly

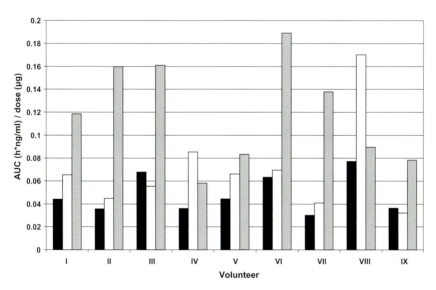

Figure 2.3 Dose-corrected plasma AUC of nine volunteers after single random doses of pasteurized and fermented milk fortified with (6S)-5-methyltetrahydrofolate with and without folate binding protein [67]. Black columns, pasteurized milk with 542 nmol (6S)-5-methyltetrahydrofolate and 262 nmol folate binding protein per portion; white columns, fermented milk with 392–445 nmol (6S)-5-methyltetrahydrofolate and 156–442 nmol folate binding protein per portion; gray columns, fermented milk with 450 nmol (6S)-5-methyltetrahydrofolate per portion.

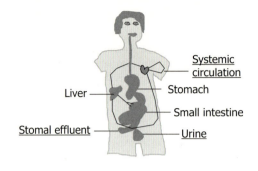

Non-absorbed folate:
Stomal folate excretion

Absorbed folate:
AUC and urinary folate

Figure 2.4 The human AUC–ileostomy model for the assessment of folate bioavailability [53, 67]. AUC, area under the plasma (concentration) curve.

standardized assay conditions, where subjects serve as their own controls, are a prerequisite for the assessment of relative absorption from single doses of dietary or supplemental folate. The number and time points of multiple post-prandial blood samples can affect the accuracy (as the size of the AUC will be underestimated when the maximum concentration (C_{max}) is missed) and sensitivity of the model. Therefore, data derived from studies simply comparing absorption by folate C_{max} or by (extrapolated) AUCs derived from post-dose collection periods shorter than 6 h must be interpreted with care, especially when different modes of dose application (orally in the form of food versus bolus versus intravenous) were used.

Using ileostomy volunteers, who lack a colon and its microflora affecting fecal folate excretion, the AUC approach can be extended by accounting for non-absorbed folate using post-dose ileostomal effluent samples [53, 67, 68] (Figure 2.4). Non-absorbed folate from oral doses can be estimated by quantifying post-dose folate excretion in ileostomal effluent samples collected in regular intervals up to 24 h post-dose. Effluent samples either can be pooled [67, 68] for estimation of total folate excretion, or can be analyzed individually, providing information on the intestinal passage time of different food matrices and folate excretion over time [53]. By the AUC or AUC–ileostomy approach, equivalent relative folate absorption from pharmaceutical preparations of (6S)-5-methyltetrahydrofolate and folic acid was determined [67, 69]. Short-term absorption from endogenous food folates from cereal products [70], vegetables and fruit [53, 68, 71, 72], and fortified dairy and cereal products [67] was also assessed. McKillop *et al.* [73] reported that folate absorption was not affected by the extent of conjugation (monoglutamyl versus polyglutamyl folates). Green and black tea were suggested to lower the absorption of supplemental folic acid [74], and dairy folate binding protein lowered the absorption of (6S)-5-methyltetrahydrofolate fortificant from pasteurized and fermented milk [67].

In recent years, several stable isotope protocols have been developed, in many cases as dual-label protocols with simultaneous application of an intravenous reference dose to a single oral dose [54, 55, 75], and short-term folate absorption has been determined from the isotope ratios in urine [54, 55] and plasma [75]. The use of folate doses labeled with stable isotopes is a superior tool, allowing discrimination of endogenous folate from body pools and labeled folate from the dose in post-dose body fluids. Therefore, interpretation of data is unambiguous, as all label in excess of the natural abundance is derived from the absorbed fraction. However, stable isotope studies demand advanced analytical equipment and are expensive. It is difficult to prepare labeled test foods, and the labeled dose is either given as a supplement [57] or incorporated into the test food as (labeled) fortificant [54, 55, 75]. Although intrinsic labeling of endogenous food folate would be desirable, it has so far been reported only once for ^{13}C- and ^{15}N-labeled spinach using a hydroponics system [76]. Choosing a suitable reference dose for the estimation of folate absorption is of great importance. It was observed that intravenous and oral doses of folic acid undergo different metabolic handling [55], and the same was assumed for folic acid versus reduced folates [77]. Therefore, folic acid is not considered an appropriate reference dose for bioavailability studies [55, 77]. Studies with stable isotope tracer design with multiple dosing [56, 78], where isotopic enrichment of, for example, plasma folate, urinary folate, and urinary folate catabolites is determined, revealed that folate body pools and excretion do not differ in pregnant women and non-pregnant controls. Using a kinetic model with several compartments, the mass and turnover of different body pools could be estimated and the importance of catabolism as a route of folate turnover demonstrated. In another multiple dosing trial with dual isotope labeling, the relative bioavailability of pharmaceutical hexaglutamylfolic acid was estimated to be about 80% of that of monoglutamylfolic acid [57].

Obviously no "golden method" for the assessment of folate bioavailability exists, but data on folate bioavailability can be carefully interpreted with consideration of the model's strengths and limitations. Since single studies rarely provide evidence on folate bioavailability, conclusions should be drawn using data from multiple trials with consistent results after thorough evaluation of the study designs.

References

1 Eitenmiller, R.R. and Landen, W.O. (1990) Folate. in *Vitamin Analysis for the Health and Food Sciences* (eds. R.R. Eitenmiller and W.O. Landen), CRC Press, Bocca Raton, FL, pp. 411–466.

2 Hawkes, J.G. and Villota, R. (1989) Folates in foods: reactivity, stability during processing, and nutritional implications. *Critical Reviews in Food Science and Nutrition*, **28**, 439–538.

3 Witthöft, C.M., Forssén, K., Johannesson, L., and Jägerstad, M. (1999) Folates – food sources, analyses, retention and bioavailability. *Scandinavian Journal of Nutrition*, **43**, 138–146.

4 Department of Health, Scottish Office Home and Health Department, Welsh Office, Department of Health and Social Services, Northern Ireland (1992) Folic Acid and the Prevention of Neural Tube

Defects. Health Publications Unit, Heywood, Lancashire.

5 Ball, G.F.M.F. (1998) *Bioavailability and Analysis of Vitamins in Foods*, Chapman & Hall, London, pp. 439–496.

6 Gregory, J.F. (1989) Chemical and nutritional aspects of folate research: analytical procedures, methods of folate synthesis, stability, and bioavailability of dietary folates. *Advances in Food and Nutrition Research*, **33**, 1–101.

7 Medical Research Council Vitamin Study Research Group (1991) Prevention of neural tube defects: results of the Medical Research Council vitamin study. *Lancet*, **338**, 131–113.

8 Honein, M.A., Paulozzi, L.J., Mathews, T.J., Erickson, J.D., and Wong, L.-Y.C. (2001) Impact of folic acid fortification of the US food supply on the occurrence of neural tube defects. *JAMA*, **285**, 2981–2986.

9 Molloy, A.M. (2005) The role of folic acid in the prevention of neural tube defects. *Trends in Food Science and Technology*, **16**, 241–245.

10 Tamura, T. and Picciano, M.F. (2006) Folate and human reproduction. *American Journal of Clinical Nutrition*, **83**, 993–1016.

11 Boushey, C.J., Beresford, S.A., Omenn, G.S., and Motulsky, A.G. (1995) A quantitative assessment of plasma homocysteine as a risk factor for vascular disease – probable benefits of increasing folic acid intakes. *JAMA*, **274**, 1049–1057.

12 Green, R. and Jacobsen, R.H. (1995) Clinical implications of hyperhomocysteinemia, in *Folate in Health and Disease* (ed. L.B. Bailey), Marcel Dekker, New York, pp. 75–122.

13 Kim, Y.-I. (1999) Folate and carcinogenesis: evidence, mechanisms, and implications. *Journal of Nutrional Biochemistry*, **10**, 66–88.

14 Kim, Y.-I. (2004) Will mandatory fortification prevent or promote cancer? *American Journal of Clinical Nutrition*, **80**, 1123–1128.

15 Mason, J.B., Dickstein, A., Jacques, P.F., Haggarty, P., Selhub, J., Dallal, G., and Rosenberg, I.H. (2007) A temporal association between folic acid

fortification and an increase in colorectal cancer rates may be illuminating important biological principles: a hypothesis. *Cancer Epidemiology, Biomarkers and Prevention*, **16** (7), 1325–1329.

16 Peters, G.J., Hooijberg, J.H., Kaspers, G.J.L., and Jansen, G. (2005) Folates and the antifolate treatment of cancer; role of folic acid supplementation on efficacy of folate and non-folate drugs. *Trends in Food Science and Technology*, **16**, 289–297.

17 Smith, A.D., Kim, Y.-I., and Refsum, H. (2008) Is folic acid good for everyone? *American Journal of Clinical Nutrition*, **87**, 517–533.

18 Sauberlich, H.E. (1991) Relationship of vitamin B_6, B_{12}, and folate to neurological and neuropsychiatric disorders, in *Micronutrients in Health and Disease Prevention* (eds. A. Bendich and C.E. Butterworth), Marcel Dekker, New York, pp. 187–218.

19 Seshadri, S., Beiser, A., Selhub, J., Jacques, P.F., Rosenberg, I.H., D'Agostino, R.B., Wilson, P.W.F., and Wolf, P.A. (2002) Plasma homocysteine as a risk factor for dementia and Alzheimer's disease. *New England Journal of Medicine*, **346** (7), 476–483.

20 de Bree, A., van Dusseldorp, M., Brouwer, I.A., van het Hof, K.H., and Steegers-Theunissen, R.P.M. (1997) Folate intake in Europe: recommended, actual and desired intake. *European Journal of Clinical Nutrition*, **51**, 643–660.

21 Hatzis, C.M., Bertsias, G.K., Linardakis, M., Scott, J.M., and Kafatos, A.G. (2006) Dietary and other lifestyle correlates of serum folate concentrations in a healthy adult population in Crete, Greece: a cross-sectional study. *Nutrition Journal*, **5**, 5–14.

22 Pfeiffer, C.M., Johnsson, C.L., Jain, R., Yetley, E.A., Picciano, M.F., Rader, J.I., Fisher, K.D., Mulinare, J., and Osterloh, J.D. (2007) Trends in blood folate and vitamin B-12 concentrations in the United States, 1988–2004. *American Journal of Clinical Nutrition*, **86**, 718–727.

23 Yates, A.A., Schlicker, S.A., and Suitor, C.W. (1998) Dietary Reference Intakes: the new basis for recommendations for calcium, and related nutrients,

B vitamins, and choline. *Journal of the American Dietetic Association*, **98**, 699–706.

24 Ashwell, M., Lambert, J.P., Alles, M.S., *et al.* (2008) How we will produce the evidence-based EURRECA toolkit to support nutrition and food policy. *European Journal of Nutrition*, **47** (1), 2–16.

25 Nordic Council of Ministers (2004) *Nordic Nutrition Recommendations NNR 2004*. 4th edn, Norden, Uppsala.

26 Berry, R.J., Li, Z., Erickson, J.D., *et al.* (1999) Prevention of neural-tube defects with folic acid in China. *New England Journal of Medicine*, **341**, 1485–1490.

27 Hertrampf, E. and Cortés, F. (2004) Folic acid fortification in wheat flour: Chile. *Nutrition Reviews*, **62**, S44–S48.

28 Centers for Disease Control and Prevention (2004) Spina bifida and anencephaly before and after folic acid mandate–United States, 1995–1996 and 1999–2002. *Morbidity and Mortality Weekly Report*, **53**, 362–365.

29 Gregory, J.F. (1995) The bioavailability of folate, in *Folate in Health and Disease* (ed. L.B. Bailey), Marcel Dekker, New York, pp. 195–235.

30 Pfeiffer, C.M., Caudill, S.P., Gunter, E.W., Osterloh, J., and Sampson, E.J. (2005) Biochemical indicators of B vitamin status in the US population after folic acid fortification: results from the National Health and Nutrition Examination Survey 1999–2000. *American Journal of Clinical Nutrition*, **82**, 442–450.

31 Pfeiffer, C.M., Fazili, Z., McCoy, L., Zhang, M., and Gunter, E.W. (2004) Determination of folate vitamers in human serum by stable-isotope-dilution tandem mass spectrometry and comparison with radioassay and microbiological assay. *Clinical Chemistry*, **50** (2), 423–432.

32 Kelly, P., McPartlin, J., Gogins, M., Weir, D.G., and Scott, J.M. (1997) Unmetabolized folic acid in serum: acute studies in subjects consuming fortified food and supplements. *American Journal of Clinical Nutrition*, **65**, 1790–1795.

33 Sweeney, M.R., McPartlin, J., Weir, D.G., Daly, L., and Scott, J.M. (2006) Postprandium serum folic acid response to multiple dose of folic acid in fortified bread. *British Journal of Nutrition*, **95**, 145–151.

34 Sweeney, M.R., McPartlin, J., and Scott, J.M. (2007) Folic acid fortification and public health: report on the threshold doses above which unmetabolised folic acid appear in serum. *Public Health*, **7**, 41–47.

35 Wright, A.J.A., Finglas, P.M., and Southon, S. (2001) Proposed mandatory fortification of the UK diet with folic acid: have potential risks been underestimated? *Trends in Food Science and Technology*, **16**, 313–321.

36 Powers, H.J. (2007) Folic acid under scrutiny. *British Journal of Nutrition*, **98** (4), 556–666.

37 VLAG (Advanced Studies in Food Technology, Agrobiotechnology, Nutrition and Health Sciences) (1997) Bioavailability '97 Symposium, May 25–28, 1997: Book of Abstracts. VLAG, Wageningen.

38 Schlemmer, U. (ed.) (1993) Bioavailability '93: Nutritional, Chemical, and Food Processing Implications of Nutrient Availability, May 9–12, 1993, Ettlingen, Germany, Proceedings, Parts 1 and 2. Bundesforschungsanstalt für Ernährung, Karlsruhe.

39 Scott, J.M., Weir, D.G., and Kirke, P.N. (1995) Folate and neural tube defects, in *Folate in Health and Disease* (ed. L.B. Bailey), Marcel Dekker, New York, pp. 329–360.

40 Tamura, T. (1990) Microbiological assay of folates, in *Folic Acid Metabolism in Health and Disease* (eds. M.F. Picciano, E.L.R. Stokstad, and J.F. Gregory), Wiley-Liss, New York, pp. 121–137.

41 Freisleben, A., Schieberle, P., and Rychlik, M. (2003) Specific and sensitive quantification of folate vitamers in foods by stable isotope dilution assays using high-performance liquid chromatography–tandem mass spetrometry. *Analytical and Bioanalytical Chemistry*, **376**, 149–156.

42 Fazili, Z., Pfeiffer, C.M., Zhang, M., Jain, R.B., and Koontz, D. (2008) Influence of 5,10-methylenetetrahydrofolate reductase

polymorphism on whole-blood folate concentrations measured by LC–MS/MS, microbiological assay, and Bio-Rad radioassay. *Clinical Chemistry*, **54** (1), 197–201.

43 Fazili, Z., Pfeiffer, C.M., and Zhang, M. (2006) Comparison of serum folate species analyzed by LC–MS/MS with total folate measured by microbiologic assay and Bio-Rad radioassay. *Clinical Chemistry*, **53** (4), 781–784.

44 Nelson, B.C., Pfeiffer, C.M., Margolis, S.A., and Nelson, C.P. (2003) Affinity extraction combined with stable isotope dilution LC/MS for the determination of 5-methyltetrahydrofolate in human plasma. *Analytical Biochemistry*, **313**, 117–127.

45 Garratt, L.C., Ortori, C.A., Tucker, G.A., Sablinsky, F., Bennett, M.J., and Barrett, D.A. (2005) Comprehensive metabolic profiling of mono- and polyglutamated folates and their precursors in plant and animal tissue using liquid chromatography/negative ion electrospray ionisation tandem mass sprectrometry. *Rapid Communications in Mass Spectrometry*, **19**, 2390–2398.

46 Indyk, H.E., Evans, E.A., Boström Caselunghe, M.C., Persson, B.S., Finglas, P.M., Woollard, D.C., and Filonzi, E.L. (2000) Determination of biotin and folate in infant formula and milk by optical biosensor-based immunoassay. *Journal of AOAC International*, **83** (5), 1141–1148.

47 Boström Caselunghe, M. and Lindeberg, J. (2000) Biosensor-based determination of folic acid in fortified food. *Food Chemistry*, **70**, 523–532.

48 Koontz, J.L., Phillips, K.M., Wunderlich, K.M., Exler, J., Holden, J.M., Gebhardt, S.E., and Haytowitz, D.B. (2005) Comparison of total folate concentrations in food determined by microbiological assay at several experienced U.S. commercial laboratories. *Journal of AOAC International*, **88** (3), 805–813.

49 Puwastien, P., Pinprapai, N., Judprasong, K., and Tamura, T. (2005) International inter-laboratory analyses of food folate. *Journal of Food Composition and Analysis*, **18**, 387–397.

50 Verwei, M., Arkbåge, K., Havenaar, R., Van den Berg, H., Witthöft, C., and Schaafsma, G. (2003) Folic acid and 5-methyltetrahydrofolic acid in fortified milk are bioaccessible as determined in a dynamic *in vitro* gastrointestinal model. *Journal of Nutrition*, **133**, 2377–2383.

51 Öhrvik, V. and Witthöft, C. (2008) Orange juice is a good folate source in respect to folate content and stability during storage and simulated digestion. *European Journal of Nutrition*, **2**, 92–98.

52 Keagy, P.M. (1990) Animal assays for folate bioavailability: a critical evaluation, in *Folic Acid Metabolism in Health and Disease* (eds. M.F. Picciano, E.L.R. Stokstad, and J.F. Gregory), Wiley-Liss, New York, pp. 139–150.

53 Witthöft, C.M., Strålsjö, L., Berglund, G., and Lundin, E. (2003) A human model to determine folate bioavailability from food – a pilot study for evaluation. *Scandinavian Journal of Nutrition*, **47**, 6–18.

54 Pfeiffer, C.M., Rogers, L.M., Bailey, L.B., and Gregory, J.F. III (1997) Absorption of folate from fortified cereal-grain products and of supplemental folate consumed with or without food determined by using a dual-label stable-isotope protocol. *American Journal of Clinical Nutrition*, **66**, 1388–1397.

55 Finglas, P.M., Witthöft, C.M., Vahteristo, L., Wright, A.J.A., Southon, S., Mellon, F., Ridge, B., and Maunder, P. (2002) Use of an oral/intravenous dual-label stable-isotope protocol to determine folic acid bioavailability from fortified cereal-grain foods in women. *Journal of Nutrition*, **132**, 936–939.

56 Gregory, J.F., Caudill, M.A., Opalko, F.J., and Bailey, L.B. (2001) Kinetics of folate turnover in pregnant women (second trimester) and nonpregnant controls during folic acid supplementation: stable-isotopic labeling of plasma folate, urinary folate and folate catabolites shows subtle effects of pregnancy on turnover of folate pools. *Journal of Nutrition*, **131**, 1928–1937.

57 Melse-Boonstra, A., Verhof, P., West, C., van Rhijn, J.A., van Breemen, R.B., Lasaroms, J.J.P., Garbis, S.D., Katan, M.B., and Kok, F.J. (2006) A dual-isotope-labeling method of studying the bioavailability of hexaglutamyl folic acid

relative to that of monoglutamyl folic acid in humans by multiple orally administered low doses. *American Journal of Clinical Nutrition*, **84**, 1128–1133.

58 deJong, R.J., Verwei, M., West, C.E., van Vliet, T., Siebelink, E., van den Berg, H., and Castenmiller, J.J.M. (2005) Bioavailability of folic acid from fortified pasteurised and UHT-treated milk in humans. *European Journal of Clinical Nutrition*, **59**, 906–913.

59 Brouwer, I.A., van Dusseldorp, M., West, C.E., Meyboom, S., Thomas, C.M.G., Duran, M., van het Hof, K.H., Eskes, T.K.A.B., Hautvast, J.G.A.J., and Steegers-Theunissen, R.P.M. (1999) Dietary folate from vegetables and citrus fruit decreases plasma homocysteine concentrations in humans in a dietary controlled trial. *Journal of Nutrition*, **129**, 1135–1139.

60 Broekmans, W.M.R., Klöpping-Ketelaars, I.A.A., Schuurman, C.R.W.C., Verhagen, H., van den Berg, H., Kok, F.J., and van Poppel, G. (2000) Fruits and vegetables increase plasma carotenoids and vitamins and decrease homocysteine in humans. *Journal of Nutrition*, **130**, 1578–1583.

61 Vahteristo, L., Kariluoto, S., Bärlund, S., Kärkkäinen, M., Lamberg-Allardt, C., Salovaara, H., and Piironen, V. (2002) Functionality of endogenous folates from rye and orange juice using human *in vivo* model. *European Journal of Nutrition*, **41**, 271–278.

62 Green, T.J., Murray Skeaff, C., Rockell, J.E.P., and Venn, B.J. (2005) Folic acid fortified milk increases blood folate and lowers homoysteine concentration in women of childbearing age. *Asia Pacific Journal of Clinical Nutrition*, **14** (2), 173–178.

63 Johansson, M., Witthöft, C.M., Bruce, Å., and Jägerstad, M. (2002) Study of wheat breakfast rolls fortified with folic acid. The effect on folate status in women during a 3-month intervention. *European Journal of Nutrition*, **41**, 279–286.

64 Hannon-Fletcher, M.P., Armstrong, N.C., Scott, J.M., Pentieva, K., Bradbury, I., Ward, M., Strain, J.J., Dunn, A.A., Molloy, A.M., Kerr, M.A., and McNulty, H. (2004) Determining bioavailability of folates in a controlled intervention study. *American Journal of Clinical Nutrition*, **80**, 911–918.

65 Houghton, L.A., Sherwood, K.L., Pawlosky, R., Ito, S., and O'Connor, D.L. (2006) [6S]-5-methyltetrahydrofolate is at least as effective as folic acid in preventing a decline in blood folate concentrations during lactation. *American Journal of Clinical Nutrition*, **83**, 842–850.

66 Pintó, X., Vilaseca, M.A., Balcells, S., Artuch, R., Corbella, E., Meco, J.F., Vila, R., Pujol, R., and Grinberg, D. (2005) A folate-rich diet is as effective as folic acid from supplements in decreasing plasma homocysteine concentrations. *International Journal of Medical Sciences*, **2**, 58–63.

67 Witthöft, C.M., Arkbåge, K., Johansson, M., Lundin, E., Berglund, G., Zhang, J.X., Lennernäs, H., and Dainty, J.R. (2006) Folate absorption from folate-fortified and processed foods using a human ileostomy model. *British Journal of Nutrition*, **95**, 181–187.

68 Konings, E.J.M., Troost, F.J., Castenmiller, J.J.M., Roomans, H.H.S., van der Brandt, P., and Saris, W. (2002) Intestinal absorption of different types of folate in healthy subjects with an ileostomy. *British Journal of Nutrition*, **88**, 235–242.

69 Pentieva, K., McNulty, H., Reichert, R., Ward, M., Strain, J.J., McKillop, D.J., McPartlin, J.M., Conolly, E., Molloy, A., Krämer, K., and Scott, J.M. (2004) The short-term bioavailabilities of [6S]-5-methyltetrahydrofolate and folic acid are equivalent in men. *Journal of Nutrition*, **134**, 580–585.

70 Fenech, M., Noakes, M., Clifton, P., and Topping, D. (1999) Aleurone flour is a rich source of bioavailable folate in humans. *Journal of Nutrition*, **129**, 1114–1119.

71 Prinz-Langenohl, R., Brönstrup, A., Thorand, B., Hages, M., and Pietrzik, K. (1999) Availability of food folate in humans. *Journal of Nutrition*, **129**, 913–916.

72 Rychlik, M., Netzel, M., Pfannebecker, I., Frank, T., and Bitsch, I. (2003)

Application of stable isotope dilution assays based on liquid chromatography–tandem mass spectrometry for the assessment of folate bioavailability. *Journal of Chromatography B*, **792**, 167–176.

73 McKillop, D.J.M., McNulty, H., Scott, J.M., McPartlin, J.M., Strain, J.J., Bradbury, I., Girvan, J., Hoey, L., McCreedy, R., Alexander, J., Patterson, B.K., Hannon-Fletcher, M., and Pentieva, K. (2006) The rate of intestinal absorption of natural food folates is not related to the extent of folate conjugation. *American Journal of Clinical Nutrition*, **84**, 167–173.

74 Alemdaroglu, N.C., Dietz, U., Wolffram, S., Spahn-Langguth, H., and Langguth, P. (2008) Influence of green and black tea on folic acid pharmacokinetics in healthy volunteers: potential risk of diminished folic acid bioavailability. *Biopharmaceutics and Drugs Disposition*, **29** (6), 335–348.

75 Rogers, L.M., Pfeiffer, C.M., Bailey, L.B., and Gregory, J.F. (1997) A dual-label stable isotopic protocol is suitable for determination of folate bioavailability in humans: evaluation of urinary excretion and plasma folate kinetics of intravenous and oral doses of [$^{13}C_5$]- and [$^2H_2$]folic acid. *Journal of Nutrition*, **127**, 2321–2327.

76 Hart, D.J., Wright, A.J.A., Wolfe, C.A., Dainty, J., Perkins, L.R., and Finglas, P.M. (2006) Production of intrinsically labelled spinach using stable isotopes (^{13}C or ^{15}N) for the study of folate absorption. *Innovative Food Science and Emerging Technologies*, **7**, 147–151.

77 Wright, A.J.A., Finglas, P.M., Dainty, J.R., Wolfe, C.A., Hart, D.J., Wright, D.M., and Gregory, J.F. (2005) Differential kinetic behavior and distribution for pteroylglutamic acid and reduced folates: a revised hypothesis of the primary site of PteGlu metabolism in humans. *Journal of Nutrition*, **135**, 619–623.

78 Gregory, J.F., Williamson, J., Liao, J.-F., Bailey, L.B., and Toth, J.P. (1998) Kinetic model of folate metabolism in nonpregnant women consuming [$^2H_2$] folic acid: isotopic labeling of urinary folate and the catabolite *para*-acetamidobenzoylglutamate indicates slow, intake-dependent, turnover of folate pools. *Journal of Nutrition*, **128**, 1896–1906.

3
Quantitation of Vitamins Using Microbiological Assays in Microtiter Formats

Wolfgang Weber, Sabine Mönch, Michael Rychlik, and Sylvia Stengl

3.1
Introduction

3.1.1
History

Before the development of specific chemical methods, bioassays were the methods of choice in vitamin analysis due to their specificity and sensitivity. In the first half of the last century, animal assays using vertebrates such as rats, chicks, pigeons, and guinea pigs were commonly applied methods [1, 2]. The increase in body weight in curative growth assays of animals deprived of a vitamin were the most frequently employed methods for vitamin B_1 [3] and vitamin A [4]. Alternative designs were (i) monitoring of the time-dependent appearance or disappearance of vitamin-specific deficiency symptoms such as the black-tongue disease in dogs for niacin [5] and (ii) evaluation of the weight gain of specific organs such as the ash of leg bones of rats or chicks in the case of vitamin D [6]. For vitamin E, the fertility-enhancing effect on rats was quantified by counting the number of successfully mated females [7].

During the mid-1950s, ethical concerns and animal protection issues led to the abolition of the latter assays. However, bioassays using microorganisms are not affected by these restrictions and have been widely used or are still in use today for specific compounds. Starting with the first report on riboflavin determination using *Lactobacillus casei* in 1939 by Snell and Strong [8], particularly B vitamins were determined by microbiological assays (MAs). Further early examples of MAs for quantitations of B vitamins are the following

- thiamin by *Lactobacillus* spp. [9]
- vitamin B_6 by *Saccharomyces carlsbergensis* [10]
- vitamin B_{12} by *Ochromonas malhamensis* [11],
- biotin by *Allescheria boydii* [12],
- folic acid by *Streptococcus faecalis* [13],
- niacin by *Lactobacillus arabinosus* [14], and
- pantothenic acid by *Lactobacillus arabinosus* [15].

3.1.2
Principle of Microbiological Assays

As generally for bioassays, the growth of an organism for which the analyte to be determined is essential is measured in a specifically deficient nutrient solution. The medium is then supplemented with a sample or a vitamin standard and the resulting growth is measured at intervals, for example by titration, gravimetrically, by turbidity, or nephelometrically. Most commonly, growth is reflected by turbidity of the assay, which conveniently can be measured in an ultraviolet–visible spectrophotometer.

3.1.3
Requirements for Classic Microbiological Analysis

In traditional microbiology, colonies of the target microorganisms must first be cultured and subsequently must be maintained by regular inoculation. Before the actual assay procedure can begin, the cultures must be freshly prepared, and the number of microorganisms must be adjusted before the organisms are transferred to the medium.

To perform accurate MAs, a clean and easy to clean room with sterile working conditions is required. Preferably, the room is equipped with a sterile working bench. Further equipment needed is an autoclave for sterilizing tubes and culture media, an incubator capable of maintaining constant temperatures in the range 27–40 °C, and a nephelometer or spectrophotometer for measuring the microorganisms' growth. A suitable water supply is crucial – deionized water usually is not sufficient and, therefore, glass-distilled water is preferred.

3.1.4
**Partial Replacement of Vitamin Microbiological Assays
by Chromatographic Methods**

However, MAs are difficult to carry out routinely in the general setting of a modern laboratory for food analysis. In particular, maintaining inoculum cultures, sterile conditions and a suitable water supply is a decisive restriction.

Generally, microorganisms must be regulated before the organisms are transferred to the medium, which is tedious and labor intensive. Moreover, studies of accuracy and precision have shown that MAs often do not produce satisfactory results.

Additionally, the measurement of a total vitamin figure in many cases no longer meets actual requirements. In modern food and nutrition sciences, differentiation of vitamers, conjugates, and pro-vitamins is important due to different bioavailabilities or due to different responses of the microorganisms. Therefore, chromatographic techniques coupled to specific detectors have gained increasing importance in vitamin analysis. Starting with ultraviolet (UV) and fluorescence detection, the versatility of modern mass detectors has spurred many applications in vitamer differentiation and multivitamin detection [16].

Chapter 14 describes all the difficulties encountered in the simultaneous analysis of various vitamins and how liquid chromatography–mass spectrometry (LC–MS) has proved itself as a promising technique in this analytical area. In addition to performing multi-analyte analysis and providing unambiguous evidence of identification, the sensitivity and selectivity of LC–MS allow sample pretreatment to be simplified and labile vitamins to be preserved.

Despite the continually increasing importance of chromatographic methods, MAs still retain their specific place in vitamin analysis. Due to their often superior sensitivity, MAs are commonly used for vitamin B_{12}, folic acid, and biotin analyses. For vitamin groups such as the folates, where the detection, separation, and quantitation of all vitamers is difficult, MA offers the advantage of obtaining a sum value without the need for extensive method development. Moreover, discussions about problems in vitamer differentiation and still potentially unknown vitamers have not yet been finally settled. In this respect, MA may detect more vitamin active compounds than its chromatographic alternatives. Because of these considerations, for vitamins such as the folates and vitamin B_{12}, MAs are the "gold standard" and are still in use as reference methods in many countries [17–19].

3.1.5
New Developments in Microbiological Assay Technology

3.1.5.1 Coupling with Chromatography
On comparing the obvious advantages of the two methodologies, the combination of chromatography and MA appears promising. If a sample contains different vitamin-active compounds, chromatography enables them to be separated and the individual components to be evaluated separately. When MA is then applied to these single fractions, it allows their bioactivity to be assessed, even if reference compounds are not available or even if the identity of the assessed compounds is unknown.

These combinations of liquid chromatography (LC) and MA may be run in an off-line mode, such that the chromatographically resolved fractions are collected and subsequently assessed separately by MA. As this approach is fairly labor intensive and time consuming, modern applications favor the on-line mode by applying the MA automatically to all single fractions in one step without separate manual handling.

The first on-line combination was the so-called bioautography, which includes separation by silica gel thin-layer chromatography (TLC) and subsequent spraying of the microorganism on the complete plate. The spray solution also contains the nutrition medium except for the vitamer to be detected. Bioactivity of the single spots on the plate is then evaluated visually from the growth of bacterial colonies or by spraying of a tetrazolium salt that reacts to give a deep-blue formazan, which results from the growing bacteria [20].

An example of this technique is outlined in Chapter 13, which describes a sensitive application of bioautography using a vitamin B_{12}-dependent *Escherichia coli* mutant. By application of bioautography, loss of vitamin B_{12} is readily evaluated during storage of vitamin B_{12}-fortified foods or during food processing. This technique is advantageous with regard to simplicity, flexibility, speed, and relative cheapness for the analysis of vitamin B_{12} compounds in foods.

3.1.5.2 Coupling with Liquid Chromatography

In the case of vitamins, for which the TLC resolution is not sufficient, or if a quantitation of bioactivity is required, high-performance liquid chromatography (HPLC) has to be coupled with the MA.

The first off-line combinations for folates were introduced by Chapman *et al.* [21], who separated serum folates by ion chromatography and detected folic acid in one of the separated fractions using *Lactobacillus rhamnosus*.

In a similar study, Kelly *et al.* [22] applied reversed-phase HPLC to detect folates in serum.

In both studies, the sensitivity of the MA was necessary as unmetabolized folic acid occurred in only miniscule amounts in the serum. Hence the studies proved the limitation of the gastrointestinal tract and the liver to convert folic acid into natural folates after dosage of folic acid.

It was only in 1998 that Belz and Nau [23] presented the first on-line application to mouse plasma, erythrocytes, and embryos. From the same laboratory, Kehlenbach [24] extended the same assay to folate quantitation in foods. The author was able to use ion-pair chromatography and to avoid an acidic mobile phase, which in common folate HPLC often leads to interconversions between folate vitamers. In validation studies using certified reference materials, the developed LC–MA method revealed a bias of less than 18% with respect to the certified reference values and proved the accuracy of this method.

3.1.5.3 Microtiter Formats and Standardized Assays

The microbiological determination of vitamins is fairly labor intensive and time demanding, in addition to requiring considerable laboratory organization. Moreover, it is always a matter of uncertainty whether the inoculation suspension is suitable or the microorganism has the desired sensitivity and specificity.

In contrast to maintaining active inoculum cultures, freezing and thawing before the day of use are possible alternatives.

In the case of the folate MA, cryopreservation was introduced by Horne and Patterson [25]. However, the authors retained the traditional organism (ATCC 7469), which necessitated sterilization of reagents by filtration and stringent aseptic precautions. In further developments, the use of a chloramphenicol-resistant strain of *L. casei* allowed it to be run completely open on the laboratory bench [26].

Advances in microtiter plate technology and improved optical qualities of the plates enabled MAs for vitamins to be performed on microtiter plates [27]. The obvious advantages of such miniaturization of the assays not only include speed of reading and reduced reagent costs, but also these simplified assays have become less labor intensive and are incorporated more readily into the modern clinical or research laboratory.

Therefore, the first aim of this study was to develop a standardized, easy to use and ready to use MA on microtiter plates to be performed without the need for special expertise in microbiological techniques. Moreover, a further goal was to apply the new technology to the determination of folates, vitamin B_{12}, riboflavin, and pantothenic acid along with a demonstration of its validity.

3.2
Methods and Materials

(6S)-Tetrahydrofolic acid [(6S)-H_4folate], (6S)-5-methyltetrahydrofolic acid [(6S)-5-methyl-H_4folate] and 10-formylfolic acid were obtained from Dr. Schircks Laboratories (Jona, Switzerland), (6S)-5-formyltetrahydrofolic acid [(6S)-5-formyl-H_4folate] was purchased from Sigma (Deisenhofen, Germany), and folic acid was obtained from Fluka (Neu-Ulm, Germany).

Takadiastase and acid phosphatase from potato were obtained from Sigma. Chicken pancreatin is available from R-Biopharm (Darmstadt, Germany).

[$^2H_4$]-5-Methyl-H_4folate, [$^2H_4$]-5-formyl-H_4folate, [$^2H_4$]-10-formylfolic acid, [$^2H_4$]-H_4folate, and [$^2H_4$]-folic acid were synthesized as reported recently [28].

All glassware for the dilution series was washed with 1% Tween 80 solution, then treated with hydrochloric acid (0.1 mol l^{-1}), and with aqueous sodium hydroxide (0.1 mol l^{-1}), rinsed hot and cold three times and finally washed with distilled water.

The microorganisms used were as follows:

- **vitamin B_{12}** *Lactobacillus delbrueckii* subsp. *lactis* (*L. leichmanii*) ATCC 7830 for cyanocobalamin and hydroxycobalamin,
- **folates** *Lactobacillus rhamnosus* ATCC 7469 for folic acid, monoglutamyl folates, and their polyglutamic forms,
- **pantothenic acid** *Lactobacillus plantarum* ATTC 8014 for pantothenic acid and panthenol,
- **vitamin B_2** *Lactobacillus rhamnosus* ATCC 7469 for riboflavin and riboflavin phosphate.

Briefly, the microorganisms were preserved in 96-well microtiter plates by freeze-drying in nutrient solution and addition of carbohydrates to the freezing and storage medium.

3.2.1
Extraction of Vitamins

Vitamins were extracted according to the test description of the commercial Vita-Fast® assays (R-Biopharm).

3.2.1.1 Extraction of Total Folate Content (Native Folates and Added Folic Acid)
To extract the bound, native folate and also to determine it in non-fortified samples, the samples were treated with different enzymes.

Exactly 1 g (ml) of homogenized sample and 10 mg of chicken pancreatin were weighed into a 50 ml centrifuge vial, about 30 ml of phosphate buffer (0.05 mol l⁻¹; 0.1% ascorbate; pH 7.2) were added and the suspension was shaken. Subsequently, the vial was filled to exactly 40 ml with phosphate buffer and incubated for 2 h at 37 °C in the dark (shaken occasionally). Extracts of cereal and yeast products and liver were incubated for at least 12 h or overnight. Thereafter, the extract was heated for 30 min at 95 °C in a water-bath. During extraction, the tightly closed vials were shaken well at least five times. After being chilled quickly to below 30 °C, the vials were centrifuged and, depending on the concentration range, the clear supernatant was further diluted in 1.5 ml (or 2.0 ml) sterile reaction vials with sterile water.

3.2.1.2 Extraction of Total Vitamin B_{12} Content (Native and Added Vitamin B_{12})

To extract the bound, native vitamin B_{12} and also to determine it in non-fortified samples, the sample has to be extracted with NaCN/KCN and treated with the enzyme takadiastase.

Exactly 1 g (ml) of homogenized sample was weighed into a 50 ml centrifuge vial and 20 ml of redistilled or deionized water and 250 µl of NaCN solution (1%, freshly prepared) were added. The suspension was shaken and its pH adjusted to 4.5 with HCl.

Alternatively, instead of water an acetate buffer was used for the extraction (no pH adjustment is necessary); to 1 g of sample, 20 ml of acetate buffer (pH 4.5) were added and the mixture was shaken; thereafter, 250 µl of NaCN solution (1%, freshly prepared) were added.

Subsequently, 300 mg of takadiastase were added and the mixture was shaken well and incubated for 1 h at 37 °C in the dark (shaken occasionally). Then, the vial was filled to exactly 40 ml with redistilled or deionized water. Thereafter, the extract was heated for 30 min at 95 °C in a water-bath. During extraction, the tightly closed vials were shaken well at least five times. After being chilled quickly to below 30 °C, the vials were centrifuged and, depending on the concentration range, the clear supernatant was further diluted in 1.5 ml (or 2.0 ml) sterile reaction vials with sterile water.

3.2.1.3 Extraction of Total Pantothenic Acid or Riboflavin Content (Native and Added Vitamins) in Foods

To extract the bound, native vitamins or to determine them in non-fortified samples, the samples were treated with different enzymes.

Exactly 1 g (ml) of homogenized sample was weighed into a 50 ml centrifuge vial and 20 ml of redistilled or deionized water were added. The suspension was shaken and its pH adjusted to 4.5 with HCl.

Alternatively, instead of water a citrate buffer can be used for the extraction (no pH adjustment is necessary); to 1 g (ml) of sample, 20 ml of citrate buffer (pH 4.5) were added and the vial was shaken.

Subsequently, 300 mg of takadiastase and, for extraction of total riboflavin, 10 mg of acid phosphatase from potato were added and the suspension was shaken well and incubated for 1 h at 37 °C in the dark (shake occasionally). Then, the vial was filled to exactly 40 ml with redistilled or deionized water. Thereafter, the extract was heated for 30 min at 95 °C in a water-bath. During extraction, the tightly closed vials were shaken well at least five times. After being chilled quickly to below 30 °C, the vials were centrifuged and, depending on the concentration range, the clear supernatant was further diluted in 1.5 ml (or 2.0 ml) sterile reaction vials with sterile water.

3.2.1.4 Extraction of Total Riboflavin Content (Native and Added Vitamin B$_2$ in Yeasts and Yeast Products)

Exactly 1 g (ml) of homogenized sample was weighed into a 50 ml centrifuge vial and about 10 ml of H_2SO_4 (0.1 mol l^{-1}) were added. After being shaken well, the vial was autoclaved for 30 min at 121 °C (with the screw-cap not totally closed). Thereafter, the samples were chilled quickly to below 30 °C, 3 ml of sodium acetate solution (2.5 mol l^{-1}) were added and and the mixture was shaken. Subsequently, 300 mg of takadiastase and, for extraction of total riboflavin, 10 mg of acid phosphatase from potato were added and the suspension was shaken well and incubated overnight for 12–16 h at 37 °C in the dark. Then, the vial was filled to exactly 40 ml with redistilled or deionized water. Thereafter, the extract was heated for 30 min at 95 °C in a water-bath. During extraction, the tightly closed vials were shaken well at least five times. After being chilled quickly to below 30 °C, the vials were centrifuged and, depending on the concentration range, the clear supernatant was further diluted in 1.5 ml (or 2.0 ml) sterile reaction vials with sterile water.

3.2.1.5 Incubation of the Microbiological Assay

The assay medium and the diluted sample extract or standard were pipetted into the wells of a microtiter plate, which was coated with the microorganisms detailed earlier.

The assays were run according to the test description of the commercial VitaFast assays (R-Biopharm).

3.2.1.6 Calculation of the Limits of Detection and Quantitation for the Microbiological Assay

The limit of detection (LOD) was calculated per definition as the mean absorbance of the blank from 10 different lots plus three times the standard deviation (SD) between lots. A calibration curve was generated from the mean absorbance of the standards from these lots using RIDA®SOFT Win (R-Biopharm) four-parameter software. The LOD was extrapolated from this curve.

The limit of quantitation (LOQ) was assigned the value of standard 1 of the calibration curves. The diluted sample extract must fall on the curve between standard 1 and the highest standard 5. Results outside this range are not valid, which renders standard 1 the LOQ.

3.2.1.7 Stable Isotope Dilution Assay for Folates

Folates were analyzed according to Gutzeit *et al.* [29]. Briefly, [$^2H_4$]-5-methyl-H$_4$folate, [$^2H_4$]-5-formyl-H$_4$folate, [$^2H_4$]-10-formylfolic acid, [$^2H_4$]-H$_4$folate, and [$^2H_4$]-folic acid were used as the internal standards and added to the samples in defined amounts during extraction. Deconjugation was performed using chicken pancreatin and rat plasma. The extracts were purified by solid-phase extraction on a strong anion exchanger and subjected to LC–tandem mass spectrometry (MS/MS).

3.3
Results

For easy to handle tests, suitable inoculum cultures had to be stabilized in microtiter plates. For decades, microorganisms have been kept in glycerin and other freezing solutions or in liquid nitrogen. An alternative to freeze storage is freeze-drying, which is also mild and allows the microorganisms to be stored at refrigerator or room temperature. To keep the microorganisms alive, trehalose or sucrose was added to the freezing and storage medium.

This procedure mimics their natural environment, as microorganisms survive in that they protect themselves from complete drying with a glass-like sugar layer. The carbohydrates form with the proteins of the cell hydrogen bond connections so that the cells are preserved for months in the natural dry condition.

The storage buffer has to provide these cells with alkali and alkaline earth metal ions such as magnesium, calcium, potassium, and sodium. However, these metal ions have the disadvantage that, promoted by the stress of the freeze-drying and dry storage, some microorganisms adapt to a deficient environment, mutate, or otherwise realign, so that on the addition of water or sample there are microorganisms present which can grow without vitamin. Therefore, it is advantageous to shock freeze and freeze-dry the microorganism in assay medium.

In contrast to frozen storage and in accordance with carbohydrate dry storage, the cells are immediately active again upon addition of liquid or water. The usual freezing shock to the cells does not arise.

Hence the microorganism for the vitamin determination no longer needs to be newly grown, because it is present in the wells of the microtiter plate in defined number and constitution. Only exact quantities of vitamin-deficient nutrient medium and sample or vitamin standard concentration need to be added to the wells. The microorganisms in the wells have a storage life of at least 12 months.

For microbiological determination, a plurality of dilution series has to be applied so that at end of the incubation time a growth or metabolic value falls in the measurement range of the standard concentration series run in parallel. For each test, a calibration curve valid only for this attempt has to be prepared. Further, for reasons of safety and precision, each concentration stage of the standard series and the sample series has to be applied at least in triplicate. The vitamin content of the sample is then determined by comparison with the known vitamin content of the parallel standard series. Generally, valid precision indications are not possible; the coefficient of variation should, however, lie at around 10% or below.

For test preparation, it is only necessary to add assay medium, incremental concentrations of the target vitamin, and sample extracts to the wells coated with the specific microorganisms.

3.3.1
Principle of Microbiological Assay in Microtiter Plate Format

The vitamin is extracted from the sample and the extract is diluted. The assay–medium and the diluted extract and are pipetted into the wells of a microtiter plate, which are coated with the respective microorganism. Following the addition of the vitamin as a standard or as a compound of the sample, the bacteria grow until the vitamin is consumed. The incubation is performed in the dark at 37 °C (98.6 °F). The intensity of metabolism or growth in relation to the extracted vitamin is measured as turbidity and compared with a calibration curve. The measurement is carried out using a microtiter plate reader at 610–630 nm (alternatively at 540–550 nm).

3.3.2
Description of the Microorganisms Used in Microbiological Assays

L. rhamnosus, L. delbrueckii, and L. plantarum belong to the lactic acid bacteria, and have an absolute growth requirement for the vitamins folic acid (L. rhamnosus), riboflavin (L. rhamnosus), vitamin B_{12} (L. delbrueckii), and pantothenic acid (L. plantarum). As a group, the lactic acid bacteria are good candidates for vitamin analysis owing to their genetic stability (i.e., resistance to undergoing mutation), simple metabolic requirements, ease of maintenance, and general safety as a non-pathogenic organism [30].

3.3.3
Assays for Single Vitamins

3.3.3.1 Folic Acid

3.3.3.1.1 Response of the Microorganism to Different Folate Vitamers
The microorganisms applied for folate MA range from Streptococcus faecalis, through Enterococcus hirae to Pediococcus acidilactici, along with the most popular L. rhamnosus. However, these organisms show different responses to food folates, which is mainly due to the different utilization of single folate vitamers. Of these, L. rhamnosus is reported to respond to most vitamers. However, application of higher polyglutamates results in less growth than with the monoglutamates. Several researchers [31–33] quantified the response of L. rhamnosus to various pteroylpolyglutamates and found for pteroyldiglutamate almost 100% response compared with folic acid, whereas the response decreased to 60% for pteroyltri- and -tetraglutamates and further to 20%, 4% and 2% for pteroylpenta-, -hexa-, and -heptaglutamate, respectively. Therefore, the use of conjugase treatment for poly-glutamate conversion to diglutamate or monoglutamate is essential in folate analysis. As chromatographic assays use monoglutamates as the reference compounds, deconjugation down to the monoglutamates is required for these methods. This is most efficiently achieved by a combination of rat plasma and chicken pancreas

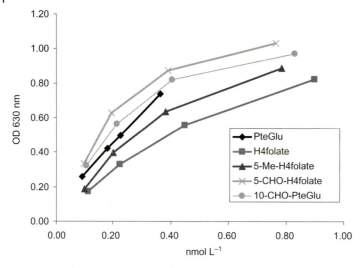

Figure 3.1 Response curves of *Lactobacillus rhamnosus* to different folate monoglutamates.

deconjugase [34]. The former degrades the polyglutamates to monoglutamates and is supported by the pancreas enzyme, which converts polyglutamates most efficiently into the diglutamates. However, as *L. rhamnosus* also shows responses of almost 100% to diglutamates, for MA the single use of rat plasma or pancreatin is sufficient.

As the range of monoglutamates may vary significantly between the foods to be analyzed, we studied the response of folate MA to different monoglutamates applied as pure reference compounds. The results for the most important folate vitamers in foods are shown in Figure 3.1. The highest response was obtained for 5-CHO-H$_4$folate followed by those for 10-CHO-folate, folic acid, 5-CH$_3$-H$_4$folate and H$_4$folate. The response of the last folate was significantly lower than that of the other vitamers and differences in the responses of up to 30% were found. As the response curve for folic acid was intermediate between all curves, its use as a reference compound for all folates can be recommended.

For further folates with less abundance in foods, such as 5,10-methenyl-H$_4$folate, 10-CHO-H$_2$folate, and 5-CH$_3$-H$_2$folate, responses are shown in Figure 3.2. Whereas the last two vitamers showed similar curves to those of the folates shown in Figure 3.1, 5,10-methenyl-H$_4$folate revealed the lowest response of all folates.

These results are in good agreement with the study of Kehlenbach [24], who developed a, LC–MA combination and calibrated the MA with different folate monoglutamates for quantitation of the single folates in complex extracts. Similar responses were found for folic acid, 5,10 methenyl-H$_4$folate, 5-CHO-H$_4$folate, and 5-CH$_3$-H$_4$folate, whereas smaller responses were found for 5,10-methenyl-H$_4$folate, H$_2$folate, and H$_4$folate.

From these results, it is obvious that MA of samples containing a variety of different folate vitamers may result in a growth response, which has to be evaluated

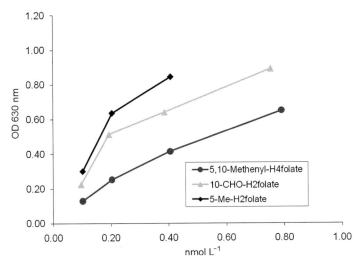

Figure 3.2 Response curves of *Lactobacillus rhamnosus* to different folate monoglutamates.

carefully when folic acid is used as the single calibrant. However, to a first approximation, the most important food folates are folic acid, 5-CHO-H$_4$folate, and 5-CH$_3$-H$_4$folate, the responses of all of which are rather similar. Therefore, the results of a folic acid calibration can be considered as valid for common vitamer distributions. For some foods containing significant amounts of H$_4$folate, such as camembert cheese and spinach, the additional use of H$_4$folate as a calibrant and subsequent calculation of the possible range of the folate content may be reasonable.

Analysis of Foods by Microbiological Assay Using Different Folate Calibrants In order to prove the dependence of the result of MA on the calibrant, we analyzed three foods high in folates for their total folate content and compared it with the results of a stable isotope dilution assay (SIDA) based on LC–MS/MS. The latter included the use of deuterated folyl monoglutamates as internal standards and revealed the folate distribution and also the total folate content calculated as the sum of all vitamers. For the two foods examined, that is, wheat germ and broccoli, SIDA resulted in a total folate content of 336 g and 394 µg per 100 g dry mass, respectively. As the folate distribution revealed 5-CHO-H$_4$folate to comprise 75% of the wheat germ folates and 5-CH$_3$-H$_4$folate to comprise 85% of the broccoli folates, these vitamers appear to be the best calibrants for the respective MAs. The results from these calibrants are listed in Table 3.1.

For broccoli, SIDA gave a total folate content that was 10% higher than that of the respective MA based on 5-CH$_3$-H$_4$folate calibration, which might be attributed to the superior compensation for losses by using stable isotope-labeled internal standards. However, this small difference is not statistically significant compared with the standard error of repeatability in folate assays, which often exceeds 10%. In the case of wheat germ, the SIDA value differed from the MA data based on

Table 3.1 Total folate content calculated as PteGlu from different calibrations of the MA and from SIDA.

Sample	Time of deconjugation	MA with calibrant (µg per 100 g dry mass)		SIDA (µg per 100 g dry mass)
		5-CH$_3$-H$_4$folate	5-CHO-H$_4$folate	
Wheat germ	2 h	449	357	–
	Overnight	764	673	336
Broccoli	2 h	337	245	–
	Overnight	357	266	394

5-CHO-H$_4$folate calibration and deconjugation over 2 h by less than 10%. When deconjugation was performed overnight, the MA value exceeded that from SIDA by 100%. A possible cause might have been liberated compounds, which stimulate the growth of the microorganisms but cannot be assigned to known folate structures. Therefore, the duration of deconjugation prior to MA should be restricted to 2 h.

3.3.3.1.2 Basis of the Extraction Procedure

Analysis of the added folic acid content in a product requires a hot water extraction from a solid matrix. For liquids, a sample can be taken directly after thermal or filter sterilization. Analyzing the native folate content of solid and liquid samples requires an enzymatic extraction in order to liberate the analyte from the matrix and deconjugate higher folyl polyglutamates. The enzyme chicken pancreatin is recommended for most matrices and incubation for 2 h with this enzyme is sufficient to render all folates detectable by *L. rhamnosus*.

3.3.3.1.3 Validation of the Assay

Calibration   As outlined earlier, folate-dependent growth of *L. rhamnosus* is reflected by the turbidity in wells, which can easily be quantified by measuring the absorbance of visible light at 630 nm. For calibration of the assay, folic acid at concentrations ranging between 0.16 and 1.28 µg per 100 g (ml) was applied and the microorganisms showed the response detailed in Figure 3.3. An almost linear curve was obtained and allowed to a first approximation calculation of the total folate in samples by simple linear regression. For more exact data processing, an algorithm for calculations of nonlinear relationships was employed (RIDA SOFT Win four-parameter software). From Figure 3.3, it is also obvious that absorption of the standards does not show an SD higher than 5% within one lot of lyophilized microtiter wells.

Stability   The stability of the test was routinely checked after defined storage intervals. Test kits were stored in a cool room at temperatures of 2–8 °C according

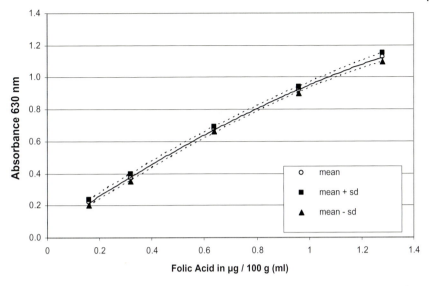

Figure 3.3 Typical calibration curve for the folic acid MA: mean of five stability tests of one lot; additional mean ± SD and mean – SD curves are shown.

to the recommended storage condition. The absorbance values of the standards were always found to remain at constant levels during the observation time over a period of more than 1 year.

3.3.3.1.4 Accuracy of the Folate MA

Recovery of Folic Acid from Spiked Samples First, the accuracy of the test kit was examined in spike recovery studies. One commercial breakfast cereal and one FAPAS® (Food Analysis Performance Assessment Scheme) cereal were spiked with defined amounts of folic acid, which was added to samples in volumes that delivered approximately 100% of the endogenous folate content, thus doubling the final content. The results are presented in Table 3.2.

The expected concentration of the FAPAS reference material is given as a target concentration within an acceptable range. The other products have expected concentrations corresponding to their label declarations for folic acid. The spiking was carried out prior to the extraction, and all spiked samples were submitted to the same extraction protocol as the unspiked samples. The recovery rates, ranging between 100.6 and 109.3%, show that the entire amount of folic acid was recovered from all samples with a high degree of accuracy.

Precision in Various Food Matrices Results from various commercial samples were compared with the declared folic acid content on the package labels. However, the "true" concentration of the respective food products is not known. Very often the concentration in food samples is higher than that indicated on the label,

Table 3.2 Recovery of spiked folic acid in solid matrices.

Sample	Expected concentration (μg per 100 g)	Unspiked result (μg per 100 g)	Spiked amount (μg per 100 g)	Spiked result (μg per 100 g)	Recovery (%)
FAPAS Test 2141 Breakfast Cereal	458 (342–575)	509	500	1012	100.6
Cornflakes	167	131	150	294	108.7
Choco Crispies	167	198	150	362	109.3
Day Vita	250	247	250	507	104.0

because the manufacturers take into account degradation of the vitamin during storage [35].

All samples in Table 3.2 were treated according to the enzymatic extraction for total folates. Three dilutions were prepared from the extract of each sample matrix. The mean result of the different dilutions had a coefficient of variation of less than 10% (Table 3.3). The hospital milk drink and the baby formula were also analyzed by an external reference laboratory (classic microbiological assay for B vitamins, AOAC Official Method 960.46) [17], and the results of 80 μg per 100 ml and 90 μg per 100 g, respectively, correlate very well with the microtiter MA results.

Analysis of Reference Materials Reference materials from the American Association of Cereal Chemists (AACC), National Institute of Standards and Technology (NIST), FAPAS, and Bureau Communautaire de Référence (BCR) were extracted enzymatically according to the procedure given in Section 3.2.1.1 for the analysis of total folates, and diluted according to the given target concentration such that the concentration of the diluted extract fell within the range of the calibration curve. Results from the reference materials were within 11% of the target, and fell well within the acceptable range for all tests (Table 3.4).

Limits of Detection and Quantitation The LOD and LOQ were found to be 0.018 μg per 100 g (ml) and 0.16 μg per 100 g (ml), respectively.

3.3.3.2 Microbiological Assay for Vitamin B_{12}

3.3.3.2.1 Basis of the Extraction Procedure

Analysis of the added vitamin B_{12} content in a product requires a hot water extraction from a solid matrix. For liquids, a sample can be taken directly after thermal or filter sterilization. Analyzing the native vitamin B_{12} content of solid and liquid samples requires an enzymatic extraction with takadiastase, an amylase, in order to liberate the analyte from the matrix. NaCN/KCN is added to the extraction solvent to convert all cobalamin forms into cyanocobalamin, which is the most stable form of vitamin B_{12}.

Table 3.3 Intra-assay variance of the folic acid MA in food samples (triplicate analyses per sample dilution).

Sample	Concentration indicated on label [μg per 100 g (ml)]	Dilution factor	Mean result [μg per 100 g (ml)]	Mean result of dilutions [μg per 100 g (ml)]	Coefficient of variation (%)
Cereal flakes with fruits	261	2000	254.6	275.7	7.6
		1500	296.6		
		1000	276.0		
Diet potato soup	100	600	117.4	123.2	6.5
		400	132.4		
		200	119.9		
Tomato soup	200	1200	108.0	111.8	9.6
		800	103,4		
		400	123.9		
Infant formula	15	100	29.1	28.2	4.1
		60	28.6		
		40	26.9		
Semolina with milk	62	100	101	104	6.9
		150	96		
		200	104		
		300	113		
Hospital milk drink	60	100	81	80	1.8
		200	79		
Baby formula	140	200	91	90	1.6
		400	89		

3.3.3.2.2 Validation of the Assay

Calibration   The calibration curve for the test covers a range between 0.03–0.18 μg per 100 g of sample. Results for samples read from the calibration curve have to be multiplied by the dilution factor corresponding to the sample preparation in order to obtain the vitamin B_{12} content of the samples analyzed. A typical curve is shown in Figure 3.4.

Table 3.4 Accuracy of the folic acid MA as determined by using NIST, AACC, BCR, and FAPAS reference materials (from different lots).

Reference material[a]	Target concentration (µg per 100 g)	Dilution factor	Concentration measured (µg per 100 g)	Mean result of dilutions (µg per 100 g)	Coefficient of variation (%)	Proportion of target recovered (%)
NIST 1846 Infant Formula	129 (101–157)	100	123 ($n = 2$)	133	6.4	103.1
		200	133 ($n = 20$)			
		300	134 ($n = 8$)			
		400	131 ($n = 27$)			
		500	137 ($n = 3$)			
		800	138 ($n = 3$)			
AACC VMA 399 Cereal	1395 (1160–1620)	2000	1348 ($n = 30$)	1363	5.0	97.7
		4000	1378 ($n = 30$)			
BCR CRM 121 Wholemeal Flour	50 (43–57)	40	46 ($n = 6$)	48.5	7.7	97.0
		280	51 ($n = 3$)			
BCR CRM 421 Milk Powder	142 (128–156)	200	131 ($n = 2$)	136	5.4	96.0
		300	132 ($n = 3$)			
		400	133 ($n = 2$)			
		500	142 ($n = 3$)			
		1500	143 ($n = 6$)			
FAPAS Test 2141 Breakfast Cereal	458 (342–575)	400	497 ($n = 2$)	509	4.2	111.0
		800	495 ($n = 6$)			
		1600	507 ($n = 6$)			
		3200	530 ($n = 6$)			
		6400	518 ($n = 2$)			

a)   CRM, certified reference material.

3.3.3.2.3   Accuracy of the Assay

Recovery of Vitamin B$_{12}$ from Spiked Samples   The accuracy of the test kit was examined in spike recovery studies on two different study designs. In the first trial, solid and liquid matrices were spiked with a vitamin B$_{12}$ solution prepared from solid vitamin B$_{12}$ purchased from a commercial source. The spike solution was added to samples of milk, infant formula, cheese, and oats (Table 3.5) in volumes

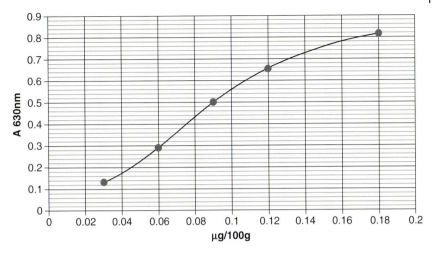

Figure 3.4 Typical calibration curve for the vitamin B$_{12}$ microtiter MA.

Table 3.5 Recovery of spiked vitamin B$_{12}$ in solid and liquid matrices.

Matrix type	Sample	Recovery (%)
Liquid	Milk	91.2
	Infant formula	119.2
Solid	Cheese	100.2
	Oats	106.0
Mean Recovery		104.2

that delivered approximately 100% of the original vitamin B$_{12}$ content, thus doubling the final content.

The concentration of the spike solution was verified in the same assay. All samples in this trial were tested in duplicate in two trials, and the percentage recovery reported in Table 3.5 is the average of all trials. The mean recovery of 104.2% of the total spiked content shows a high degree of accuracy for the microtiter MA.

The second spiking trial similarly showed excellent recovery rates (Table 3.6). The products' expected concentrations correspond to their label declarations for vitamin B$_{12}$. The spiking was carried out prior to the extraction, and all spiked samples were submitted to the same extraction protocol as the unspiked samples. The recovery rates show that the entire amount of vitamin B$_{12}$ was recovered from all samples with a high degree of precision.

Method Comparisons Samples were analyzed in-house with the microtiter MA and HPLC methods in parallel. The same samples were sent to external laboratories for analysis by HPLC or the classic microbiological turbidimetric method. Excellent agreement between results from these methods were achieved in all

Table 3.6 Recovery of spiked vitamin B_{12} in solid matrices.

Sample	Expected concentration (µg per 100 g)	Unspiked result (µg per 100 g)	Spiked amount (µg per 100 g)	Spiked result (µg per 100 g)	Recovery (%)
Cornflakes	0.85	0.65	0.85	1.48	97.7
Choco Krispies	0.85	0.76	0.85	1.63	102.4
Day Vita	0.65	0.76	0.85	1.66	105.9

Table 3.7 Comparison of results for vitamin B_{12} from various analytical methods performed internally and externally.

Sample	Concentration indicated on label [µg per 100 g (ml)]	VitaFast: mean result of dilutions [µg per 100 g (ml)]	Other methods [µg per 100 g (ml)]
Drink food without lactose	0.70	0.74 ($n = 4$)	Internal HPLC: 0.71
Drink food extra	0.47	0.29 ($n = 6$)	Internal HPLC: 0.29
			External HPLC: 0.34
Hospital milk drink standard	0.21	0.21 ($n = 4$)	Internal HPLC: 0.22
			External HPLC: 0.20
Hospital milk drink extra	0.21	0.32 ($n = 6$)	Internal HPLC: 0.26
			External HPLC: 0.31
Cereals	1.0	1.29 ($n = 6$)	Internal HPLC: 1.33
Fruit mix	1.2	1.61 ($n = 6$)	Internal HPLC: 1.56
Juice milk drink	0.20	0.20 ($n = 4$)	External microtiter MA: 0.17
Banana milk pudding	0.65	0.82 ($n = 6$)	Classic microbiological assay: AOAC 960.46 [17]
			0.90 (external lab. 1)
			0.76 (external lab. 2)
			0.89 (external lab. 3)

cases. The differences (i) between internal analyses were below 20%, (ii) between MA and external HPLC below 18%, and (iii) between MA and external classical MA below 10% (Table 3.7).

Analysis of Reference Materials   Reference materials from AACC, NIST, FAPAS, and BCR were extracted enzymatically according to the procedure given in Section

Table 3.8 Accuracy of the vitamin B_{12} assay determined with AACC, NIST, BCR, and FAPAS reference materials.

Reference material	Target concentration (µg per 100 g)	Dilution factor	Concentration measured (µg per 100 g)	Mean result of dilutions (µg per 100 g)	Coefficient of variation (%)	Proportion of target recovered (%)
NIST 1846 Infant Formula	3.9 (3.6–4.2)	20	4.1 ($n = 2$)	4.0	5.2	102.6
		40	4.2 ($n = 3$)			
		100	3.8 ($n = 4$)			
AACC VMA 399 Cereal	21.2 (12.2–25.0)	70	18.1 ($n = 2$)	20.8	6.1	98.1
		140	18.9 ($n = 2$)			
		200	21.5 ($n = 8$)			
		240	21.0 ($n = 10$)			
		280	19.9 ($n = 2$)			
		300	24.8 ($n = 3$)			
		400	20.4 ($n = 4$)			
		480	21.5 ($n = 10$)			
FAPAS T2130 Baby Food	1.46 (0.82–2.11)	10	1.56 ($n = 3$)	1.69	6.8	115.8
		20	1.80 ($n = 3$)			
		30	1.58 ($n = 6$)			
		40	1.77 ($n = 6$)			
		50	1.76 ($n = 3$)			
BCR CRM 421 Milk Powder	3.4 (2.9–3.9)	20	3.4 ($n = 3$)	3.2	7.9	94.0
		40	3.3 ($n = 9$)			
		80	2.9 ($n = 3$)			
FAPAS T2143 Powdered Baby Food	1.73 (0.97–2.50)	16	1.68 ($n = 3$)	1.76	4.9	101.7

3.2.1.2 for the analysis of total vitamin B_{12}. Sample extracts were diluted according to the given target concentration such that the concentration of the diluted extract fell within the range of the calibration curve. Results from the reference materials were within 16% of the target, and fell well within the acceptable range for all tests (Table 3.8). The FAPAS T2143 Powdered Baby Food was also analyzed by an external reference laboratory using the VitaFast Vitamin B_{12} kit, and the result of 1.69 µg per 100 g correlates very well.

Limits of Detection and Quantitation   The LOD and LOQ were found to be 0.021 µg per 100 g (ml) and 0.03 µg per 100 g (ml), respectively.

3.3.3.3 Microbiological Assay for Pantothenic Acid

3.3.3.3.1 Basis of the Extraction Procedure
Analysis of the added pantothenic acid content in a product requires a hot water extraction from a solid matrix. For liquids, a sample can be taken directly after thermal or filter sterilization. Analyzing the native pantothenic acid content of solid and liquid samples requires an enzymatic extraction in order to liberate the analyte from the matrix. The enzyme takadiastase (an amylase) from the organism *Aspergillus oryzae* is recommended for most matrices.

3.3.3.3.2 Validation of the Assay

Calibration   The calibration curve of the test covers the range 0.04–0.24 mg per 100 g (ml) of sample and reveals typically a slight sigmoidal shape (Figure 3.5). The software RIDA SOFT Win includes the dilution factor in its final calculations, and outputs the result in mg of Ca pantothenate per 100 g of sample. This result can be multiplied by 0.92 to obtain the concentration of the active vitamin, pantothenic acid.

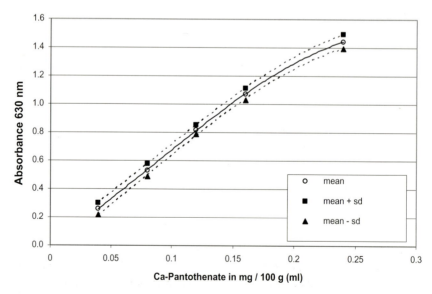

Figure 3.5 Typical calibration curve for the pantothenic acid MA: mean of five stability tests of one lot; additional mean ± SD and mean – SD curves are shown.

Table 3.9 Recovery of spiked Ca-pantothenate in solid and liquid matrices.

Matrix type	Sample	Recovery (%)
Liquid	Milk	90.1
	Infant formula	104.6
Solid	Wheat cereal	92.4
	Oats	93.0
Mean recovery		95.0

3.3.3.3.3 Accuracy of the Assay

Recovery of Pantothenic Acid from Spiked Samples According to our general validation procedure, the accuracy of the MA was first examined using spike recovery
studies. Different solid and liquid matrices from commercial sources were spiked
with a pantothenic acid solution prepared from solid Ca pantothenate. The prepared solution had a calcium pantothenate concentration of $120 \mu g\, ml^{-1}$, which
corresponds to an active pantothenic acid content of $110.4 \mu g\ ml^{-1}$. The spike
solution was added to samples of milk, infant formula, and cereals (Table 3.9)
in volumes that delivered approximately 100% of the original pantothenic acid
content, thus doubling the final content.

The concentration of the spike solution was verified in the same assay. All
samples were tested in duplicate on two occasions, and the percentage recovery
reported in Table 3.9 is the average of all trials. The mean recovery of 95% of the
total spiked content reveals excellent accuracy for the MA.

Accuracy and Precision in Various Food Matrices Results from various samples
were compared with the declared pantothenic acid content on package labels.
However, except for the two reference materials, the "true" concentration of the
respective food products was not known. Very often the concentration in food
samples is higher than that indicated on the label, because the manufacturers tend
to over-fortify their products with the vitamins to compensate for degradation
during storage. In a recent report on pantothenic acid, the fortification often was
found to exceed 30% of the labeled value [36]. The results of the latter study were
confirmed by our analyses as the mean result in all cases except for the cereals is
greater than that indicated on the product label.

Each sample was prepared according to the relevant extraction protocol for that
matrix, as outlined in Section 3.2.1.3: infant formula, cereal, sausage samples,
dextrose, and milk powder were extracted enzymatically, multivitamins were dissolved and diluted by a factor of at least 10, and liquid samples were simply sterilized without further treatment. Three dilutions were prepared from the extract of
each sample matrix. The mean result for the different dilutions had a coefficient
of variation of less than 6% (Table 3.10).

Table 3.10 Intra-assay variance of the pantothenic acid MA for food samples.

Sample[a]	Concentration indicated on label (mg per 100 g)	Mean result of dilutions (mg per 100 g) ($n = 3$)	Coefficient of variation (%)
Infant formula	1.76	2.1	4.1
Multivitamin sweet (bonbon)	26.6	33.6	2.0
Cereals	5.55	5.5	1.5
Multivitamin pills	400	465	1.4
Ham sausage 1	6.95	7.82	1.5
Ham sausage 2	6.95	6.89	2.2
Dextrose powder	6	9.2	4.6
RM: Vit001 internal			
Milk powder	3.8	4.4	5.4
RM: Vit002 internal			
Fruit juice	0.9	1.07	2.5
Energy drink	2	1.83	1.0
Multivitamin juice drink	3	3.48	0.8

a) RM, reference material.

Analysis of Reference Materials   Reference materials from AACC and NIST were extracted enzymatically and diluted according to the given target concentration. Results from both reference materials were within 11% of the target, and were well in line with the acceptable range for all tests (Table 3.11).

Limits of Detection and Quantitation   The LOD and LOQ were found to be 0.0035 µg per 100 g (ml) and 0.04 µg per 100 g (ml), respectively.

3.3.3.4   Microbiological Assay for Riboflavin

3.3.3.4.1   Basis of the Extraction Procedure

Analysis of the added riboflavin content in a solid product requires a hot water extraction. For liquids, a sample can be taken directly after thermal or filter steri-lization. Analyzing the native riboflavin content of solid and liquid samples requires an enzymatic extraction in order to liberate the analyte from the matrix. The enzyme takadiastase (an amylase) from *A. oryzae* is recommended for most matrices. In addition to this starch-degrading enzyme, acid phosphatase is also used for most matrices, and functions to cleave phosphate groups from flavin adenine dinucleotide (FAD) and flavin mononucleotide (FMN), thus releasing riboflavin.

3.3.3.4.2   Validation of the Assay

Calibration   The calibration curve for the test covers the range 0.04–0.24 mg per 100 g. The RIDA SOFT Win software includes the dilution factor in its final cal-

Table 3.11 Accuracy of the pantothenic acid MA as determined with AACC and NIST reference materials.

Reference material	Target concentration (range) (mg per 100 g)	Concentration measured (mg per 100 g)		Proportion of target recovered (%)
AACC VMA 399 Cereal	37.35 (31.01–41.9)	40.3 (dilution 300) 42.3 (dilution 500)	41.3	110.5
		39.5 (dilution 500) 38.5 (dilution 300)	39	104.4
		37.4 (dilution 400) 38.2 (dilution 600)	37.8	101.2
		33 (dilution 150) 36.1 (dilution 300) 35.4 (dilution 600)	34.8	93.2
NIST 1846 Infant Formula	4.87 (4.14–5.60)	4.3 (dilution 50) 4.6 (dilution 25)	4.5	92.4
		4.6 (dilution 50) 4.5 (dilution 30)	4.6	94.5
		4.8 (dilution 40) 4.4 (dilution 20)	4.6	94.5
		4.8 (dilution 40)	4.8	98.6

culations, and outputs the result in mg of riboflavin per 100 g of sample. A typical curve is shown in Figure 3.6.

3.3.3.4.3 Accuracy of the Assay

Recovery of Riboflavin from Spiked Samples Spike recovery studies were first performed to examine the accuracy of the MA. Commercial solid and liquid samples were spiked with a riboflavin solution prepared from solid riboflavin. The spike solution was added to samples of milk and two cereals (Table 3.12) in amounts that doubled the original riboflavin content of the samples.

The concentration of the spike solution was verified in the same assay. All samples were tested in duplicate on two occasions, and the percentage recovery reported in Table 3.11 is the average of all trials. The mean recovery of 103.9% of the total spiked content shows a high degree of accuracy for the microtiter MA.

Method Comparisons Samples were analyzed in-house in parallel with the microtiter MA and HPLC with fluorescence detection. The same samples were sent to external laboratories for analysis by HPLC or the classic microbiological turbidimetric method. Excellent agreement between results from these methods were achieved in all cases. The differences (i) between internal analyses were below

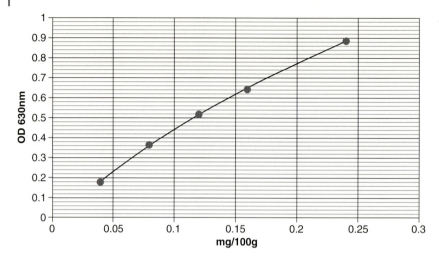

Figure 3.6 Typical calibration curve for the MA for riboflavin. Five standards were measured in triplicate and the average absorbance from each standard across the 10 lots was plotted on the curve.

Table 3.12 Recovery of spiked riboflavin in solid and liquid matrices.

Matrix type	Sample	Recovery (%)
Liquid	Milk	110.0
Solid	Wheat cereal	96.0
	Oats	105.6
		Mean: 103.9

10%, (ii) between MA and external HPLC below 20%, and (iii) between MA and external classical MA below 10% (Table 3.13).

Analysis of Reference Materials   Reference materials from AACC, NIST, FAPAS, and BCR were extracted in the presence of takadiastase and phosphatase for analysis of total riboflavin. The FAPAS liquid supplement was prepared without using takadiastase during extraction. Results from all reference materials revealed biases below 5% compared with the certified reference values and proved the excellent accuracy of the assay (Table 3.14).

Limits of Detection and Quantitation   The LOD and LOQ were found to be 0.0018 mg per 100 g (ml) and 0.04 mg per 100 g (ml), respectively.

Table 3.13 Comparison of the riboflavin MA with results from various analytical methods performed internally and externally.

Sample	Concentration indicated on label [mg per 100 g (ml)]	Internal MA result: mean result of dilutions [mg per 100 g (ml)]	Other methods [mg per 100 g (ml)]
Cereal drink Day Vita	1.00	1.21 ($n = 4$)	Internal HPLC: 1.15
Cornflakes	1.30	1.61 ($n = 4$)	Internal HPLC: 1.54
Baby milk drink 1	1.00	1.09 ($n = 4$)	Internal HPLC: 1.00
			External HPLC: 1.07
			Classic microbiological assay: AOAC 960.46 [17]: 1.12 (external lab. 1)
Baby milk drink 2	0.47	0.70 ($n = 4$)	External HPLC: 0.63
			Classic microbiological assay: AOAC 960.46: 0.72 (external lab. 1)
Hospital milk drink	0.70	0.77 ($n = 4$)	External HPLC: 0.62
			Classic microbiological assay: AOAC 960.46: 0.70 (external lab. 1)
Banana milk pudding	0.70	0.70 ($n = 4$)	External HPLC: 0.65
			Classic microbiological assay: AOAC 960.46
			0.71 (external lab. 1)
			0.66 (external lab. 2)
			0.76 (external lab. 3)

3.4
Conclusion

The presented MA with a miniaturized microtiter design, distributed under the trade name VitaFast by R-Biopharm, proved to be an excellent tool for applications that require the analysis of total vitamin contents. It does not need a classical microbiological laboratory setup and it is robust and easy to handle. Due to its excellent precision and accuracy compared with classical MAs, the assay is suitable for use in routine vitamin analysis. Therefore, the MAs presented as commercial kits have been awarded Performance Tested Method (PTM) status from the Association of Analytical Communities Research Institute after intense evaluation. In

Table 3.14 Accuracy of the riboflavin MA as determined with AACC, NIST, BCR, and FAPAS reference materials.

Reference material	Target concentration (mg per 100 g)	Dilution factor	Concentration measured (mg per 100 g)	Mean result of dilutions (mg per 100 g)	Coefficient of variation (%)	Proportion of target recovered (%)
NIST 1846 Infant formula	1.74 (1.64–1.84)	14	1.81 ($n = 5$)	1.74	4.4	100.2
		20	1.66 ($n = 5$)			
		28	1.76 ($n = 5$)			
AACC VMA 399 Cereal	5.97 (4.89–7.60)	50	5.80 ($n = 10$)	5.79	3.8	97.0
		100	5.78 ($n = 10$)			
FAPAS T2139 Liquid supplement	8.86 (5.32–12.40)	60	8.71 ($n = 4$)	8.43	6.3	95.2
		120	8.14 ($n = 3$)	(internal HPLC 8.34)		
FAPAS T2141 Breakfast cereal	2.07 (1.65–2.49)	16	2.00 ($n = 5$)	2.07	4.7	100.0
		32	2.14 ($n = 5$)	(internal HPLC 2.03)		
FAPAS T2148 Breakfast cereal	1.99 (1.58–2.40)	16	2.07 ($n = 2$)	2.17	5.7	109.0
		32	2.27 ($n = 2$)	(internal HPLC 2.00)		
BCR CRM 421 Milk Powder	1.45 (1.39–1.51)	12	1.46 ($n = 2$)	1.44	1.5	99.3
		15	1.41 ($n = 9$)			
		20	1.43 ($n = 6$)			
		30	1.45 ($n = 3$)			
		32	1.75 ($n = 9$)			
		64	1.85 ($n = 3$)			

addition to its routine use, it is also a promising tool in research, if the identification of unknown vitamin forms requires an easy to use screening assay for bioactivity-guided fractionations. In addition to the assays considered in this study, VitaFast assays for further water-soluble B-vitamins are also available.

References

1 Roche (1976) *Vitamin Compendium*, Hoffmann-La Roche, Basel.

2 US Pharmacopeal Convention (1939) *United States Pharmacopeia*, US Pharmacopeal Convention, Rockville, MD.

3 Birch, T.W. and Harris, L.J. (1934) Bradycardia in the vitamin B1-deficient rat and its use in vitamin B_1 determinations. *Biochemical Journal*, **28**, 602–621.

4 Richards, M.B. and Simpson, B.W. (1934) Curative method of vitamin A assay. *Biochemical Journal*, **28**, 1274–1292.

5 Koehn, C.J. and Elvehjem, C.A. (1937) Further studies on the concentration of the antipellagra factor. *Journal of Biological Chemistry*, **118**, 693–699.

6 Schultzer, P. (1931) Investigations on the determination of vitamin D. Comparison between the preventive and the curative method. *Biochemical Journal*, **25**, 1745–1754.

7 Bacharach, A.L., Allchorne, E., and Glynn, H.E. (1937) The method of estimating vitamin E. I. The influence of vitamin E deficiency on implantation. *Biochemical Journal*, **31**, 2287–2292.

8 Snell, E.E. and Strong, F.M. (1939) A microbiological assay for riboflavin. *Industrial and Engineering Chemistry, Analytical Edition*, **11** (6), 346–350.

9 Sarett, H.P. and Cheldelin, V.H. (1944) Use of *Lactobacillus fermentum* 36 for thiamine assay. *Journal of Biological Chemistry*, **155**, 153–160.

10 Hopkins, R.H. and Pennington, R.J. (1947) Assay of vitamin B_6 complex. *Biochemical Journal*, **41**, 110–114.

11 Ford, J.E. (1958) B_{12}-vitamins and growth of the flagellate *Ochromonas malhamensis*. *Journal of General Microbiology*, **19**, 161–172.

12 Villela, G.G. and Cury, A. (1951) A microbiological assay method for biotin. *Proceedings of the Society for Experimental Biology and Medicine*, **76**, 341–343.

13 Teply, L.J. and Elvehjem, C.A. (1945) Titrimetric determination of "*Lactobacillus casei* factor" and "folic acid". *Journal of Biological Chemistry*, **157**, 303–309.

14 Barton-Wright, E.C. (1944) The microbiological assay of nicotinic acid in cereals and other products. *Biochemical Journal*, **38**, 314–319.

15 Skeggs, H.R. and Wright, L.D. (1944) Use of *Lactobacillus arabinosus* in the microbiological determination of pantothenic acid. *Journal of Biological Chemistry*, **156**, 21–26.

16 Gentili, A., Caretti, F., D'Ascenzo, G., Marchese, S., Perret, D., Di Corcia, D., and Rocca, L.M. (2008) Simultaneous determination of water-soluble vitamins in selected food matrices by liquid chromatography/electrospray ionization tandem mass spectrometry. *Rapid Communications in Mass Spectrometry*, **22**, 2029–2043.

17 AOAC International (2006) AOAC Official Method 960.46: Vitamin Assays: Microbiological Methods, in *Official Methods of Analysis*, 18th edn., AOAC International, Philadelphia, PA.

18 Bundesamt für Verbraucherschutz und Lebensmittelsicherheit (2002) Amtliche Sammlung Untersuchungsverfahren (ASU), L 00.00-87, Behr's Verlag, Hamburg.

19 Bundesamtes für Gesundheit (2010) Vitaminibestimmungen in Lebensmitteln und Kosmetika, *Schweizer Lebensmittelbuch*, BAG, Bern, Chapter 62.

20 Usdin, E., Shockman, G.D., and Toennies, G. (1954) Bacterimetric studies. IX. Tetrazolium bioautography. *Applied Microbiology*, **2**, 29–33.

21 Chapman, S.K., Greene, B.C., and Streiff, R.R. (1978) A study of serum folate by high-performance ion-exchange and ion-pair partition chromatography. *Journal of Chromatography B*, **145**, 302–306.

22 Kelly, P., McPartlin, J., and Scott, J. (1996) A combined high-performance liquid chromatographic-microbiological assay for serum folic acid. *Analytical Biochemistry*, **238**, 179–183.

23 Belz, S. and Nau, H. (1998) Determination of folate patterns in mouse plasma, erythrocytes, and embryos by HPLC coupled with a microbiological assay. *Analytical Biochemistry*, **265**, 157–166.

24 Kehlenbach, U. (2004) Optimierung und Anwendung einer Analysemethode zur selektiven und sensitiven Bestimmung von Folatmustern in klinischen Proben und Lebensmitteln. PhD thesis, Tierärztliche Hochschule Hannover.

25 Horne, D.W. and Patterson, D. (1988) *Lactobacillus casei* microbiological assay of folic acid derivatives in 96-well microtiter plates. *Clinical Chemistry*, **34**, 2357–2359.

26 O'Broin, S. and Kelleher, B. (1992) Microbiological assay on microtitre plates of folate in serum and red cells. *Journal of Clinical Pathology*, **45**, 344–347.

27 Newman, E.M. and Tsai, J.F. (1986) Microbiological analysis of 5-formyltetrahydrofolic acid and other folates using an automatic 96-well plate reader. *Analytical Biochemistry*, **154**, 509–515.

28 Freisleben, A., Schieberle, P., and Rychlik, M. (2002) Syntheses of labelled vitamers of folic acid to be used as internal standards in stable isotope dilution assays. *Journal of Agricultural and Food Chemistry*, **50**, 4760–4768.

29 Gutzeit, D., Moench, S., Jerz, G., Winterhalter, P., and Rychlik, M. (2008) Folate content in sea buckthorn berries and related products (*Hippophae rhamnoides* L. ssp. *rhamnoides*): LC–MS/MS determination of folate vitamer stability influenced by processing and storage assessed by stable isotope dilution assay. *Analytical and Bioanalytical Chemistry*, **391**, 211–219.

30 Ball, G.F.M. (1998) *Bioavailability and Analysis of Vitamins in Foods*, Chapman & Hall, London.

31 Goli, D.M. and Vanderslice, J.T. (1991) Investigation of the conjugase treatment procedure in the microbiological assay of folate. *Food Chemistry*, **43**, 57–64.

32 Keagy, P.M. (1985) Folacin. Microbiological and animal assay, in *Methods of Vitamin Assay* (eds. J. Augustin, P.B. Klein, D.A. Becker, and P.B. Venugopal), John Wiley & Sons, Inc., New York, p. 445–471.

33 Krumdieck, C.L., Tamura, T., and Eto, I. (1983) Synthesis and analysis of the pteroylpolyglutamates. *Vitamins and Hormones*, **40**, 45–104.

34 Rychlik, M., Englert, K., Kapfer, S., and Kirchhoff, E. (2007) Folate contents of legumes determined by optimized enzyme treatment and stable isotope dilution assays. *Journal of Food Composition and Analysis*, **20**, 411–419.

35 Allen, L.H. (2003) B vitamins: proposed fortification levels for complementary foods for young children. *Journal of Nutrition*, **133**, 3000S–3007S.

36 Rychlik, M. (2003) Simultaneous analysis of folic acid and pantothenic acid in foods enriched with vitamins by stable isotope dilution assays. *Analytica Chimica Acta*, **495**, 133–141.

4
Biosensors in Vitamin Analysis of Foods

Anthony O'Kane and Lennart Wahlström

4.1
Introduction

Although surface plasmon resonance (SPR)-based biosensors are most commonly associated with protein interaction analysis in the life sciences and drug discovery, the technology is now also well established in the food industry [1–6].

Final product analysis levels of many water-soluble vitamins in fortified foods is routinely carried out via microbiological analysis (MA) in the majority of food processor quality control laboratories and contract laboratories. As these methods commonly require a 2–3 day turnaround for results, alternative robust and reliable methods for many of these vitamins, in a range of matrices, are in high demand [7–9]. High-performance liquid chromatography (HPLC) methods are useful in the analysis of simple pharmaceutical products and raw ingredient premixes, but the complexities of sample preparation and sensitivity in the range of samples to which vitamins are now added have limited the adoption of this approach across the industry. Traditional immunological methods, such as enzyme-linked immunosorbent assay (ELISA) and radioimmunoassay (RIA), can be used to good effect although they have been slow to gain acceptance due to concerns over poor reproducibility [10–14].

The main advantages of SPR-based detection over alternative analytical techniques such as MA include ease of use, simpler and faster sample purification or preparation, and reduced assay time (from days to minutes in some cases) in the measurement of vitamins and also a whole range of endogenous and xenobiotic food components.

4.2
Technology

Biacore™ Q is a fully automated SPR-based instrument from GE Healthcare (Uppsala, Sweden), designed for routine analysis of food. To make the analysis

Fortification of Foods with Vitamins, First Edition. Edited by Michael Rychlik.
© 2011 Wiley-VCH Verlag GmbH & Co. KGaA. Published 2011 by
Wiley-VCH Verlag GmbH & Co. KGaA.

process as simple and consistent as possible, an extensive range of Qflex® Kits has been developed for use with Biacore Q. These kits are designed for screening or quantifying specific, commercially important food additives and potential contaminants. Analyte levels in the low picomolar range can typically be measured, exploiting the sensitivity generated by the inhibition assay format used in these assays. Samples and reagents are injected automatically and the analysis procedure is monitored in real time. Reproducibility and reliability of results are assured by elimination of the variability inherent in manual procedures and the absence of additional labeling reagents.

Frequently, minimal sample preparation is involved and accurate measurements may be carried out in complex matrices. This is made possible because matrix effects (interference due to components other than the analyte itself in the sample or sample extract) are minimized because the contact time between sample and specific binding protein within the continuous flow system of the instrument is brief. High-affinity interactions are thus favored over low-affinity matrix interactions. The longer contact times between proteins and sample in a typical ELISA (approximately 1 h) significantly increase matrix effects, causing a loss of sensitivity and an increase in the frequency of false positives [5, 6, 9, 15].

Assays performed using Biacore Q are rapid and it is simple to alternate between assays, particularly when using Qflex Kits. Biacore Q includes a surface preparation unit that allows easy immobilization of small molecules on the sensor surface allowing the end-user the ability to develop their own in-house assay systems in addition to the range provided of Qflex Kits. Several Qflex Kits are also delivered with preimmobilized sensor chips, further reducing the manual input necessary for analysis.

4.3
Surface Plasmon Resonance (SPR)

SPR is a phenomenon based on the transfer of energy (photons) from an incident beam of light to a group of electrons (a plasmon) on a metal layer between two transparent media with different refractive indices [16]. The light beam strikes the metal layer at an angle greater than the critical angle, that is, under total internal reflection. The energy transfer occurs at the resonance angle of the plasmons, which depends on the mass of bound material on the metal surface. At this angle, the intensity of the reflected light is a minimum or "dip," which is detected by a photodetector array. Hence the sensor chip consists of a layer of gold on a glass slide, to which a layer of carboxymethyldextran is bound. The sensor responds directly to changes in the refractive index caused by binding events and the buffer environment in the microfluidic system. The change in signal is directly proportional to the molecular weight and concentration of bound species. In concentration analysis of small molecules, the target molecule is bound to the surface of a sensor chip. Once a target molecule has been immobilized, solution from the test material is passed over the sensor chip. Any binding to the target molecule can be detected in real time. Molecules such as vitamins can therefore be easily identified and quantified if a suitable binding partner is available.

The Biacore instrument uses a sensor chip as the interface between the SPR detection system and the fluidic system which delivers the reagents. The reflection of a specific wavelength of light off the gold layer in the chip produces an evanescent wave which penetrates the chip surface and fluid environment. At the resonance angle of the incident light with this wavelength, a change in the reflected light is produced, which is detected by the instrument. When molecules in the test solution bind to a target molecule at the sensor chip surface, the refractive index increases; when they dissociate, the refractive index decreases. These changes form the basis of a Biacore sensorgram, a continuous, real-time monitoring of the association and dissociation of interacting molecules. The sensorgram permits the determination in real time of the specificity of binding, binding kinetics, and affinity and, most importantly, the concentration of biologically active molecules, such as vitamins, in a sample (see Figure 4.1).

SPR technology forms the basis of a rapid and reliable automated biosensor system, Biacore Q. This meets the specific needs of the food industry for analyzing additives, residues, and contaminants in a variety of foodstuffs, without the need for radiolabels or fluorescent tags, but with comparable sensitivity [11–13, 17–21]. To achieve this, the Biacore Q system combines SPR detection with two other key technologies, surface chemistry (sensor chips) and microfluidics.

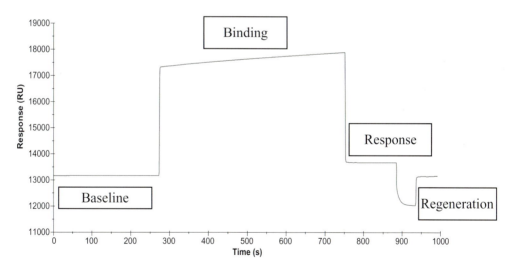

Figure 4.1 The sensorgram provides real-time information about an entire interaction, with binding responses measured in resonance units (RU). Binding responses at specific times during the interaction can also be selected as report points. Depending on the interactions under investigation, any remaining bound sample molecules may be removed in a regeneration step that prepares the surface for the next sample injection.

4.4
Biosensor Assay

Assays performed on Biacore Q using Qflex Kits are based on inhibition or surface competition, as these methods alleviate the problem of accurately interpreting the small changes in SPR response elicited by interactions involving low molecular weight molecules in solution. Typically, an active derivative of the sample under analysis is immobilized covalently on the sensor surface of a chip. Then a fixed concentration of specific binding protein is mixed with the sample and injected over the sensor surface. Free binding protein that has not bound to analyte can bind to the surface and in this way give rise to a response. The SPR response is inversely proportional to the concentration of the sample, which may be estimated by interpolation of the response on a calibration curve. The binding to the surface takes place under conditions that are limited by mass transport, which means that the degree of response is directly proportional to the time of contact of the sample on the surface, that is, the injection volume. The response is measured as the difference in absolute response obtained immediately before and immediately after the injection of the sample [17].

Analysis is fully automated and the sensor surface is regenerated after each sample injection, ready for the next sample to be tested. The chip surfaces are stable to many hundreds of measurement and regeneration cycles.

The principle is illustrated in Figure 4.2.

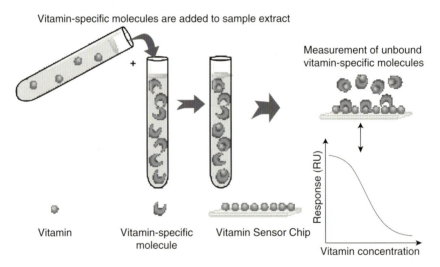

Figure 4.2 An excess of specific antibodies is mixed by an automated procedure with sample or calibration solution containing the free ligand, and the antibody and ligand form a complex. When injected into the detection unit, non-complexed antibodies are measured by the biosensor system when they bind to the ligand immobilized on the sensor chip. The biosensor response will be inversely proportional to the ligand concentration.

4.5
Water-Soluble Vitamin Analysis by Inhibition Protein Binding Assay on Biacore Q

4.5.1
Biotin

Biotin has traditionally been measured in foods using a microbiological assay which relies on the growth of a biotin sensitive strain of *Lactobacillus*. With some refinement, the method has remained essentially unchanged for over 70 years, requiring extensive sample-specific preparation, careful maintenance of cultures, and a 2–3 day turnaround. Biotin is also difficult to analyze via HPLC due to its poor absorbance spectra, making common methods too insensitive for routine users [2]. The biotin assay using Biacore Q involves a simple core sample preparation method, of dilution in water, optional autoclaving to release protein-bound biotin forms, and filtration. Some variations have also been adopted for difficult samples, including the use of amylase to break up starch. However, the method demonstrates a cross-reactivity with lumichrome of >40%. This is a degradation product of riboflavin when exposed to ultraviolet (UV) light and has the potential to interfere with the accuracy of the assay in products co-fortified with riboflavin. A validated protocol involving the addition of riboflavin-binding protein to the samples to sequester lumichrome is provided by the manufacturer [5, 6, 15]. Regeneration and conditioning solutions are also provided in the kit.

The biotin assay can be applied to a range of sample types, including infant formulas, cereals, vitamin supplement tablets, and premixes, with only minor changes in sample preparation. The Performance Tested Method (PTM) study by the Association of Analytical Communities (AOAC) Research Institute (RI) [22] generated a repeatability value of as low as 1.7% in independent testing (eight replicates of single sample); the results of the evaluation are shown in Table 4.1 in more detail [23].

Table 4.1 Summary of results from independent evaluation of Qflex Kit Biotin as part of the AOAC RI PTM program.

Detection limit	$0.5\,\mathrm{ng\,ml^{-1}}$
Quantification limit	$2.0\,\mathrm{ng\,ml^{-1}}$
Accuracy, compared with MA	96%
Precision within one run	4.5%
Precision between runs	5.9%
Repeatability	2.3–15%
Reproducibility	5.9–22.7%
Recovery	72–95%

4.5.2
Folic Acid

Folic acid, in common with most of the B vitamins, is also normally analyzed using MA. However, recent suggestions from the US Food and Drug Administration (FDA) and other bodies have questioned the suitability of these methods in terms of their accurate response to all endogenous forms of folate. In particular, the natural variable polyglutamation of folates has been singled out as a source of potential error [7, 18].

The folic acid assay is based on a very similar template to the biotin assay, with identical regeneration, conditioning, and antibody dilution buffers. The sample preparation also involves dilution into water and autoclaving followed by filtration.

The assay recognizes pteroylglutamic acid (PGA), the monoglutamated synthetic form of folic acid, as the primary analyte; 100% cross-reactivity is also shown towards 5'-methyltetrahydrofolate (5-CH$_3$-H$_4$folate), which is the predominant bio-available form of folate present naturally in milk based products. While recognized by the assay, 5-CH$_3$-H$_4$folate will rapidly degrade during sample preparation; to prevent this, a 1% ascorbate solution is used in place of water to prepare samples. The results shown in Table 4.2 represent the findings of the AOAC RI PTM evaluation and validation data for the folic acid assay. Studies carried out at the Swedish University of Agricultural Sciences, Uppsala, further support the validity of the method [2, 23].

4.5.3
Pantothenic Acid

Pantothenic acid (vitamin B$_5$) is fairly ubiquitous in most foodstuffs. It has been very difficult to develop alternative spectroscopic methods to MA for pantothenic

Table 4.2 Summary of results from independent evaluation of Qflex Kit Folic Acid as part of the AOAC RI PTM program.

Accuracy, compared with MA	97–100%
Repeatability	3–5%
Reproducibility	5–10%
Precision within one run	1%
Precision between runs	2%
Precision between kits, same calibration solutions	2%
Precision between kits, different calibration solutions	3%
Recovery	87–96%
Detection limit	<1 ng ml^{-1}
Quantifiable region	2.0–70 ng ml^{-1}
Dilution of sample extracts	2–5% CV[a]
Different amounts of sample	2–14% CV

a) CV, coefficient of variation.

Table 4.3 Summary of results from independent evaluation of Qflex Kit Pantothenic Acid as part of the AOAC RI PTM program.

	SPM 1[a)	SPM 2[b)
Accuracy, compared with LC–MS analysis	103%	104%
Accuracy, compared with MA	104%	118%
Repeatability	5.4%	0.9–6.9%
Limited laboratory reproducibility	3.2–17.4	6.3–26.4%
Recovery	>95%	>97%
Precision within one run	1.5–4.6%	
Precision between runs	5.3–7.1%	
Quantification range	100–1500 ng ml^{-1}	

a) SPM 1 = sample preparation method 1, acetic acid precipitation method.
b) SPM 2 = sample preparation method 2, simple dilution method.

acid, owing to the lack of a unique chromophore, meaning that only expensive mass spectrometry (MS) coupled detection is feasible, restricting these methods to research rather than routine analysis.

Qflex Kit Pantothenic Acid PI contains the core reagents to perform the assay, including a preimmobilized sensor chip and stock calibration and antibody solutions. Regeneration solution is prepared in the laboratory according to the handbook instructions.

The pantothenic acid assay uses two sample preparation methods: a universal method, which uses an acetic acid protein-precipitation step (Sample Preparation Method 1), and a simpler method that can also be applied to a wide range of samples involving simple dilution into phosphate–citrate buffer (pH 7.0) and filtration (Sample Preparation Method 2). The validation data from the PTM study, shown in Table 4.3, contain results for both sample preparation methods, and comparisons with a published liquid chromatography (LC)–MS method, developed by Fonterra (Auckland, New Zealand), and also the MA [4, 9, 23, 24].

4.5.4
Vitamin B$_2$

Vitamin B$_2$ or riboflavin plays an important role in energy metabolism and nerve function [25]. In food and other preparations, the molecule is very quickly degraded by UV light to non-biologically active compounds. It is normally analyzed by a combined HPLC–fluorescence method to detect vitamin B$_1$ (thiamine by conversion to the fluorescent derivative thiochrome) and vitamin B$_2$, which has a natural strong fluorescent character. Relatively, this method does not present the difficulties, in terms of analysis time and reproducibility, seen in the MA methods used for the other vitamins. However, the SPR method still represents a more rapid, reliable, and solvent-free solution to the analysis of vitamin B$_2$ in a wide range of materials.

Table 4.4 Summary of results from independent evaluation of Qflex Kit Vitamin B$_2$ as part of the AOAC RI PTM program.

Accuracy, compared with HPLC analysis	112%
Repeatability (samples)	1.2–5.5%
Limited laboratory reproducibility 1	14.1% (between 2 laboratories)
Limited laboratory reproducibility 2	2.7% (between samples)
Precision within one run (run 1, buffer)	2.3–3.4%
Precision between runs (buffer)	1.1–4.4%
Recovery	>87%
Detection limit	17.1 ng ml^{-1}
Quantifiable region	50–500 ng ml^{-1}

Sample preparation for the analysis of the majority of sample types for vitamin B$_2$ involves dilution into phosphate–citrate buffer at pH 7.0, stirring for 30 min, and filtration. It is important, of course, to protect samples from UV light during this preparation. The assay was assessed independently during the AOAC PTM study, being compared directly with the official fluorimetric method described above; the results are given in Table 4.4 [3, 23].

4.5.5
Vitamin B$_{12}$

Vitamin B$_{12}$ is found only in animal-derived or yeast-based foods, and normally at levels much lower than those of the other B vitamins. It is routinely added, however, to a wide range of foodstuffs, such as cereals and infant formulas, in the form of cyanocobalamin, in order to add value and fulfill particular nutritional needs [25].

Some success has been achieved in the development of chromatographic methods for vitamin B$_{12}$, but lack of sensitivity and interference from a range of food components have limited their application to pharmaceutical preparations and premixes. MA is still the most widely used method for vitamin B$_{12}$ quantification across the industry.

The development of a biosensor method for vitamin B$_{12}$ had to overcome the additional technical challenges presented by the relatively lower levels of this vitamin, but still resulted in a very straightforward, robust, and sensitive method.

Sample preparation for vitamin B$_{12}$ analysis uses a pH 4.5 phosphate–citrate buffer containing 0.005% sodium cyanide. Samples are dispersed in this buffer, autoclaved, and finally filtered before analysis. Under these conditions, all vitamin B$_{12}$ forms are converted to the most stable cyanocobalamin form, which is detected by the assay allowing a measurement of the total vitamin B$_{12}$ content [5, 6, 10, 15, 23, 24].

A summary of findings of the AOAC PTM is shown in Table 4.5.

Table 4.5 Summary of results from independent evaluation of Qflex Kit Vitamin B_{12} as part of the AOAC RI PTM program.

Detection limit	$0.06\,ng\,ml^{-1}$
Quantification limit	$0.21\,ng\,ml^{-1}$
Accuracy, compared with MA	113%
Precision within one run (run 1)	1.1–2.4%
Precision between runs	1.7–12.5%
Repeatability	1.3–6.2%
Reproducibility	5.1–22%
Recovery	85–117%

4.6
Validation Considerations

Although the Biacore instrument utilizes a sophisticated physico-chemical phenomenon to measure the interactions in these assays, it is important to bear in mind that the actual quantification is based on the interaction of biochemical components, rather than direct physical detection. Hence standard validation protocols intended for physico-chemical methods such as HPLC and LC–MS must be adapted in order to be applicable to these methods. In particular, it should be understood that the calibration curve generated by the instrument uses a four-parameter equation fitting function; a linear fitting is not appropriate, even when the line appears straight. The precise mechanics of the interactions which form the calibration curve are dependent upon a myriad of variables affecting affinity and binding kinetics, including pH, ionic strength, and so on. Calculation of linearity of measurement should therefore be performed on the relationship between measured concentration and externally spiked reference samples, rather than on the internal calibration. Calibration curves for all the vitamins will trend towards horizontal maximum and minimum response values if extended, so extrapolation is not advised. The accuracy of the fit can also be affected by the calibration points used, so it is not recommended to employ any calibration parameters other than those recommended in the kit instructions, as this can reduce the assay performance. Shortening of the calibration range to focus on a particular routine sample can, in particular, have the effect of reducing the accuracy and increasing the variability of the assay [7, 9, 22, 26–28].

4.7
Conclusions

The quantification assays described in this chapter using Biacore Q and Qflex Kits are all characterized by a set of common advantages: the assays are highly automated, they are easily controlled and evaluated via wizard-driven software, results are presented in real time, no fluorescent or radioactive labels are required, sample preparation is minimal, and the samples do not require clean-up or solvent extraction. In addition, the advent of kits providing preimmobilized sensor surfaces further reduces the manual input required by the user and provides a standardized sensor surface. Lastly, the possibility of easily regenerating and reusing sensor surfaces with no significant decrease in assay performance is a major attraction in the quest for an assay that delivers consistent data over several hundred runs, allowing for valid comparative studies between geographically disparate locations or at different times.

Together with the inherent robustness of the technology and the proven resistance of Biacore Q to stressful industrial environments, these features add to the speed with which screening and quantification can be pursued, meaning that analyses can be in line with the demands of a modern production facility.

Although SPR-based biosensors remain to make significant inroads into many areas of food safety testing and analysis, all comparative tests to date demonstrate that they can deliver data of a quality at least as high as, if not superior to, those of the main competing technologies. The technology and assays are also relatively safe and easy to use, with clear advantages of speed and throughput.

References

1 Homola, J. (2003) *Analytical and Bioanalytical Chemistry*, **377**, 528–539.

2 Boström Caselunghe, M. and Lindeberg, J. (2000) Biosensor-based determination of folic acid in fortified food. *Food Chemistry*, **70**, 523–532.

3 Caelen, I., Kalman, A., and Wahlström, L. (2004) Biosensor-based determination of riboflavin in milk samples. *Analytical Chemistry*, **76**, 137–143.

4 Haughey, S.A., O'Kane, A.A., Baxter, G.A., Kalman, A., Trisconi, M.-J., Indyk, H.E., and Watene, G.A. (2005) *Journal of AOAC International*, **88**, 1008–1014.

5 Indyk, H.E., Evans, E.E., Boström Caselunghe, M.C., Persson, B., Finglas, P.M., Woollard, D.C., and Filonzi, E.L. (2000) *Journal of AOAC International*, **83**, 1141–1148.

6 Indyk, H., Persson, B., Boström Caselunghe, M., Moberg, A., Filonzi, E., and Woollard, D. (2002) *Journal of AOAC International*, **85**, 72–81.

7 Rader, J.I. (1998) Folic acid: considerations regarding food values in databases. Presented at Emerging Issues for the Next Generation of Databases: 22nd National Nutrient Data Bank Conference, 1998.

8 Stenberg, E., Persson, B., Roos, H., and Urbaniczky, C. (1991) Quantitative determination of surface concentration of protein with surface plasmon resonance using radiolabeled proteins. *Journal of Colloid and Interface Science*, **143**, 513–526.

9 Woolard, D.C., Indyk, H.A., and Christiansen, S.K. (2000) The analysis of pantothenic acid in milk and infant formulas by HPLC. *Food Chemistry*, **69**, 201–208.

10 Eitenmiller, R.R. and Landen, W.O. Jr. (1999) Vitamin B-12, biotin, and pantothenic acid, in *Vitamin Analysis for the Health and Food Sciences* (eds. R.R. Eitenmiller and W.O. Landen Jr.), CRC Press, Boca Raton, FL, pp. 467–501.

11 Mittermayr, R., Kalman, K., Trisconi, M.-J., and Heudi, O. (2004) *Journal of Chromatography A*, **1032**, 1–6.

12 Rychlik, M. (2000) *Journal of Agricultural and Food Chemistry*, **48**, 1175–1181.

13 Rychlik, M. and Freisleben, A. (2002) *Journal of Food Composition and Analysis*, **15**, 399–409.

14 Rychlik, M. (2003) *Analyst*, **128**, 832–837.

15 Indyk, H.E. (2006) Optical biosensors: making sense of interactions. *Chemistry in New Zealand*, July, 42–46.

16 GE Healthcare (2005) *Biacore Q Instrument Handbook*, GE Healthcare, Little Chalfont.

17 Karlsson, R. and Stahlberg, R. (1995) *Analytical Biochemistry*, **228**, 274–280.

18 Koontz, J.L., Phillips, K.M., Wunderlich, K.M., Exler, J., Holden, J.M., Gebhardt, S.E., and Haytowitz, D.B. (2005) Comparison of total folate concentrations in foods determined by microbiological assay at several experienced U.S. commercial laboratories. *Journal of AOAC International*, **88** (3), 805–813.

19 Lofas, S. and Johnsson, B. (1990) *Journal of the Chemical Society, Chemical Communications*, 1526–1528.

20 Liedberg, B., Nylander, C., and Lundstrom, I. (1983) *Sensors and Actuators*, **4**, 299–304.

21 Malmqvist, M. (1993) Biospecific interaction analysis using biosensor technology. *Nature*, **361**, 186–187.

22 AOAC International (2002) AOAC Guidleines for Single Laboratory Validation of Chemical Methods for Dietary Supplements and Botanicals, http://www.aoac.org (accessed 14 August 2009).

23 AOAC International (2006) *Official Methods of Analysis*, 18th edn., AOAC International, Philadelphia, PA, http://www.eoma.aoac.org (accessed 21 September 2009).

24 Kalman, A., O'Kane, A., Caelen, I., Trisconi, M.-J., and Wahlstrom, L. (2003) *Biacore Journal*, **3**, 18–21.

25 Food Standards Agency (2002) Draft Report of the Expert Group on Vitamins and Minerals. http://www.food.gov.uk (accessed 22 September 2008).

26 AOAC International (2002) Interlaboratory collaborative study. Appendix D, in *Official Methods of Analysis*, 18th edn., AOAC International, Philadelphia, PA, http://www.aoac.org.

27 AOAC International (2004) Definitions and Calculations of HORRAT Values from Intralaboratory Data, http://www.aoac.org (accessed 21 September 2009).

28 Commission of the European Communities (2003) Preliminary Draft Proposal for a Regulation of the European Parliament and of the Council on the Addition of Vitamins and Minerals and of Certain Other Substances to Foods, SANCO/329/03. Commission of the European Communities, Brussels.

5
International Perspectives in Vitamin Analysis and Legislation in Vitamin Fortification

Michael Rychlik

5.1
Introduction

The particular features of vitamins, which are the need for alimentary supply by humans and the physiological requirements in minute amounts, have always had an impact on analytical methods. The traces to be quantified are too small even for some highly sensitive, instrumental analytical methods, which makes bioassays still important. For the determination of vitamins, those organisms can be used which do not produce these compounds by themselves, but need them for their growth.

In former times, vertebrates such as chicks for vitamin K were used, but nowadays only bacteria are applied. In addition, for evaluation of a food's nutritive value, the measurement of all active species of a vitamin is necessary. In this regard, vitamins may occur as stereoisomers or as bound forms showing different bioactivities and, therefore, have to be quantified separately. An additional complication was introduced with synthetic vitamins, with increased variety and analytical challenges with new derivatives or lacking isomeric purity of the synthetic forms. With improved separation efficiency of chromatographic stationary phases, high-performance liquid chromatography (HPLC) methods came to the fore due to their ability to differentiate the individual vitamins, along with their improved accuracy and decreased run times, and they often showed significant advantages over bioassays. The classic division into water- and fat-soluble vitamins, however, is still reasonable.

5.2
General Requirements for Modern and Future Vitamin Assays

In many countries, regulations for the fortification of foods with vitamins are in place and maximum regulation levels (MRLs) for the European Union (EU) are currently under discussion (see below). Keeping in mind the number of vitamins and upcoming regulation levels, several requirements for future analytical methods can be deduced:

5.2.1
Multimethods

The term "multi" may stand for several approaches: First, actual dietary recommendations have to consider different bioavailabilities of the various vitamers of a single group of vitamins. Therefore, new methods aim at differentiating all vitamers within one sample preparation scheme and one chromatographic or electrophoretic run. The transfer of all vitamers into one major form such as for vitamin B_6 into pyridoxol [1] or folates into 5-methyltetrahydrofolates [2] no longer meets modern requirements.

With the expected EU regulations with upper fortification limits in place, for each vitamin group the number of vitamers may be limited, but it becomes desirable to quantitate all vitamins that have been added. The most versatile analytical method to achieve this task is modern liquid chromatography–tandem mass spectrometry (LC–MS/MS). Applications to water- and fat-soluble vitamins are presented Chapters 14 and 19, respectively, dealing with the requirements for and modern approaches to quantitate simultaneously all vitamins of the respective group.

When endogenous vitamins have to be quantified simultaneously in foods, the variety of compounds to be analyzed increases dramatically as conjugated forms may occur. The conjugates include phosphates for many water-soluble vitamins, nucleotides for vitamin B_2, niacin, and pantothenic acid, and polyglutamates for folates.

Moreover, vitamins might be bound to or encapsulated in the food matrix, which requires liberation procedures, preferably by enzymes. However, optimum deconjugation of all these compounds is a demanding task and several approaches have been reported. For fortified foods such as infant formulas, a fairly simple digestion with hydrochloric acid followed by treatment with takadiastase proved to be sufficient to extract completely added thiamine, riboflavin, nicotinamide, nicotinic acid, pyridoxine, pyridoxal, folic acid, and vitamin B_{12} [3]. However, if endogenous contents of phosphorylated and protein-bound vitamins B_1, B_2, and B_6 have to be analyzed, a mixture of α-amylase, papain, and acid phosphatase was necessary to determine the sum of all vitamers. Comparison with certified reference materials proved the completeness of extraction [4].

As these treatments may run for many hours, stability of the vitamins might be critical. For this challenge, the use of stable isotope-labeled standards is recommended.

The term "multimethod" may apply not only to various analytes, but also to various matrices. In this regard, application of the assays to different foods and to clinical matrices such as blood plasma or urine may be valuable. For instance, bioavailability studies require the analysis of one vitamin group in these kinds of samples to allow biokinetic calculation of applied doses and the response of the organism. The use of the same assay for both kinds of samples significantly improves the validity of the conclusions.

5.2.2
Assays of Bioactivity

A complementary approach to multivitamer methods is to quantify the bioactivity of all members of one whole vitamin group. Given an analogous response of the biotest system to vitamin activity in humans, this alternative is still attractive as the equipment is often less complicated. As already indicated in the Introduction, the use of vertebrates is not longer acceptable for ethical reasons and, therefore, bacteria are still suitable test organisms. New developments are directed towards easier microbiological handling. For example, 96-well microtiter assays based on precoating with bacteria (VitaFast®) can be operated even in the imperfectly sterile environment of a chemical laboratory. These kinds of assays are available for riboflavin, thiamine, pyridoxine, niacin, folic acid, biotin, pantothenic acid, and vitamin B_{12}. A practice report on this methodology is presented in Chapter 3.

5.2.3
Fast Assays

In contrast to microbiological assays requiring analysis times of around 48 h, the need for high-throughput methods spurred the development of fast alternatives.

Very promising in this regard is the biosensor technology based on surface plasmon resonance (SPR) measurement. The SPR technique involves a mass-sensitive instrument that transduces with high sensitivity the amount of bound molecules to surfaces by measuring the intensity of reflected polarized light at the resonance angle [5]. The specific binding of vitamins to the layer can be conveyed by interaction with antibodies or specific proteins such as folate-binding protein (FBP). This technology was commercialized as the Biacore® system and has been reported for biotin [6], folates [6], pantothenic acid [7], riboflavin [8], and vitamin B_{12} [9]. A collaborative study on vitamin B_{12}, biotin, folic acid, and pantothenic acid in fortified foods under AOAC criteria has recently been performed successfully and the report is in progress [10]. Current applications of the Biacore system are also presented in Chapter 4.

5.2.4
Accurate Methods

Accuracy is a central parameter for each analytical method and the need to determine the "true" value of the analyte is particularly important for candidate reference methods. For vitamin assays, there is a huge variety of official reference methods ranging from microbiological assays to HPLC-based assays for folates to vitamin B_6, respectively. However, a tendency towards LC–MS/MS can be recognized, particularly with the use of isotopically labeled internal standards. An actual review of stable isotope dilution assays (SIDAs) is presented in Chapter 1. Although the accuracy of the latter method is outwith the present discussion, the limited availability of the standards restricts its use and the highly variable mass spectrometric settings depending on the respective instruments hindered collaborative trials under a stringent official protocol. However, first attempts towards isotope dilution assays as reference methods in the area of mycotoxins point the way to implementing SIDAs as official methods. Furthermore, quality assurance in accredited laboratories is hindered by the lack of stable certified reference materials due to the lability of vitamins.

5.2.5
Sensitive Methods

For vitamins occurring in trace contents, the sensitivity of common assays still is not sufficient. Particularly for vitamin B_{12}, modern liquid chromatography–mass spectrometry (LC–MS) instrumentation with improved sensitivity offers the prospect of permitting the quantitation even of endogenous contents. This topic and the coupling of HPLC to inductively coupled plasma mass spectrometry (ICP-MS) is covered in Chapter 12. Moreover, if limited sample size, such as for human blood studies, requires a further increase in sensitivity, liquid chromatography (LC) coupled to accelerator mass spectrometry offers the chance to detect even attomolar concentrations, as shown for 5-methyltetrahydrofolate in blood plasma [11].

5.2.6
Developments for Single Vitamins or Vitamin Groups

5.2.6.1 Fat-Soluble Vitamins
This group of vitamins is naturally dissolved in edible fats. For complete measurement in foods and separation from fats, the latter typically are saponified, using thermal and alkali treatments of the sample, which may lead to losses. These can be compensated for in different ways.

For hydrophobic analytes, there is principally the choice in HPLC between a normal phase and an organic mobile phase or a reversed phase (RP) with an aqueous eluent. However, organic solvents are often not compatible with mass spectrometric applications. Additionally, for RP phases new materials with higher selectivities and improved separation efficiency are increasingly being developed, so these have become more important than normal-phase applications. Vitamins D and E are comprehensibly covered in Chapters 17 and 16, respectively. Vitamins A and K are considered below.

5.2.6.1.1 Vitamin A and Carotenoids
For vitamin A, the base compound all-*trans*-retinol and some of its metabolites along with a series of carotenoids acting as pro-vitamins have to be quantified separately. Chapter 15 is dedicated to the analysis of carotenoids and presents a procedure for the effective determination of 14 carotenoids. Moreover, carotenoids with new structures are still being identified [12] by using hyphenated techniques such as HPLC–NMR spectroscopy [13].

5.2.6.1.2 Vitamin K

In addition to its function in blood clotting, vitamin K has been recognized to be essential for calcium homeostasis and bone mineralization. The current analytical methods for this vitamin are reviewed in Chapter 19 including a survey of its biological role, dietary sources, stability, bioavailability, and markers for deficiency.

5.2.6.2 Water-Soluble Vitamins

This group of compounds are analyzed by RP-HPLC. New studies and developments are discussed below.

5.2.6.2.1 Vitamin B_1 and B_2

The sensitive analysis of thiamin is generally achieved by fluorimetric detection of thiochrome, which is generated from thiamin by post- or pre-column derivatization.

Due to its native fluorescence, riboflavin is preferably quantified by HPLC coupled to fluorescence detection. The simultaneous detection of riboflavin along with its conjugates flavin mononucleotide (FMN) and flavin adenine dinucleotide (FAD) by HPLC using an amide stationary phase, thus omitting the use of an ion-pair reagent, has been reported [14].

Reviews on thiamin and riboflavin analysis are presented in Chapters 6 and 7, respectively.

5.2.6.2.2 Folates

For folates, a new commercial microtiter assay is presented in Chapter 3. Calibration functions for different vitamers were recorded photometrically and proved the linearity of the assay. Thus the assay appeared to be very suitable for quantifying fortified folic acid in foods. However, if endogenous folates have to be analyzed, different slopes for the various vitamers restrict the use of the assay. 5-Methyltetrahydrofolate revealed a significantly lower and 5-formyltetrahydrofolate a distinctly higher slope of the calibration curve. These results were in agreement with a survey by Koontz *et al.* [15], whose review clearly indicates significant differences in bacterial response to single folate vitamers.

If fortified foods including folic acid are heated, the potential formation of new folate derivatives may occur, as described in Chapter 11, which includes SIDAs of new folates in commercial products and provides a preliminary risk evaluation after oral application of the new compounds in a human study.

5.2.6.2.3 Vitamin B_{12}

Due to low fortification levels, analysts consider vitamin B_{12} to be the most challenging vitamin. Therefore, two chapters are dedicated to modern analytical methods for this compound. Chapter 12 deals with HPLC methods including modern couplings to ICP-MS and electrospray ionization mass spectrometry, and Chapter 13 describes a sensitive application of bioautography using bacteria. As vitamin B_{12} contents are very low, the use of microbiological assays and radiobinding assays prospects will remain the choice for quantitating endogenous contents in foods in the medium term.

5.3
Fortification with Vitamins – The International Perspective

Malnutrition is not solely a phenomenon in developing countries. However, in these areas, deficiency is caused by inadequate intake of energy, whereas the population in Western countries is at risk of being deficient in certain micronutrients such as vitamins and minerals. These concerns spurred the propagation of vitamin-containing supplements and foods fortified with vitamins. However, the impact on the general nutrition status of the population in the EU and North America has been restricted as the groups of people with low vitamin intake are not aware of it and, therefore, do not voluntarily consume fortified products. For vitamins, in the 1990s the folates came into focus, as the incidence of neural tube defects appeared that were clearly associated with a deficient folate intake. Therefore, over 50 countries throughout the world have introduced mandatory folate fortification. In Table 5.1, a list of these countries is given and reveals that mandatory folate fortification is in force mainly on the American continent, but also in Africa, Asia, and the Pacific Region. This measure was implemented in 1998 in the USA and Canada and most recently, in September 2009, in Australia.

In the EU, however, the governments of the Member States did not agree on mandatory fortification and it is remains the responsibility of each individual Member State to allow voluntary fortification. In some Member States, for example The Netherlands, even voluntary fortification was prohibited some years ago by law, but in 2004 this was concluded not to be compliant with European regulations (see below) and, therefore, the law was withdrawn. Moreover, the Codex Alimentarius Commission of the Food and Agriculture Organization of the United Nations (FAO) and the World Health Organization (WHO) recognized the need to provide consumers with a broader choice of foods and also to provide the industry with more flexibility along with reducing trade barriers [16].

Under regulation 1925/2006/EC on the addition of vitamins and minerals, foods voluntarily fortified with folic acid are widely available in the EU. The range of products that are fortified covers dairy products, breakfast cereals, cereal bars, fruit juices, fat spreads, bread, and beverages.

The most controversially discussed aspect is the minimization of adverse health effects associated with over-exposure to vitamins. For some of these, for example, β-carotene, in particular smokers would have to face a higher risk of developing lung cancer. A second example is vitamin K, which is known to interfere with drugs for anticoagulant treatment. Moreover, the discussions suffer from a lack of knowledge about tolerable upper tolerable intake levels (ULs) for many other vitamins.

The first EU regulation on added micronutrients was EU Directive 2002/46/EC, which initiated harmonization of the legislation for food supplements in the indi-

Table 5.1 Countries with mandatory fortification of flour with folic acid.

North America

Canada, United States

Caribbean

Barbados, Cuba, Dominican Republic, Grenada, Guadeloupe, Guyana, Haiti, Jamaica, Puerto Rico, Saint Vincent, Trinidad and Tobago

South and Central America

Argentina, Belize, Bolivia, Brazil, Chile, Colombia, Costa Rica, Ecuador, El Salvador, Guatemala, Honduras, Mexico, Nicaragua, Panama, Paraguay, Peru, Surinam, Uruguay, Venezuela

Middle East/North Africa

Bahrain, Egypt, Iran, Iraq, Israel, Jordan, Kuwait, Morocco, Oman, Palestine, Occupied Territory, Qatar, Saudi Arabia, Yemen

Sub-Saharan Africa

Côte d'Ivoire, Ghana, Guinea, Malawi, Nigeria, South Africa, Zambia

Central Asia

Kazakhstan, Kyrgyzstan, Turkmenistan

South-East Asia

Indonesia

Oceania

Australia, New Zealand, Fiji, Vanuatu

---

vidual Member States. It included the definitions of micronutrients with a list of possible substances with which supplements may be fortified.

However, specific maximum and minimum regulation limits had not been set and a discussion paper from the EC DG Health and Consumer Protection was sent to the Member States and other stakeholders to formulate their opinion on addition levels in foods and supplements [17]. Answers to this discussion paper were published from 13 Member States and over 40 stakeholders from all over the world, including vitamin producers, consumer protection associations, and manufacturers' associations.

As a result, in 2006 regulation 1925/2006/EC came into force, which names 13 vitamins that may be added to processed foods excluding beverages containing more than 1.2% by volume of alcohol. In order to harmonize the regulation for fortified foods and food supplements, regulation EC 1170/2009 entered into force and included updated admissions for new vitamin and mineral forms.

However, general MRLs for the whole of the EU were still not set as there is a lack of agreement centered on (i) unknown ULs for some vitamins, (ii) different food intakes between population groups and Member States, and (iii) the question of whether minimum levels also have to be set. The issues in detail are as follows:

1) **Upper limits**

   In regulations 2002/46/EC and 1925/2006/EC, ULs have been considered as the basis for upcoming MRLs. Although the Scientific Committee on Food(SCF) has established ULs for vitamins A, D, E, and B_6, and also for niacin and folic acid, given in Table 5.2, and further TGLs have been proposed (Table 5.3) [24], vitamins K, B_1, B_2, B_{12}, and C and biotin are still under current consideration.

2) **Intake of foods and supplements**

   The intake of foods, whether fortified or not, is different between sub-populations of a given Member State and differs between the Member States.

Table 5.2 Tolerable upper intake levels (ULs) of vitamins and minerals for adults, established by the Scientific Committee on Food.

Vitamin	UL	Ref.
Vitamin A	3000 µg[a)]	[18]
Vitamin D	50 µg	[19]
Vitamin E	300 mg	[20]
Niacin[b)]	900 mg[c)]	[21]
Vitamin B_6	25 mg	[22]
Folic acid[d)]	1000 µg	[23]

a) The UL for adults does not apply for postmenopausal women.
b) The ULs apply only for nicotinamide.
c) The UL for adults does not apply during pregnancy or lactation.
d) The UL does not include dietary folate from natural sources.

Table 5.3 Suggested guidance levels (GLs) or temporary guidance levels (TGLs) for adult intake of vitamins and minerals, for which no tolerable upper intake levels (ULs) have been established by the Scientific Committee on Food.

Micronutrient	Classification	Adults	Ref.
β-Carotene	TGL	5 mg	[24]
Vitamin K	GL[a)]	1000 µg	[25]
Thiamin	TGL	50 mg	[24]
Riboflavin	GL[a)]	43 mg	[25]
Vitamin B_{12}	GL[a)]	2000 µg	[25]
Pantothenic acid	GL[a)]	200 mg	[25]
Biotin	GL[a)]	1000 µg	[25]
Vitamin C	TGL	1000 mg	[24, 26]

a) The GL established by the UK Expert Group on Vitamins and Minerals (EMV).

Therefore, general levels, on the one hand, may set some of these at risk and, on the other, may render addition unreasonable for others. In addition, the consumption of supplements containing vitamins renders the vitamin intake of the population hardly predictable and further hampers the setting of MRLs.

3) **Minimum levels**

In addition to maximum levels, the question is not settled as to whether minimum levels will be required for some nutrients to be present in a significant amount, which is the amount in 100 g or 100 ml of the food being equivalent to 15% of the recommended allowance specified in the Annex to Directive 90/496/EEC.

Further regulations will have to consider the vitamin overages in products that are applied by manufacturers to anticipate losses during storage and processing. A general guideline for this aspect has been suggested, for example, by the Working Group on Nutritional Issues of the German Chemical Society [27], which specified the tolerable ranges of vitamin contents in fortified products.

References

1 Reitzer-Bergaentzle, M., Marchioni, E., and Hasselmann, C. (1993) HPLC determination of vitamin B_6 in foods after pre-column derivatization of free and phosphorylated vitamers into pyridoxol. *Food Chemistry*, **48**, 321–324.

2 Ndaw, S., Bergaentzle, M., Aoude-Werner, D., Lahely, S., and Hasselmann, C. (2001) Determination of folates in foods by high-performance liquid chromatography with fluorescence detection after precolumn conversion to 5-methyltetrahydrofolates. *Journal of Chromatography A*, **928**, 77–90.

3 Vinas, P., Lopez-Erroz, C., Balsalobre, N., and Hernandez-Cordoba, M. (2003) Reversed-phase liquid chromatography on an amide stationary phase for the determination of the B group vitamins in baby foods. *Journal of Chromatography A*, **1007**, 77–84.

4 Ndaw, S., Bergaentzle, M., Aoude-Werner, D., and Hasselmann, C. (2000) Extraction procedures for the liquid chromatographic determination of thiamin, riboflavin and vitamin B_6 in foodstuffs. *Food Chemistry*, **71**, 129–138.

5 Blake, C.J. (2007) Analytical procedures for water-soluble vitamins in foods and dietary supplements: a review. *Analytical and Bioanalytical Chemistry*, **389**, 63–76.

6 Indyk, H.E., Evans, E.A., Caselunghe, M.C.B., Persson, B.S., Finglas, P.M., Woollard, D.C., and Filonzi, E.L. (2000) Determination of biotin and folate in infant formula and milk by optical biosensor-based immunoassay. *Journal of AOAC International*, **83**, 1141–1148.

7 Haughey, S.A., O'Kane, A.A., Baxter, G.A., Kalman, A., Trisconi, M.-J., Indyk, H.E., and Watene, G. (2005) Determination of pantothenic acid in foods by optical biosensor immunoassay. *Journal of AOAC International*, **88**, 1008–1014.

8 Caelen, I., Kalman, A., and Wahlstrom, L. (2004) Biosensor-based determination of riboflavin in milk samples. *Analytical Chemistry*, **76**, 137–143.

9 Indyk, H.E., Persson, B.S., Caselunghe, M.C.B., Moberg, A., Filonzi, E.L., and Woollard, D.C. (2002) Determination of vitamin B_{12} in milk products and selected foods by optical biosensor protein-binding assay: method comparison. *Journal of AOAC International*, **85**, 72–81.

10 Committee on Food Nutrition (2008) Methods Committee Reports. *Journal of AOAC International*, **91**, 42B–43B.

11 Buchholz, B.A., Arjomand, A., Dueker, S.R., Schneider, P.D., Clifford, A.J., and Vogel, J.S. (1999) Intrinsic erythrocyte labeling and attomole pharmacokinetic tracing of ^{14}C-labeled folic acid with accelerator mass spectrometry. *Analytical Biochemistry*, **269**, 348–352.

12 Pott, I., Breithaupt, D.E., and Carle, R. (2003) Detection of unusual carotenoid esters in fresh mango (*Mangifera indica* L. cv. Kent). *Phytochemistry*, **64**, 825–829.

13 Glaser, T., Lienau, A., Zeeb, D., Krucker, M., Dachtler, M., and Albert, K. (2003) Qualitative and quantitative determination of carotenoid stereoisomers in a variety of spinach samples by use of MSPD before HPLC–UV, HPLC–APCI-MS, and HPLC–NMR on-line coupling. *Chromatographia*, **57**, S/19–S/25.

14 Chase, G.W., Landen, W.O., Eitenmiller, R.R., and Soliman, A.M. (1992) Liquid chromatographic determination of thiamine, riboflavin, and pyridoxine in infant formula. *Journal of AOAC International*, **75**, 561–565.

15 Koontz, J.L., Phillips, K.M., Wunderlich, K.M., Exler, J., Holden, J.M., Gebhardt, S.E., and Haytowitz, D.B. (2005) Comparison of total folate concentrations in foods determined by microbiological assay at several experienced U.S. commercial laboratories. *Journal of AOAC International*, **88**, 805–813.

16 Codex Alimentarius Commission, Codex Committee on Nutrition and Foods for Special Dietary Uses (2008) Discussion paper on the proposal for new work to amend the Codex general principles for the addition of essential nutrients to foods (CAC/GL 09-1987).

17 European Commission (2006) Discussion paper on the setting of maximum and minimum amounts for vitamins and minerals in foodstuffs. http://ec.europa.eu/food/food/labellingnutrition/supplements/discus_paper_amount_vitamins.pdf (accessed 09-12-2010).

18 SCF (Scientific Committee for Food) (2002) Opinion of the Scientific Committee on Food on the tolerable upper intake level of preformed vitamin A (retinol and retinyl esters). http://ec.europa.eu/food/fs/sc/scf/out 145_en.pdf (accessed 09-12-2010).

19 SCF (Scientific Committee for Food) (2002) Opinion of the Scientific Committee on Food on the tolerable upper intake level of vitamin D. http://ec.europa.eu/food/fs/sc/scf/out 157_en.pdf (accessed 09-12-2010).

20 SCF (Scientific Committee for Food) (2003) Opinion of the Scientific Committee on Food on the tolerable upper intake level of vitamin E. http://ec.europa.eu/food/fs/sc/scf/out195_en (accessed 09-12-2010).

21 SCF (Scientific Committee on Food) (2002) Opinion of the Scientific Committee on Food on the tolerable upper intake levels of nicotinic acid and nicotinamide (niacin). http://ec.europa.eu/food/fs/sc/scf/out80f_ev (accessed 09-12-2010).

22 Scientific Committee on Food (2000) Opinion of the Scientific Committee on Food on the tolerable upper intake level of vitamin B_6. http://ec.europa.eu/food/fs/sc/scf/out80c_en (accessed 09-12-2010).

23 Scientific Committee on Food (2000) Opinion of the Scientific Committee on Food on the tolerable upper intake level of folate. http://ec.europa.eu/food/fs/sc/scf/out80e_en.pdf (accessed 09-12-2010).

24 Rasmussen, S.E., Andersen, N.L., Dragsted, L.O., and Larsen, J.C. (2006) A safe strategy for addition of vitamins and minerals to foods. *European Journal of Nutrition*, **45**, 123–135.

25 Expert Group on Vitamins and Minerals (2003) *Safe Upper Levels for Vitamins and Minerals. Report of the Expert Group on Vitamins and Minerals*, Food Standards Agency Publications, London.

26 NDA Panel (Scientific Panel on Dietetic Products, Nutrition and Allergies of the European Food Safety Authority) (2004) Opinion of the Scientific Panel on Dietetic Products,

Nutrition and Allergies on a request from the Commission related to the tolerable upper intake level of vitamin C (L-ascorbic acid, its calcium, potassium and sodium salts, and L-ascorbyl-6-palmitate. *EFSA Journal*, **59**, 1–21.

27 Working Group on Nutritional Issues of the German Chemical Society GDCh (2009) Recommendations for tolerances for variances in the contents of nutrients and their labelling. *Lebensmittelchemie*, **63**, 98 (in German).

Part II
Analysis of Water-Soluble Vitamins

6

HPLC Determination of Thiamin in Fortified Foods

Roland Bitsch and Irmgard Bitsch

6.1
Introduction

In accordance with IUPAC nomenclature, thiamin or vitamin B_1 (Figure 6.1) is named as 2-{3-[(4-amino-2-methylpyrimidin-5-yl) methyl]-4-methylthiazol-5-yl} ethanol, with a molecular weight of $265.354\,58\,g\,mol^{-1}$ and a molecular formula of $C_{12}H_{17}N_4OS^+Cl^-.HCl$. This vitamin is found in a variety of animal and vegetable foods at a relatively low level (<0.5 mg per 100 g). The most important dietary sources for humans are whole-grain products, which provide ~40% of the thiamin requirement. Other foods that contribute to thiamin intake are meat products (~27%), especially pork skeletal muscle, heart, liver, brain, and kidneys, vegetables (~12%), milk and milk products (~8%), legumes (~5%), fruits (~4%), and eggs (~2%) [1]. A list of the thiamin contents of various foods is available at the United States Department of Agriculture (USDA) food database [2].

In addition to free thiamin, animal tissues contain thiamin monophosphate (TMP), thiamin diphosphate (TDP), and thiamin triphosphate (TTP) in different amounts. Recently, Poel [3] used a special extrapolation method for the semiquantitative determination of those thiamin phosphates which are not available as standard substances. Using this method, the author was able to detect in pork muscle TTP along with thiamin tetraphosphate for the first time. Another thiamin derivative in some foods is hydroxyethylthiamin pyrophosphate (HET), which functions as active acetaldehyde in metabolism. Compared with thiamin, its bioactivity is nearly 80% in microbiological assays. Even though this relative bioactivity was only determined in bacteria, Jakobsen [4] pointed to the need to determine the concentration of HET in food as part of the total thiamin complex. Compared with the content of thiamin, the author found 2% HET in broccoli, 7–24% in liver and meat, and 37% in dried yeast.

In plant and animal tissues, thiamin in the form of its diphosphate ester functions as the prosthetic group of several key enzymes involved primarily in energy metabolism. The binding affinity of TDP to the apoenzymes is tight but noncovalent. The most important enzymatic systems which utilize TDP as a cofactor are the mitochondrial multienzyme complexes pyruvate dehydrogenase

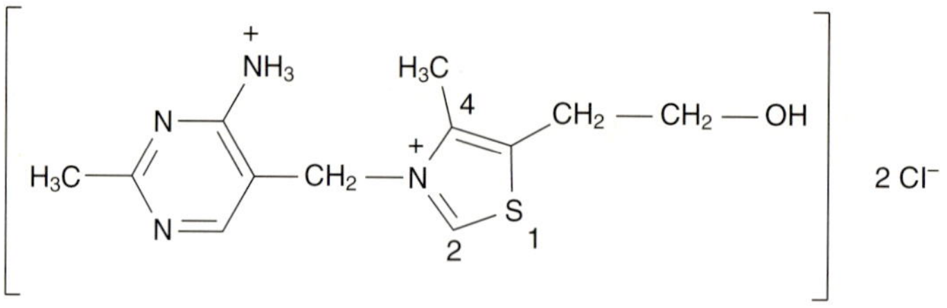

Figure 6.1 Structure of thiamin chloride hydrochloride.

(PDH), α-ketoglutarate dehydrogenase (KGDH), and branched-chain α-keto acid dehydrogenase (BCKDH), and the cytosolic transketolase. The thiamin-dependent mitochondrial dehydrogenase complexes convert 2-keto acids to acyl-CoA, CO_2, and reducing equivalents [NADH (reduced nicotinamide adenine dinucleotide)] and are therefore essential for mitochondrial energy production: the PDH in glycolysis, the KGDH in the citric acid cycle, and the BCKDH in amino acid catabolism [5]. In addition to their main metabolic pathways, mitochondrial dehydrogenase complexes, especially the KGDH complex, produce oxygen free radicals and reactive oxygen species in side reactions (paracatalytic reactions) and are sensitive to oxidative stress [6, 7]. In thiamin deficiency, the activities of the thiamin-dependent enzymes decrease, especially in the brain. This causes reversible and irreversible brain lesions due to impaired mitochondrial oxidative metabolism, resulting in apoptosis and severe cellular deficits [8, 9]. Studies of proteins, cells, animal models, and humans suggest that treatments to diminish or to bypass the reductions in KGDH activity might be beneficial in age-related neurodegenerative disorders [10].

The transketolase participates in the pentose phosphate pathway, which generates NADPH (nicotinamide adenine dinucleotide phosphate) equivalents. These are required for a variety of processes that are related to oxidative stress, including nitric oxide synthesis and keeping glutathione and protein sulfhydryl groups in a reduced state. The pentose pathway also supplies ribose for the production of nucleotides, nucleic acids, coenzymes, and polysaccharides and is involved in various cellular biosynthetic reactions, including lipids and neurotransmitters [11].

The overall reduction of thiamin-dependent processes in thiamin deficiency increases oxidative stress in the organism by affecting the intramitochondrial and cytosolic redox state. This can partly be overcome by application of substances with reducing activity, such as caffeic acid or sodium bisulfite [12]. In addition to metabolic consequences of thiamin deficiency, several structural consequences have been described in the literature, such as cellular membrane damage and irregular and ectopic cells [8, 13].

Inadequate consumption of thiamin is the main cause of thiamin deficiency in developing countries. Different reports of endemic thiamin deficiency in human populations at the end of the twentieth century were compiled by Bates [14, 15]. His reviews dealt with the populations of Indonesia, the Seychelles, the Gambia, Cuba, Thailand, and Mozambique and also the Amazonian Indians of South America. Affected persons exhibited either wet (cardiomyopathy with cardiac hypertrophy and dilatation) or dry (peripheral neuropathy, polyneuritis) beriberi, or a mixture of both. In Western countries, alcohol misuse, often combined with an inadequate nutrient intake, is the most frequent cause of insufficient thiamin supply. Alcohol affects cellular thiamin uptake, which is mediated by a saturable high-affinity transport system, and its utilization in metabolism. Studies have demonstrated that between 30 and 80% of alcoholics are thiamin deficient, and this puts them at risk of developing the Wernicke–Korsakoff syndrome (WKS) [16]. The findings of WKS and cerebellar degeneration, both consequences of severe thiamin deficiency, are present at autopsy in approximately 13 and 42% of alcoholics, respectively [11]. Further patients at risk of developing WKS include people with eating disorders or on radical diets, patients on dialysis, women with hyperemesis gravidarum, anorectic cocaine users, and HIV-positive patients [17]. Several cases of WKS after bariatric surgery were described by Aasheim [18]. Another major clinical manifestation of thiamin deficiency in humans involves the cardiovascular system. Heart failure was diagnosed in elderly persons because of inadequate nutrient intake or the use of loop diuretics causing hyperexcretion of thiamin [19, 20]. Other patients at high risk of thiamin deficiency are those with cancer [21], end-stage chronic liver failure [22] or renal diseases [23, 24]. Low plasma thiamin concentrations are also prevalent in patients with type 1 and type 2 diabetes and are associated with increased thiamin clearance [25]. From animal studies, a link between thiamin deficiency and colon carcinogenesis was suggested [26]. In 2003, an epidemic of Wernicke encephalopathy developed in Israeli infants fed a thiamin-deficient soy-based formula. Approximately 20 infants were affected out of an estimated 3500 fed the thiamin-deficient formula. Early diagnosis and treatment with parenteral thiamin led to complete neurologic recovery [27].

6.2
Fortification of Foods with Thiamin

Taken together, the above-described cases of thiamin deficiency throughout the world underline the need to improve the supply of this important vitamin, primarily because very little thiamin can be stored in the body and depletion can occur as quickly as within 14 days. Particularly the chemical reactivity of the thiamin molecule makes it very vulnerable to losses during food processing and storage. In model systems, the thermal decay of thiamin was characterized as a first-order reaction [28]. In contrast to this early study, Kessler and Fink [29] demonstrated that the thermal losses of thiamin are best described by a second-order reaction. Using the temperature-dependent rate constant and the corresponding reaction law, changes in the concentration of thiamin in milk can be determined over a prolonged period (from 1 s to 1 year) and wide temperature range (from 4 to 150 °C) range. The reaction kinetic data obtained for milk were also applicable to calculating thiamin losses after heating other foods such as meat and vegetables. In this respect, the thermal destruction of thiamin in milk after home-based food preparation was 10–30%, in meat 60–70%, and in vegetables 20–30%.

Table 6.1 US recommended dietary allowances (RDAs) and dietary reference intakes (DRIs) for thiamine (mg per day).

Age group	Age	RDA		DRI	
		Males	Females	Males	Females
Infants	0–6 months	0.2[a]	0.2[a]	0.2[a]	0.2[a]
	7–12 months	0.3[a]	0.3[a]	0.3[a]	0.3[a]
Children	1–3 years	0.5	0.5	0.5	0.5
	4–8 years	0.6	0.6	0.6	0.6
	9–13 years	0.9	0.9	0.9	0.9
Adolescents	14–18 years	1.2	1.0	1.2	1.0
Adults	>19 years	1.2	1.1	1.2	1.1
Pregnant women	All ages	–	1.4	–	1.4
Breastfeeding women	All ages	–	1.4	–	1.5

a) In these cases an RDA cannot be set, and an AI (Adequate Intake) is given.

Thiamin supplementation of staple food products is an effective, simple, and safe public health measure that can improve the thiamin status of all population groups [17]. The level of addition of thiamin to food for normal consumption should comply with the recommended intakes. These are the RDAs (recommended dietary allowances) and the DRIs (dietary reference intakes), which are established in the USA and Canada [30] (Table 6.1). In Europe, most countries tend to use their own nationally derived values, which can vary considerably [31]. On this account the EURRECA Network of Excellence started in 2007 to harmonize and standardize up-to-date micronutrient recommendations for the European Union (http://www.eurreca.org).

For food supplementation purposes, thiamin is chiefly used in the form of thiamine chloride hydrochloride, known commercially as thiamin hydrochloride. This salt is hygroscopic and usually exists as the hemihydrate containing the equivalent of about 4% water. Another commercial form of the vitamin, thiamin mononitrate, is practically non-hygroscopic and, therefore, is especially recommended for the enrichment of flour mixes. Both salts are stable to heat below pH 5.0 and destroyed rapidly at pH 7.0 or above by boiling. The most distinguishing difference between the hydrochloride salt and the mononitrate is water solubility. The hydrochloride shows good solubility in water $(1\,g\,ml^{-1})$ and the mononitrate is only slightly water soluble $(0,027\,g\,ml^{-1})$. This solubility difference leads to differentiation in industrial uses of the two thiamin forms: the hydrochloride salt is preferred for foods requiring solubility and the mononitrate for dry products [32]. Other possible sources of vitamin B_1 added for nutritional purposes to food and food supplements are benfotiamine (a synthetic *S*-acyl derivative of thiamine), thiamin monophosphate chloride and thiamin pyrophosphate chloride. The literature on the safety of these compounds and the bioavailability of vitamin B_1 from them was assembled in a Scientific Opinion of the Panel on Food Additives and Nutrient Sources Added to

Food (ANS), because some petitioners wanted to use this substances in food supplements. However, so far only thiamin hydrochloride and thiamin mononitrate may be added to food in the European Community [33].

6.3
Analytical Principles

Liquid chromatographic methods for the determination of thiamin in certain foods have been applied for nearly 30 years. One of the first papers in this respect was published in 1979 [34]. The authors compared a high-performance liquid chromatography (HPLC) method with the official wet chemistry method of the American Association of Cereal Chemists and analyzed several rice products. The results showed no significant differences between the two methods of thiamin evaluation for all rice samples analyzed, with the exception of paddy rice. The method involved the following steps: (i) grinding of the sample matrix, (ii) autoclaving, (iii) extraction with dilute acid, (iv) adjustment of the pH and incubation with enzymes, (v) extract cleanup, and (vi) analysis for thiamin by HPLC with ultraviolet (UV) detection.

Several methods developed in the past 30 years consist principally of analogous steps. The combination of steps depends on which particular thiamin derivatives have to be analyzed. For the assay of total thiamin, the extraction must liberate thiamin from the sample matrix and hydrolyze the phosphate esters. Such extractions use acid and enzyme hydrolysis that closely follow AOAC International Method 942.23 [35]. In fortified foods, such as infant formula, milk products, cereals, flour products, fruit juices, soybean milk, and polyvitaminated premixes, for which only the declared values of added thiamin and not the natural content are to be controlled, no enzyme hydrolysis is necessary. Furthermore, most vitamin-enriched foods contain higher thiamin content than unenriched samples because they are often manufactured with an overaddition of vitamins so that the labeled amounts are still present in the food at the expiry date.

For food samples with high thiamin levels, the method of analysis does not require the sensitivity of fluorescence detection and some workers have successfully used UV or diode-array detectors [34, 36–41]. For estimating the thiamin content of most biological materials, the more selective and sensitive fluorescence detection after oxidation of thiamin to thiochrome with alkaline potassium ferricyanide is widely used. However, it has to be kept in mind that the reaction is particularly susceptible to redox interference with polyphenols and other reducing components, resulting in a low thiochrome yield. This can be prevented by using cyanogen bromide as oxidizing agent. In the cyanogen bromide reaction, the production of thiochrome from thiamin is not affected by the presence of any reducing component [42]. Another possibility is the removal of interfering substances prior to the oxidation step or by using postcolumn instead of precolumn oxidation.

For the simultaneous determination of most water-soluble vitamins, including thiamin, in the same extract of foods some combined liquid chromatography

(LC)–mass spectrometry (MS) methods have recently been developed, such as LC–tandem mass spectrometry (MS/MS) and LC–electrospray ionization (ESI)-MS/MS [43, 44].

6.4
Extraction Procedures

For the determination of the total thiamin content of food products, extraction is normally performed by autoclaving with hydrochloric acid (0.1 mol l^{-1}) or sulfuric acid (0.05 mol l^{-1}) for 30 min at 121 °C or heating for 60 min at 100 °C. At this low pH, thiamin is stable and thiamin–protein complexes are hydrolyzed. Subsequently, in most cases an enzymatic incubation with dephosphorylation enzymes such as takadiastase follows. The enzyme efficiency has to be checked with each new batch for its activity to hydrolyze bound thiamin from food and for its endogenous amount of thiamin [45, 46]. In addition, some authors incubate food samples with proteases such as papain in order to free protein-bound forms of thiamin and of other B vitamins [34]. Other authors use trienzyme incubation with α-amylase, papain, and acid phosphatase for the combined extraction of thiamin, riboflavin, and vitamin B$_6$ from foods prior to LC. The digestion with three enzymes eliminates the need for acid hydrolysis to free bound forms of the vitamins [47]. For concentration and/or cleanup of the extracts before LC, several protocols for solid-phase extraction (SPE) on C$_{18}$ resin or commercial C$_{18}$ SPE columns have been developed for specific matrices [35].

Analyzing plant foods with low thiamin contents (<1 mg per 100 g) and traces of thiamin phosphates, such as Italian pasta, only acid hydrolysis without enzymatic digestion was used to extract free thiamin and to break up thiamin–protein complexes [44]. In a more recent report, free thiamin was extracted from homogenized foods, such as maize flour, tomato pulp, and kiwi flesh, using C$_1$8 cartridges and ethanol–water as eluent [45]. In both cases the highly sensitive and selective LC–ESI-MS/MS was used instead of HPLC with fluorescence or UV detection. The extraction conditions with C$_{18}$ cartridges were mild and it was not possible to estimate the thiamin forms bound to food proteins. Mild conditions were applied in order to preserve the analytes and to avoid the creation of artifacts [44].

On analyzing fortified foods and dietary supplements, incubation with enzymes is also dispensable, because thiamin and other water-soluble vitamins are typically added in a single chemical form and matrix issues are usually not as complex as with natural foods [48]. This implies a shorter time for sample preparation. The extractions used are not suitable for determining endogenous (bound) vitamins in food samples and are generally applied to enriched foods and supplements. In most cases, thiamin in fortified food is simultaneously determined with several other water-soluble vitamins. Therefore, the previous sample treatment has to consider the special chemical properties of each of them. For the quantification of nine vitamins in polyvitaminated premixes, used for the fortification of infant nutrition products, only one aqueous extraction and two different sample dilutions were necessary [39]. For the determination of thiamin with or without other water-soluble vitamins, such as riboflavin, pyridoxine, and niacin, in infant formulas, an extraction and protein precipitation with weak hydrochloric, trichloroacetic, sulfuric, or perchloric acid and filtration was used [34, 49, 50]. Some authors sonicate and concentrate the extract and/or dilute it with the mobile phase or single components of it [36, 38, 51]. For the sample preparation of multivitamin-enriched milk products, such as milk, infant nutrition, and certified reference material milk powder, an acidic solution containing zinc and tungsten salts was used for the precipitation of proteins and fat. After centrifugation and filtration, the samples were ready for chromatography [37]. The most relevant advantages of the proposed method are the simultaneous determination of the eight more common vitamins in enriched foods and a reduction in the time required for quantitative extraction [52].

The Food Composition and Methods Development Laboratory (FCMDL) of USDA has undertaken a long-term project to develop, validate, and establish analytical methods that utilize modern technologies for the simultaneous determination of water-soluble vitamins in dietary supplements and fortified and unfortified foods. An objective of this project is the development and validation of sample preparation procedures to optimize extraction, to remove interferences, and/or to concentrate vitamins that are difficult to analyze. Multiple extraction approaches with different buffers at various pH values are under systematic examination, in addition to classical and modern methods such as pressurized liquid extraction, microwave-assisted extraction, ultrasonic treatment, stirring, shaking, and Soxhlet extraction. The purpose of this project, entitled "Measurement Systems for Determination of Vitamins in Foods and Dietary Supplements," is systematic exploration to ensure complete extraction and to compare the extraction efficiencies of different procedures.

6.5
Liquid Chromatography Procedures

Several HPLC methods are available for the determination of thiamin in foodstuffs, which are mainly performed on C_{18} stationary reversed phases with isocratic mobile phases. Although C_{18} has been the most common support for thiamin chromatography, polystyrene resin provides advantages when precolumn derivatization of thiamin to thiochrome is used, because it is stable at alkaline pH. Mobile phase components of published reversed-phase systems suitable for thiamin resolution include methanol–water, methanol–buffers (phosphate and acetate), acetonitrile buffers, and buffers of various concentrations. Ion-pair reagents, usually heptane- or hexanesulfonates, are often added to the mobile phase to improve resolution. For the pre- or postcolumn derivatization of thiamin to thiochrome, alkaline hexacyanoferrate(III) is used in most cases and fluorescence is measured at wavelengths of 360–378 nm for excitation and 420–440 nm for emission [35, 45, 53, 54]. For samples with higher thiamin levels, which do not require the high sensitivity of fluorescence detection, some authors have successfully used UV detection at wavelengths of 245–254 nm [35, 36].

For the simultaneous determination of several vitamins in a single run, HPLC coupled with diode-array detection or LC–ESI-MS/MS methods are suitable. In the case of water-soluble vitamins, the application of a technique such as LC–MS, which is sensitive and selective, allows considerable simplification of the pre-analytical procedure. Leporati *et al.* [43] applied an LC–ESI-MS/MS method for the quantitative analysis of six water-soluble vitamins (B_1, B_2, three B_6 vitamers, two niacin vitamers, pantothenic acid and folic acid) in typical Italian pasta and also in fortified pasta samples produced for the US market. Gentili *et al.* [44] used an LC–ESI-MS/MS method for the simultaneous analysis of nine water soluble vitamins (B_1, B_2, five B_6 vitamers, B12, two niacin vitamers, biotin, pantothenic acid, folic acid and ascorbic acid) in various food samples, that is, maize flour, kiwi, and tomato pulp.

For the determination of thiamin in fortified foodstuffs and dietary supplements, some particular HPLC methods have been reported. Ayi *et al.* [36] and Nicolas and Pfender [35] developed very rapid and simple LC methods for the determination of nonphosphorylated thiamin in infant formulas, which are applicable to fortified and unfortified products. Both methods were validated against the AOAC manual fluorimetric method for thiamin [43]. In the majority of cases not only a single vitamin such as thiamin needs to be quantified in fortified products, but also several other water-soluble vitamins. Chase *et al.* [49] developed an LC method for estimating the vitamins B_1, B_2, and B_6 in milk- and soy-based infant formulas. The method uses *m*-hydroxybenzoic acid as an internal standard and the results compared favorably with the AOAC methods for thiamin, riboflavin, and pyridoxine [35]. Agostini *et al.* [38] used a simple HPLC method with diode-array detection, which permitted a rapid, versatile, and simultaneous determination of vitamins B_1, B_2, and B_6 and niacin in many enriched Brazilian foods. Riccio *et al.* [57] used a rapid and reliable HPLC method with diode-array detection for the simultaneous determination of vitamins B_1, B_6, and B_{12} in boiled ham and various fortified meat products. The method ensures low detection limits and good sensitivity and resolution within an analysis time of 17 min. Cooking caused considerable losses of the three essential B vitamins, indicating that fortification of meat products seems to be a useful procedure. Heudi *et al.* [39] also used a simple and rapid HPLC method with diode-array detection. They were able to quantify in a single run nine vitamins (vitamins B_1, B_2, B_6, B_{12}, and C, niacin, pantothenic acid, folic acid, and biotin) in polyvitaminated premixes used for the fortification of infant formulas. The validation experiments showed linearity of the method for the determination of thiamin concentrations between 1.25 and 50 μg ml^{-1} with acceptable repeatability, precision, and accuracy. The successful application of this method to the simultaneous determination of nine water-soluble vitamins renders it a perfect tool for the quality control of fortified foodstuffs by public and private sector analytical laboratories. Zafra-Gómez *et al.* [37] published an HPLC method with either fluorescence or UV–visible detection for the simultaneous determination of eight vitamins (vitamins B_1, B_2, B_6, B_{12}, and C, niacin, pantothenic acid, and folic acid) in two enriched UHT milk beverages marketed for children and an enriched UHT milk infant formula. The method was validated using CRM 421 milk powder, a certified reference material from the Community Bureau of Reference (BCR), Brussels, Belgium. The recoveries ranged from 90 to 100%.

In a long-term research project, the FCMDL of USDA is improving and updating procedures to measure water-soluble vitamins simultaneously in foods and dietary supplements with LC–MS and LC–IDMS (isotope dilution mass spectrometry). A method was developed for the simultaneous determination of thiamin, pyridoxine, niacin, and pantothenic acid in a single chromatographic run. The four vitamins were isotopically labeled with ^{13}C and/or ^{2}H and the method was used for the determination of the four vitamins in SRM 3280, a multivitamin/multimineral dietary supplement, which is used as a standard reference material [57]. The results by LC–IDMS were in good agreement with those given by LC–UV/fluorescence [48]. This work will be expanded to methods for these vitamins in fortified and natural levels in foods [56]. Chen *et al.* [59] used an HPLC method with different detectors for the determination of eight water-soluble vitamins (vitamins B_1, B_2, and B_6, niacin, folic acid, pantothenic acid, biotin, and ascorbic acid) in SRM 3280 dietary supplements. Analysis was performed using HPLC and different detection methods such as diode-array, fluorescence, and mass spectrometry. Following extraction, the method does not require any sample cleanup or preconcentration step except centrifugation and filtration.

6.6
Conclusion

Several validated methods for the determination of thiamin in fortified foods and dietary supplements are available. Sometimes thiamin is measured as a single vitamin, often in combination with other water-soluble vitamins. Future procedural development are directed at sample preparation with optimization of extraction, removal of interferences, and concentration without destruction of vitamins. Part of long-term projects is the effort to design methods for the simultaneous determination of multiple vitamins in extracts of fortified foods with no chromatographic separations. Of particular importance is the improvement and validation of the measurement systems and the development of new methods and essential reference materials.

References

1 Rindi, G. (1996) Thiamin. In *Present Knowledge in Nutrition*, 7th edn. (eds. E.E. Ziegler and L.J. Filer), ILSI Press, Washington, DC, pp. 160–166.

2 USDA USDA National Nutrient Database for Standard Reference, Release 22. http://www.ars.usda.gov/SP2UserFiles/Place/12354500/Data/SR22/nutrlist/sr22w404.pdf (accessed 11 January 2011).

3 Poel, C. (2008) Thiamin und Thiaminphosphate in der Muskulatur von Schweinen und in Schweinefleischprodukten. Beiträge zur Kinetik der Transformationsreaktionen von Thiamin. http://elib.tiho-hannover.de/dissertations/poelc_ss08.pdf (accessed 11 January 2011).

4 Jakobsen, J. (2008) Optimisation of the determination of thiamin, 2-(1-hydroxyethyl)thiamin, and riboflavin in food samples by use of HPLC. *Food Chemistry*, **106**, 1209–1217.

5 Bitsch, R. (2003) Thiamin: physiology, in *Encyclopaedia of Food Science and Nutrition*, 2nd edn., (eds. B. Caballero, L. Trugo, and P. Finglas), Academic Press, London, pp. 5772–5780.

6 Gibson, G.E. and Blass, J.P. (2007) Thiamin-dependent processes and treatment. *Antioxidants and Redox Signaling*, **9**, 1605–1620.

7 Bunik, V.I., Schloss, J.V., Pinto, J.T., Gibson, G.E., and Cooper, A.J.L. (2007) Enzyme-catalysed side reactions with molecular oxygen may contribute to cell signalling and neurodegenerative diseases. *Neurochemical Research*, **32**, 871–891.

8 Bâ, A. (2008) Metabolic and structural role of thiamin in nervous tissues. *Cellular and Molecular Neurobiology*, **28**, 923–931.

9 Depeint, F., Bruce, W.R., Shangari, N., Mehta, R., and OBrien, P.J. (2006) Mitochondrial function and toxicity: role of the B vitamin family on mitochondrial energy metabolism. *Chemico-Biological Interactions*, **163**, 94–112.

10 Gibson, G.G., Blass, J.P., Flint Beal, M., and Bunik, V. (2005) The α-ketoglutarate-dehydrogenase complex. A mediator between mitochondria and oxidative stress in neurodegeneration. *Molecular Neurobiology*, **31**, 43–63.

11 Singleton, C.K. and Martin, P.R. (2001) Molecular mechanisms of thiamin utilization. *Current Molecular Medicine*, **1**, 197–207.

12 Egoramaiphol, S. and Bitsch, I. (1986) The influence of orally administered caffeic acid and bisulfite upon thiamin metabolism in rats. *Journal of the Medical Association of Thailand*, **69**, 659–663.

13 Wenisch, S., Steinmetz, T., Fortmann, B., Leiser, R., and Bitsch, I. (1996) Can megadoses of thiamin prevent ethanol-induced damages of rat hippocampal CA 1 pyramidal neurones? *Zeitschrift für Ernahrungswissenschaft*, **35**, 266–272.

14 Bates, C.J. (2001) Thiamin In *Present Knowledge in Nutrition*, 8th edn. (eds. B.A. Bowman and R.M. Russell), ILSI Press, Washington, DC, pp. 184–190.

15 Bates, C.J. (2009) Thiamin.In *Present Knowledge in Nutrition*, 9th edn., vol. 1 (eds. B.A. Bowman and R.M. Russel), ILSI Press, Washington, DC, pp. 242–249.

16 Mancinelli, R. and Ceccanti, M. (2009) Biomarkers in alcohol misuse: their role in the prevention and detection of thiamin deficiency. *Alcohol and Alcoholism*, **44**, 177–182.

17 Harper, C. (2006) Thiamin (vitamin B$_1$) deficiency and associated brain damage is still common throughout the world and the prevention is simple and safe. *European Journal of Neurology*, **13**, 1078–1082.

18 Aasheim, E.T. (2008) Wernicke encephalopathy after bariatric surgery: a systematic review. *Annals of Surgery*, **248**, 714–720.

19 Wooley, J.A. (2008) Characteristics of thiamin and its relevance to the management of heart failure. *Nutrition in Clinical Practice*, **23**, 487–493.

20 Kasper, H. and Hölscher, I. (1966) Der Einfluß von Hydrochlorothiazid auf die Thiaminausscheidung im Urin. *Klinische Wochenschrift*, **44**, 568–571.

21 Kuo, S.H., Debnam, J.M., Fuller, G.N., and de Groot, J. (2009) Wernicke's encephalopathy: an underrecognized and reversible cause of confusional state in cancer patients. *Oncology*, **76**, 10–18.

22 Butterworth, R.F. (2009) Thiamin deficiency-related brain dysfunction in chronic liver failure. *Metabolic Brain Disease*, **24**, 189–196.

23 Frank, T., Bitsch, R., Maiwald, J., and Stein, G. (1999) Alteration of thiamin pharmacokinetics by end-stage renal disease (ESRD). *International Journal of Clinical Pharmacology and Therapeutics*, **37**, 449–455.

24 Frank, T., Czeche, K., Bitsch, R., and Stein, G. (2000) Assessment of thiamin status in chronic renal failure patients, transplant recipients and hemodialysis patients receiving a multivitamin supplementation. *International Journal for Vitamin and Nutrition Research*, **70**, 159–166.

25 Thornalley, P.J., Babaei-Jadidi, R., Al Ali, H., Rabbani, N., Antonysunil, A., Larkin, J., Ahmed, A., Rayman, G., and Bodmer, C.W. (2007) High prevalence of low plasma thiamin concentration in diabetes linked to a marker of vascular disease. *Diabetologia*, **50**, 2164–2170.

26 Bruce, W.R., Furrer, R., Shangari, N., O'Brien, P.J., Medline, A., and Wang, Y. (2003) Marginal dietary thiamin deficiency induces the formation of colonic aberrant crypt foci (ACF) in rats. *Cancer Letters*, **202**, 125–129.

27 Kesler, A., Stolovitch, C., Hoffmann, C., Avni, I., and Morad, Y. (2005) Acute ophthalmoplegia and nystagmus in infants fed a thiamin-deficient formula: an epidemic of Wernicke encephalopathy. *Journal of Neuro-ophthalmology*, **25**, 169–172.

28 Feliciotti, E. and Esselen, W.B. (1957) Thermal destruction rates of thiamin in puréed meats and vegetables. *Food Technology*, **11**, 77–84.

29 Kessler, H.-G. and Fink, R. (1986) Changes in heated and stored milk with an interpretation by reaction kinetics. *Journal of Food Sciences*, **51**, 1105–1111.

30 Yates, A.A., Schlicker, S.A., and Suitor, C.W. (1998) Dietary Reference Intakes: the new basis for recommendations for calcium and related nutrients, B vitamins, and choline. *Journal of the American Dietetic Association*, **98**, 699–706.

31 Doets, E.L., de Wit, L.S., Dhonukshe-Rutten, R.A.M., Cavelaars, A.E.J.M., Raats, M.M., Timotijevic, L., Brzozowska, A., Wijnhoven, T.M.A., *et al.* (2008) Current micronutrient recommendations in Europe: towards understanding their differences and similarities. *European Journal of Nutrition*, **47** (Suppl. 1), 17–40.

32 European Commission (2006) Regulation (EC) No. 1925/2006, European Commission, Brussels.

33 Eitenmiller, R.R., Ye, L., and Landen, W.O., Jr. (eds.) (2008) *Vitamin Analysis for the Health and Food Sciences*, 2nd edn,, CRC Press, Boca Raton.

34 Poma, R.B. and Tabekhia, M.M. (1979) High performance liquid chromatographic analysis of B-vitamins in rice and rice products. *Journal of Food Science*, **44**, 263–268.

35 Nicolas, E.C. and Pfender, K.A. (1990) Fast and simple liquid chromatographic determination of nonphosphorylated thiamin in infant formula, milk, and other food. *Journal of the Association of Official Analytical Chemists*, **73**, 792–798.

36 Ayi, B.K., Yuhas, D.A., Moffet, K.S., Joyce, D.M., and Deangelis, N.J. (1985) Liquid chromatographic determination of thiamin in infant formula products by using ultraviolet detection. *Journal of the Association of Official Analytical Chemists*, **68**, 1087–1092.

37 Zafra-Gómez, A., Garballo, A., Morales, J.C., and García-Ayuso, L.E. (2006) Simultaneous determination of eight water-soluble vitamins in supplemented foods by liquid chromatography. *Journal of Agricultural and Food Chemistry*, **54**, 4531–4536.

38 Agostini, A. and Godoy, H.T. (1997) Simultaneous determination of nicotinamide, nicotinic acid, riboflavin, thiamin, and pyridoxine in enriched Brazilian foods by HPLC. *Journal of High Resolution Chromatography*, **20**, 245–248.

39 Heudi, O., Kilinc, T., and Fontannaz, P. (2005) Separation of water-soluble vitamins by reversed-phase high performance liquid chromatography with ultra-violet detection: application to polyvitaminated premixes. *Journal of Chromatography A*, **1070**, 49–56.

40 Mendiola, J.A., Marin, F.R., Señoráns, F.J., Reglero, G., Martín, P.J., Cifuentes, A., and Ibáñez, E. (2008) Profiling of different bioactive compounds in functional drinks by high-performance liquid chromatography. *Journal of Chromatography A*, **1188**, 234–241.

41 Vidović, S., Stojanović, B., Veljović, J., Pražić-Arsić, L., Roglić, G., and Manojlović, D. (2008) Simultaneous determination of some water-soluble vitamins and preservatives in multivitamin syrup by validated stability-indicating high-performance

liquid chromatography method. *Journal of Chromatography A*, **1202**, 155–162.

42 Fujiwara, M. and Matsui, K. (1953) Determination of thiamin by the thiochrome reaction. Application of cyanogen bromide in place of potassium ferricyanide. *Analytical Chemistry*, **25**, 810–812.

43 Leporati, A., Catellani, D., Suman, M., Andreoli, R., Manini, P., and Niessen, W.M.A. (2005) Application of a liquid chromatography tandem mass spectrometry method to the analysis of water-soluble vitamins in Italian pasta. *Analytica Chimica Acta*, **531**, 87–95.

44 Gentili, A., Caretti, F., D'Ascenzo, G., Marchese, S., Perret, D., Di Corcia, D., and Rocca, L.M. (2008) Simultaneous determination of water-soluble vitamins in selected food matrices by liquid chromatography/electrospray ionisation tandem mass spectrometry. *Rapid Communications in Mass Spectrometry*, **22**, 2029–2043.

45 Comité Européen de Normalisation (2003) EN 14122. Foodstuffs. Determination of Vitamin B$_1$ by HPLC. Comité Européen de Normalisation (CEN), Brussels.

46 AOAC (2003) AOAC Official Method 942.23, Thiamin (Vitamin B$_1$) in Foods, Fluorimetric Method, AOAC Official Methods of Analysis, 45.1.05. AOAC International, Gaithersburg, MD

47 Ndaw, S., Bergaentzlé, M., Aoudé-Werner, D., and Hasselmann, C. (2000) Extraction procedures for the liquid chromatographic determination of thiamin, riboflavin and vitamin B$_6$ in foodstuffs. *Food Chemistry*, **71**, 129–138.

48 Wolf, W.R., Goldschmidt, R.J., and Atkinson, R.L. (2008) Simultaneous determination of water-soluble vitamins in dietary supplements and fortified foods by LC–UV/fluorescence and LC-MS. Presented at the 122nd AOAC Annual Meeting and Exposition, Dallas, TX.

49 Chase, G.W., Landen, W.O., Eitenmiller, R.R., and Soliman, A.G. (1992) Liquid-chromatographic determination of thiamin, riboflavin, and pyridoxine in infant formula. *Journal of AOAC International*, **75**, 561–565.

50 Woollard, D.C. and Indyk, H.E. (2002) Rapid determination of thiamin, riboflavin, pyridoxine, and niacinamide in infant formula by liquid chromatography. *Journal of AOAC International*, **85**, 945–951.

51 Klejdus, B., Petrlová, J., Potěšil, D., Adam, V., Mikelová, R., Vacek, J., Kizek, R., and Kubán, V. (2004) Simultaneous determination of water- and fat-soluble vitamins in pharmaceutical preparations by high-performance liquid chromatography coupled with diode array detection. *Analytica Chimica Acta*, **520**, 57–67.

52 Konings, E. (2007) Water-soluble vitamins. *Journal of AOAC International*, **90**, 21B–25B.

53 Lynch, P.L.M. and Young, I.S. (2000) Determination of thiamin by high-performance liquid chromatography. *Journal of Chromatography A*, **881**, 267–284.

54 Van den Berg, H., van Schalk, F., Finglas, P.M., and de Froidmont-Görtz, I. (1996) Third EU MAT intercomparison on methods for the determination of vitamins B$_1$, B$_2$, and B$_6$ in food. *Food Chemistry*, **57**, 101- 108.

55 Chen, P., Ozcan, M., and Wolf, R. (2007) Contents of selected B vitamins in NIST SRM 3280 multivitamin/multielement tablets by liquid chromatography isotope dilution mass spectrometry. *Analytical and Bioanalytical Chemistry*, **389**, 343–347.

56 Chen, P., Atkinson, R., and Wolf, W.R. (2009) Single-laboratory validation of a high-performance liquid chromatographic–diode array detector–fluorescence detector/mass spectrometric method for simultaneous determination of water-soluble vitamins in multivitamin dietary tablets. *Journal of AOAC International*, **92**, 680–688.

57 Riccio, F., Mennela, C., and Fogliano, V. (2006) Effect of cooking on the concentration of B vitamins in fortified meat products. *Journal of Pharmaceutical and Biomedical Analysis*, **41**, 1592–1595.

7
HPLC Determination of Riboflavin in Fortified Foods

Roland Bitsch and Irmgard Bitsch

7.1
Introduction

Riboflavin, IUPAC systematic name 7,8-dimethyl-10-(1-D-ribityl)benzo[g]pteridin-2,4-dione, is a light-sensitive molecule. Earlier names derived from its discovery and first identification are lacto-, ovo-, and uroflavin. In alkaline solution it is decomposed to lumiflavin (7,8,10-trimethylisoalloxazine); in neutral or acidic medium, the photolysis proceeds to lumichrome (7,8-dimethylisoalloxazine) within hours (Figure 7.1). This also occurs in foods, when they are exposed to daylight for longer periods, for example, in milk and milk products.

Therefore, analytical manipulations and operations with riboflavin solutions should be performed in the dark or under red light. If protected from light, riboflavin is fairly stable against heat and oxygen. Riboflavin or vitamin B_2 is the active component (prosthetic group) of the flavin nucleotides FAD (flavin adenine dinucleotide) and FMN (flavin mononucleotide = riboflavin-5′-phosphate), which fullfill as coenzymes of oxidoreductases multiple functions involving catalytic hydrogen transfer reactions in the metabolism of carbohydrates, fats, and proteins [1]. Riboflavin is readily soluble in dilute mineral or organic acids but sparingly soluble in water (<10 mg per 100 ml at pH 7 and 27 °C) and very sparingly soluble in ethanol (<5 mg per 100 ml at 27 °C). It is also readily soluble but unstable in dilute alkali (Table 4.1). In contrast. FMN and FAD are water soluble (3 g per 100 ml at pH 7 and 25 °C). Absorbance spectra reveal absorption maxima in the visible and UV regions of the spectrum. Furthermore, in weakly acidic medium at pH 6–7 the molecule exhibits a characteristic yellow–green fluorescence with excitation maximum at 444 nm and emission maximum between 530 and 565 nm (Table 7.1). The degradation products after photolysis, lumichrome and lumiflavin, also fluoresce. After irradiation (λ_{ex} = 270 nm) in alkaline solution at pH 10–12, riboflavin is converted to lumiflavin, whereby the fluorescence is shifted to the green spectral range (λ_{em} = 418 nm). This conversion to lumiflavin followed by chloroform extraction allows a very sensitive quantification of riboflavin with a detection limit of 0.02 ng of vitamin B_2 [2].

In general, riboflavin is considered as a less "critical" vitamin, because the dietary reference intakes given by international boards, which vary between 1 and

Fortification of Foods with Vitamins, First Edition. Edited by Michael Rychlik.
© 2011 Wiley-VCH Verlag GmbH & Co. KGaA. Published 2011 by
Wiley-VCH Verlag GmbH & Co. KGaA.

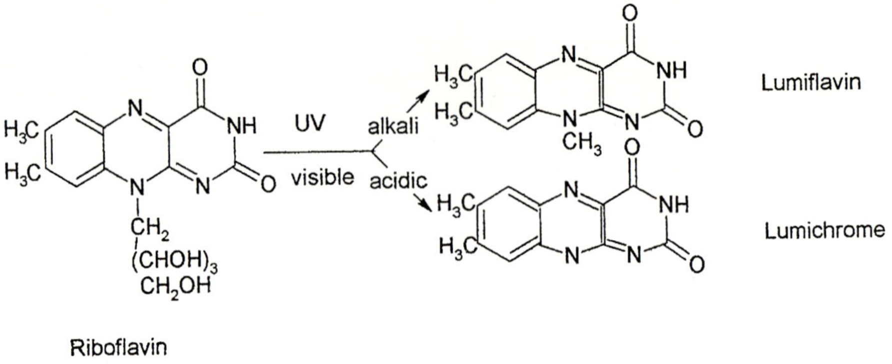

Figure 7.1 Conversion of riboflavin to lumiflavin and lumichrome.

Table 7.1 Physicochemical data for riboflavin according to [2].

Solubility	Sparingly in water (70–100 mg l^{-1}, 27 °C)
	Very sparingly in ethanol (45 mg l^{-1}, 27 °C)
	Readily in dilute alkali (decomposition) and in dilute acids
Absorbance (UV–Vis)	Maxima at 223, 267, 347, and 444 nm in HCl (0.1 mol l^{-1})
Fluorescence	Yellow–green in weakly acidic solution at pH 6–7 (λex 444 nm, λem 530–565 nm)

2 mg per day for adults, are met or even surpassed in population groups of Western countries [3]. Occasional fortifications of foods with this vitamin are restricted to multivitamin syrups, fruit juices, or breakfast cereals. Superior natural sources of this vitamin are milk and dairy products, but meat and whole meal products just as green vegetables are also relevant suppliers of this nutrient. A list of the riboflavin contents of foods is available in an online database [4]. The purpose of this chapter is to give an overview of high-performance liquid chromatography (HPLC) methods for vitamin B$_2$ quantitation.

7.2
Materials and Methods

7.2.1
Chemicals

Acetonitrile, ammonium chloride, chloroform, citrate–phosphate buffer (pH 5.5), 1,4-dioxane, formic acid, glacial acetic acid, hydrochloric acid, methanol, methylene chloride, 4-nitrophenylphosphate, potassium dihydrogenphosphate, potassium perchlorate, sodium acetate, sulfuric acid, trichloroacetic acid, and urea can be obtained from common suppliers. Claradiastase is available from Fluka (Buchs, Switzerland) and takadiastase from Pfaltz & Bauer (Waterbury, CT, USA). The standards riboflavin, FAD, FMN, 7-ethyl-8-methylriboflavin and sorboflavin can be obtained from Sigma (Deisenhofen, Germany).

7.2.2
Extraction Procedure

Methods for the extraction of riboflavin from foods depend on whether the sum of free riboflavin plus its proportion in flavoproteins FAD and FMN is required or the simultaneous determination of FAD and FMN in addition to unbound vitamin. In fortified foods, riboflavin is commonly added in its free form. Equally in milk, dairy products, and eggs, loosely bound riboflavin is by far the predominant flavin derivative. In these cases, the procedure can be simplified by acidification to remove proteins following centrifugation or filtration through cellulose filters. For protein precipitation, methanol-acetic acid mixtures, dilute trichloroacetic acid, formic acid, or perchlorate are used [5–9]. Alternatively, liquid milk or an acetate buffer extract [0.02 mol l^{-1} acetate–methanol (1:1), pH 4] of the dairy product is passed through a conditioned Sep-Pak C$_{18}$ cartridge to remove potential interferences followed by elution of riboflavin with the above acetate buffer–methanol mixture. This extraction procedure is also applicable to the analysis of foods fortified with riboflavin [10–12].

However, when milk proteins were solubilized by using formic acid–urea (6% formic acid containing urea, 2 mol l^{-1}), minor amounts of FMN and FAD can be quantified in addition to free riboflavin [13]. Obviously, these flavins were coprecipitated with milk proteins by simple acidification and escaped former analyses, thus providing an incomplete picture of the (total) riboflavin content in milk and dairy products. Older methods, based on the extraction of flavins from biological materials such as blood and urine by using acetonitrile at pH 7 or with 5% ammonium chloride solution at pH 5.5 and heating at 80 °C for 15 min to precipitate proteins, had the disadvantage that the protein precipitate may adsorb up to 20% of the riboflavin [14, 15].

Most food materials of animal and plant origin contain variable amounts of flavins in addition to free riboflavin. In a more sophisticated procedure, the individual flavins in raw and cooked meats and also in dairy products and cereals were extracted by using a methanol–methylene chloride mixture and citrate–phosphate buffer (0.1 mol l^{-1} at pH 5.5) [16].

If, on the other hand, the determination of total riboflavin is required, the extraction step should involve the release of riboflavin from the phosphorylated and protein-bound flavins by combined acid and enzymatic hydrolysis. A hydrolysis step with mineral acids followed by an effective enzyme treatment is the most important precondition for analytical accuracy, as incomplete hydrolysis is a relevant source of variations in riboflavin determination.

By autoclaving food samples at 121 °C for 15–30 min with dilute mineral acids, for example, 0.1 N HCl or H_2SO_4, protein-bound flavins are liberated and FAD is converted to FMN. Complete conversion to free riboflavin, particularly from starch- or protein-containing food matrices, requires subsequent enzymatic hydrolysis with takadiastase or clarase [17–24].

Both of these amylolytic enzymes also possess phosphatase activity and low protease activity. In view of the varying phosphatase activity in commercially available enzyme preparations, it may be necessary to determine their activity in preceding studies by using 4-nitrophenylphosphate as substrate. An estimation proved that 0.4–1.0 units of takadiastase were required in order to obtain a complete hydrolysis of up to 0.1 μmol of FAD and FMN, respectively. For practical purposes, this relation corresponds to 100–500 mg of enzyme preparation per gram of food sample. Irrespective of the enzyme preparations used, incubation periods should be at least or even exceed 18 h [19–21, 25].

7.2.3
Analytical Principles

HPLC analyses are mainly performed on C_{18} stationary reversed phases (RPs) with hydrophilic eluents based on water-methanol followed either by fluorimetric or rarely by ultraviolet (UV) detection. Less frequently used are C_{10}, amino, hydrophilic gels, and normal-phase chromatography based on silica materials.

Commonly, riboflavin, thiamin, and other B vitamins in foods and biological materials are analyzed simultaneously. In those cases, ion-pair chromatography with hexane- or heptanesulfonate or triethyl- or tetrabutylammonium phosphate is preferred to improve the resolution. Separation procedures include isocratic and gradient elution using mixtures of methanol or acetonitrile with water, acetic acid, or acetate or phosphate buffer, occasionally modified with 1,4-dioxane or formic acid. With gradient elution techniques using phosphate-buffered eluents based on methanol and/or acetonitrile, optimal resolution of FAD, FMN, and riboflavin and of simultaneously analyzed B-complex vitamins can be achieved [16, 26–30].

7.2.4
Detection Modes

Fluorescence detection as the preferred detection method avoids quantification problems due to interfering peaks that appear with UV detection. Because the fluorescence intensity is pH dependent, the mobile phases used for elution should contain buffers in the pH range 3.7–7.5 corresponding to the riboflavin fluorescence maxima (λ_{ex} = 440–450 nm, λ_{em} = 530–560 nm). The excitation maximum for fluorimetric detection of FAD and FMN is in the range 440–500 nm and the emission maximum is at 530 nm.

When employing a multistep gradient elution program based on citrate–phosphate buffer (pH 5.5) and acetonitrile on a polymer-based PLRP-S column with subsequent fluorimetric monitoring at λ_{ex} 450/λ_{em} 520 nm, detection limits of flavins in various foods as low as 0.55 pmol (0.21 ng) of riboflavin, 1.96 pmol (0.89 ng) of FMN, and 14.19 pmol (11.15 ng) of FAD are obtained per injection [16].

In food samples with low riboflavin contents, as for example in meat products and vegetables, it could, however, be advisable to convert riboflavin to lumiflavin by UV irradiation in order to enhance the sensitivity of fluorescence detection. Accordingly, after acid extraction and enzymatic hydrolysis, an aliquot of the filtrate is adjusted to pH 10–12 (15% NaOH) and irradiated under a suitable UV lamp for 30 min. Thereafter, the sample is acidified (glacial acetic acid) and lumiflavin is extracted with chloroform. After drying over anhydrous sodium sulfate, the extract is injected on to an HPLC column and the lumiflavin fluorescence is measured at $\lambda_{ex} = 270$ nm and $\lambda_{em} = 418$ nm. The detection limit is given as 0.02 ng per injection [30]. Alternatively, filtered extracts of food samples with a low riboflavin content after autoclaving and enzymatic treatment can be further concentrated and purified on disposable Sep-Pak C_{18} cartridges [31]. UV detection is about 30 times less sensitive than fluorescence detection and, therefore, is only suitable for measurements of riboflavin in foods with higher natural contents (dairy products, eggs), fortified foods, or pharmaceuticals. Absorbance maxima of riboflavin are 270 and 446 nm [6, 8, 10, 11, 30, 32–34].

7.3
HPLC Intercomparisons

In intercomparison studies of methods that were organized by the EU Community Bureau of Reference (BCR), certified reference materials such as lyophilized meat products and vegetables and wholemeal flour were analyzed for water-soluble vitamins. Accordingly, autoclaving with HCl (0.1 mol l^{-1}) at 121 °C for 15–30 min followed by incubation with takadiastase (100–500 mg g^{-1} sample), or a mixture of takadiastase and claradiastase (10:1, m/m) at pH 4.0–4.5 and 37–45 °C for at least 3 h (with a tendency to longer incubation periods up to 18 h) evolved as the optimal extraction procedure for riboflavin in foods. The enzyme should in any case be in excess by using 60–500 mg of enzyme per gram of sample. The additional application of ultrasonication to the incubation mixture shortened the enzymatic hydrolysis to 1 h. The subsequent HPLC analysis involved separation on C_{18} material with water–methanol, water–methanol–acetic acid or phosphate buffer mixtures at pH 2.9–5 or with acetonitrile substituting methanol, and occasionally ion-pair chromatography with hexane- or heptanesulfonic acid. The type of HPLC column (normal phase or RP) did not affect the results. Fluorimetric detection was employed with excitation at 422–467 nm and emission at 510–525 nm. The detection limit was 20 pg of riboflavin absolute per injection.

The coefficient of variation of the total riboflavin content between or within laboratories (milk powder and pig liver) was 10–12%, showing an acceptable variability. Higher variability of up to 40% was found for the results for vegetables and wholemeal [19, 21, 23, 24, 35].

7.4
Conclusion

HPLC analysis is established as the method of choice for the determination of riboflavin in foods. The scope of the analysis determines the mode of procedure to be used. For the determination of the flavins FAD and FMN, including free riboflavin, a nondegradative extraction with dilute organic acids can be used. Determination of total riboflavin needs combined extraction by autoclaving with mineral acids followed by enzymatic hydrolysis. Chromatography is performed mainly on RP C_{18} columns by using isocratic or less frequently gradient elution with acidified or buffered methanol–water or acetonitrile mixtures followed by fluorimetric detection as the most sensitive quantification method. In fortified foods, riboflavin can also be determined using UV detection with adequate precision.

References

1 McCormick, D.B. (2003) Riboflavin–physiology. In *Encyclopedia of Food Science and Nutrition* (eds. B. Caballero, L. Trugo, and P. Finglas), Academic Press, London, pp. 4989–4995.

2 Bitsch, R. (2003) Riboflavin–properties and determination. In *Encyclopedia of Food Science and Nutrition*, Academic Press, London, pp. 4983–4989.

3 German Society for Nutrition, Austrian Society for Nutrition, Swiss Society for Nutrition Research, Swiss Association for Nutrition (2008) *DACH Reference Values for Nutrient Intake*, 3rd edn., Umschau/Braus, Frankfurt.

4 US Department of Agriculture, Agricultural Research USDA (2010) USDA National Nutrient Database for Standard Reference, Release 22, Nutrient Lists – Riboflavin, registered under http://www.nal.usda.gov/fnic/foodcomp/Data/SR20/nutrlist/sr20a405.pdf.

5 Munari, M., Miurin, M., and Goi, G. (1991) Didactic application to riboflavin HPLC analysis. A laboratory experiment. *Journal of Chemical Education*, **68**, 78–79.

6 Albala-Hurtado, S., Veciana-Nogues, M.T., Izquierdo-Pulido, M., and Marine-Font, A. (1997) Determination of water-soluble vitamins in infant milk by high-performance liquid chromatography. *Journal of Chromatography*, **778**, 247–253.

7 Chase, G.W., Landen, W.O., Eitenmiller, R.R., and Soliman, A.M. (1992) Liquid chromatographic determination of thiamine, riboflavin, and pyridoxine in infant formula. *Journal of AOAC International*, **75**, 561–565.

8 Otles, S. and Hisil, Y. (1993) High pressure liquid chromatographic analysis of water-soluble vitamins in eggs. *Italian Journal of Food Science*, **5**, 69–75.

9 Severo-Silva, L. Jr., Trevisan, M.G., Rath, S., Poppi, R.J., and Reyes, F.G.R. (2005) Chromatographic determination of riboflavin in the presence of tetracycline in skimmed and full cream milk using fluorescence detection. *Journal of the Brazilian Chemical Society*, **16** (6A), 1174–1178.

10 Ashoor, S.H., Seperich, G.J., Monte, W.C., and Welty, J. (1983) HPLC determination of riboflavin in eggs and dairy products. *Journal of Food Science*, **48**, 92–95.

11 Ashoor, S.H., Knox, M.J., Olsen, J.R., and Deger, D.A. (1985) Improved liquid chromatographic determination of riboflavin in milk and dairy products. *Journal of AOAC International*, **68** (4), 693–696.

12 Stancher, B. and Zonta, F. (1986) High performance liquid cromatographic analysis of riboflavin (vitamin B_2) with visible absorbance detection in Italian cheese. *Journal of Food Science*, **51** (3), 857–858.

13 Bilic, N. and Sieber, R. (1990) Determination of flavins in dairy products by high-performance liquid chromatography using sorboflavin as internal standard. *Journal of Chromatography*, **511**, 359–366.

14 Speek, A.J., Van Schaik, F., Schrijver, J., and Schruers, W.H.P. (1982) Determination of the B_2 vitamer flavin-adenine- dinucleotide in whole blood with high-performance liquid chromatography with fluorimetric detection. *Journal of Chromatography*, **228**, 311–316.

15 Lopez-Anaya, A. and Mayersohn, M. (1987) Quantification of riboflavin, riboflavin 5′-phosphate and flavin adenine dinucleotide in plasma and urine by high-performance liquid chromatography. *Journal of Chromatography*, **423**, 105–113.

16 Russell, L.F. and Vanderslice, J.T. (1992) Non-degradative extraction and simultaneous quantitation of riboflavin, flavin mononucleotide and flavin adenine dinucleotide in foods by HPLC. *Food Chemistry*, **43**, 151–162.

17 Kamman, J.F., Labuza, T.R., and Warthesen, J.J. (1980) Thiamin and riboflavin analysis by high performance liquid chromatography. *Journal of Food Science*, **45**, 1497–1504.

18 Finglas, P.M. and Faulks, R.M. (1987) Critical review of HPLC methods for the determination of thiamin, riboflavin and niacin in food. *Journal of Micronutrient Analysis*, **3**, 251–283.

19 Hollman, P.C.H. and Slangen, J.H. (1993) Intercomparison of methods for the determination of vitamins in foods. *Analyst*, **118**, 481–488.

20 Arella, F., Lahély, S., Bourguignon, J.B., and Hasselmann, C. (1996) Liquid chromatographic determination of vitamins B_1 and B_2 in foods. A collaborative study. *Food Chemistry*, **56** (1), 81–86.

21 Van den Berg, H., van Schaik, F., Finglas, P.M., and Froidmont-Görtz, I. (1996) Third EU Mat intercomparison on methods for the determination of vitamins B_1, B_2 and B_6 in food. *Food Chemistry*, **57** (1), 101–108.

22 Tang, X., Cronin, D.A., and Brunton, N.P. (2006) A simplified approach to the determination of thiamine and riboflavin in meats using reverse phase HPLC. *Journal of Food Composition and Analysis*, **19**, 831–837.

23 Gentili, A., Caretti, F., D'Ascenso, G., Marchese, S., Perret, D., Di Corcia, D., and Mainero Rocca, L. (2008) Simultaneous determination of water-soluble vitamins in selected food materials by liquid chromatography/ electrospray ionization tandem mass spectrometry. *Rapid Communications in Mass Spectrometry*, **22**, 2029–2043.

24 Jakobsen, J. (2008) Optimisation of the determination of thiamin, 2-(1-hydroxyethyl)thiamin, and riboflavin in food samples by use of HPLC. *Food Chemistry*, **106**, 1209–1217.

25 Ndaw, S., Bergaentzlé, M., Aoudé-Werner, D., and Hasselmann, C. (2000) Extraction procedures for the liquid chromatographic determination of thiamin, riboflavin and vitamin B_6 in foodstuffs. *Food Chemistry*, **71**, 129–138.

26 Kanno, C., Shiharhuji, K., and Hoshi, T. (1991) Simple method for separate determination of three flavins in bovine milk by high performance liquid chromatography. *Journal of Food Science*, **56**, 678–681.

27 Greenway, G.M. and Kometa, N. (1994) On-line sample preparation for the determination of riboflavin and flavin mononucleotides in foodstuffs. *Analyst*, **119** (5), 929–935.

28 Agostini, T.S., and Godoy, H.T. (1997) Simultaneous determination of nicotinamide, nicotinic acid, riboflavin, thiamin, and pyridoxine in enriched Brazilian foods by HPLC. *Journal of High Resolution Chromatography*, **20**, 245–248.

29 Zafra-Gomez, A., Garballa, A., Morales, J.C., and García-Ayuso, L.E. (2006) Simultaneous determination of eight water-soluble vitamins in supplemented foods by liquid chromatography. *Journal of Agricultural and Food Chemistry*, **54**, 4531–4536.

30 Vidović, S., Stojanović, B., Veliković, J., and Pražić-Arsić, L. (2008) Simultaneous determination of some water-soluble vitamins and preservatives in multivitamin syrup by validated stability-indicating high-performance liquid chromatography method. *Journal of Chromatography A*, **1202**, 155–162.

31 Ang, C.Y.W. and Moseley, F.A. (1980) Determination of thiamin and riboflavin in meat and meat products by high-pressure liquid chromatography. *Journal of Agricultural and Food Chemistry*, **28**, 483–486.

32 Wimalasiri, P. and Wills, R.B.H. (1985) Simultaneous analysis of thiamin and riboflavin in foods by high-performance liquid chromatography. *Journal of Chromatography*, **318**, 412–416.

33 Blanco, D., Sanchez, L.A., and Gutierez, M.D. (1994) Determination of water soluble vitamins by liquid chromatography with ordinary and narrow-bore columns. *Journal of Liquid Chromatography*, **17**, 1525–1539.

34 Barna, E. and Dworschak, E. (1994) Determination of thiamine (vitamin B_1) and riboflavin (vitamin B_2) in meat and liver by high-performance liquid chromatography. *Journal of Chromatography A*, **668**, 359–363.

35 Ollilainen, V., Finglas, P.M., van den Berg, H., and de Froidmont-Görtz, I. (2001) Certification of B-group vitamins (B_1, B_2, B_6, and B_{12}) in four food reference materials. *Journal of Agricultural and Food Chemistry*, **49**, 315–321.

8
HPLC–MS Determination of Vitamin C in Fortified Food Products

Antonia Garrido Frenich, José Luis Martínez Vidal, Remedios Fernández Fernández, and Roberto Romero-González

8.1
Introduction

Vitamins are organic molecules that cannot be synthesized by the human organism and, therefore, must be obtained through direct ingestion from the diet. These compounds are essential nutrients, which are indispensable for proper functioning of the metabolism [1], and they can act as coenzymes and prosthetic groups of enzymes [2]. Although low concentrations of vitamins are required for human health, the lack of or an excess of these compounds can cause adverse effects.

Vitamins are usually classified according to their solubility in water or fat. Vitamin C, also known as the L-enantiomer of ascorbic acid, belongs to the former group. It is an excellent antioxidant, active against diseases such as cancer, heart disease, and various inflammatory disorders, and is essential for growth and other regulatory roles throughout the entire body [3]. Furthermore, vitamin C is very important in reproductive systems [4], and also in promoting the absorption of other micronutrients, such as iron [5, 6]. In addition, it is known that the use of multivitamin supplements, including vitamin C, may delay the progression of HIV disease [7, 8].

Vitamin C is mainly supplied by fruits and vegetables but it cannot be stored in the human body. This means that we need a continuous supply of this vitamin in our diet. Toxicity by over-ingestion of vitamin C is very rare, as it is easily eliminated via the urine. The maximum level of total chronic daily intake of vitamin C that does not impose any risk of adverse effects to human health is 2000 mg per day [9]; quantities exceeding this value are not recommended because they can lead to diarrhea and stomach upsets [10]. However, the lack of vitamin C leads to the disease scurvy, which mostly affects older and undernourished people.

The best way to preserve health and to ingest the daily needs of essential vitamins is a balanced diet, but this is not always possible. In these cases, fortified foods are a nutritional alternative, because they can provide an adequate intake of vitamins to large sectors of the population without drastic changes in their eating habits.

Fortification of Foods with Vitamins, First Edition. Edited by Michael Rychlik.
© 2011 Wiley-VCH Verlag GmbH & Co. KGaA. Published 2011 by
Wiley-VCH Verlag GmbH & Co. KGaA.

The Codex Alimentarius defines fortification or enrichment as "the addition of one or more essential nutrients to a food, whether or not it is normally contained in the food, for the purpose of preventing or correcting demonstrated deficiency of one or more essential nutrients in the population or specific population groups" [11]. Moreover, it defines restoration as "the addition to a food of essential nutrient(s), which are lost during the course of good manufacturing practice, or during normal storage and handling procedures, in amounts which will result in the presence in the food of the level(s) of the nutrient(s) present in the edible portion of the food before processing, storage, and handling" [11]. For many years, the latter procedure has been applied in Western countries to restore micronutrients lost by food processing, which can help to avoid the spread of diseases associated with vitamin deficiencies [12].

Currently, many countries permit the fortification of foods, although the variety of fortified foods depends on the country. For instance, the European Union allows voluntary fortification of spreadable fats and margarines with vitamins A and D [12]. In addition, the European Commission has established the limits of vitamins A, E, and C that can be added to meet nutritional requirements and to guarantee the stability of infant formulas and follow-on formulas [13]. Examples of the few fortified foods with vitamin C are infant formulas, breakfast cereals, apple and apricot compote, and orange and apple juice, which can readily be found in pharmacies and supermarkets.

The main problems associated with the determination of vitamins in foods are the large number of matrix interferences and the low contents to be detected [4]. Vitamin C in food is frequently determined by ultraviolet–visible (UV–Vis) spectrophotometry [14, 15], chemiluminescence [16], fluorimetry [17–20], capillary electrophoresis [21–24], titrimetry [25–28], and voltammetry [29, 30]. These methods often suffer from the presence of matrix interferences and the long analysis times required. In order to overcome these problems, high-performance liquid chromatography (HPLC) methods have been applied. HPLC combined with different detection methods, such as UV–Vis [25, 26, 31–35], fluorescence [36, 37], or electrochemical [38], has been used to quantify vitamin C in various types of foods. However, the analysis of food samples requires more selective and reliable methods, which allow the detection, identification, and determination of the analyte compounds with acceptable accuracy. These requirements can be achieved if HPLC is coupled with mass spectrometry (MS) detection. HPLC–MS offers good sensitivity, and also an improved selectivity, due to a reduction of the background, thus minimizing some of the problems associated with matrix interferents. In this way, cleanup steps may be eliminated, while still allowing reliable identification of the target analytes. However, very few papers have reported the determination of vitamin C in foods by HPLC–MS [39–41].

In the work described in this chapter, a sensitive, selective, and rapid HPLC–MS method was developed and validated for the identification and quantification of the L-enantiomer of ascorbic acid in fortified food products. It is worth pointing out that very few previous studies have focused on the determination of vitamin C in fortified foods [26], and none of them used MS as the detection technique. The performance of the method was evaluated by analyzing a large number of different fortified drinks and food products of known concentration. In addition, the HPLC results obtained in our laboratory were compared with the values indicated by the manufacturers. The proposed method has important advantages such as simplicity (no cleanup step) and high sample throughput (analysis time less than 6 min), indicating that it can be applied in routine analysis.

8.2
Materials and Methods

8.2.1
Chemicals

L-(+)-Ascorbic acid (vitamin C) and acetic acid were obtained from Panreac (Barcelona, Spain). Metaphosphoric acid (with a purity of 33.5–36.5%) was provided by Fluka (Buchs, Switzerland) and methanol (HPLC grade) was purchased from Merck (Darmstadt, Germany). Highly purified water was obtained by ultrafiltration of deionized water with a Milli-Q system (Millipore, Bedford, MA, USA). A standard solution ($1000 \, mg \, l^{-1}$) of ascorbic acid was prepared in 50 ml of methanol. A stock solution ($10 \, mg \, l^{-1}$) was prepared by dilution of the standard solution with methanol and working solutions of lower concentration were prepared in methanol. Bearing in mind that ascorbic acid is easily oxidized, all solutions were prepared daily and were stored at $-24 \, °C$.

8.2.2
Procedure

The method for the extraction of ascorbic acid from solid fortified foods was based on a standard additions methodology described previously [37]. Briefly, the procedure was as follows: 2 g of solid fortified sample was mixed with 5 ml of methanol and 25 ml of a mixture of acetic acid (8%) and metaphosphoric acid (3%). The solution was homogenized with a Polytron mixer for 1 min. Subsequently, the mixture was filtered and 10 ml of the extract were diluted to 25 ml with acetic acid (0.1%). Then 1 ml of the final solution was diluted to 2 ml by adding the appropriate amount of standard solution and mobile phase, followed by injection of 10 µl into the HPLC–MS system.

For liquid fortified foods, such as juices and royal jellies, samples were diluted with mobile phase (see below) prior to HPLC–MS analysis. The dilution depends on the type of sample; for instance, for apple juice, 10 µl were diluted to 2 ml with mobile phase prior to chromatographic analysis, whereas royal jellies, which contain higher amounts of ascorbic acid, were diluted 2000-fold.

8.2.3
Instrumentation

HPLC analysis was performed with an Alliance 2695 instrument (Waters, Milford, MA, USA) equipped with an autosampler, degasser, and column heater. To isolate and determine ascorbic acid, two coupled columns were used, a Symmetry C_{18} 75 × 4.6 mm i.d., 3.5 µm (first) and an Atlantis dC_{18} 150 × 2.0 mm i.d., 5 µm, both from Waters. The composition of the mobile phase was methanol (containing 0.005% acetic acid)–0.05% acetic acid (7:3). Isocratic elution was applied and a flow-rate of 0.3 ml min^{-1} was used. The column temperature was set at 30 °C. The mass spectrometer was a ZQ 2000 single-quadrupole instrument, provided by Waters Micromass (Manchester, UK), using electrospray ionization (ESI) in the negative ion mode. The source and probe tip temperatures were 110 and 350 °C, respectively, and the flow-rates for desolvation and cone gas were 400 and 60 l h^{-1}, respectively, from an N_2 Flo generator purchased from Claind (Lenno, Italy). The capillary voltage was set at 3.5 kV and the extractor voltage at 5 V.

A Polytron PT2100 high-speed homogenizer (Kinematica, Littan/Luzern, Switzerland) and an AB204-S analytical balance (Mettler Toledo, Greifensee, Switzerland) were also used.

8.3
Results and Discussion

8.3.1
Analysis of Ascorbic Acid by HPLC–MS

The determination of water-soluble vitamins such as ascorbic acid depends on the accuracy and sensitivity required and the interferences encountered in the sample matrix. As HPLC–MS is a suitable tool for the rapid and reliable determination of this type of compound, the chromatographic analysis was based on a previously reported HPLC–MS methodology [41]. Ascorbic acid was detected using ESI in the negative ion mode, selecting m/z 175, which corresponds to the ion $[M – H]^-$, as quantification ion, whereas another ion, m/z 115 $[M – C_2H_4O_2]^-$, was used for confirmation purposes (see Figure 8.1a). In relation to the chromatographic determination, it was found that two columns were necessary to isolate the analyte from matrix interferences and, using the conditions described previously, good separation of ascorbic acid was achieved with a convenient retention time at 4.58 min, as can be seen in Figure 8.1b.

The final conditions for sample treatment were simple and fast. For solid samples, a simple solid–liquid extraction was carried out using methanol, acetic acid, and metaphosphoric acid as extracting agents. For liquid samples, sample treatment was not necessary, and just a simple dilution with mobile phase was used, due to the sensitivity of the method and the high concentration of the analyte in the samples. It must be emphasized that sample dilution presents some advan-

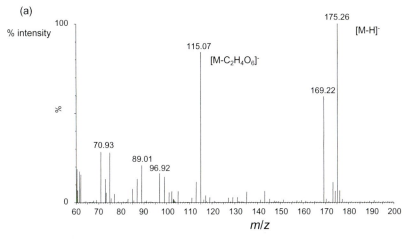

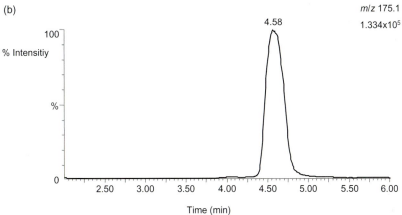

Figure 8.1 (a) Mass spectrum of ascorbic acid obtained in ESI negative ion mode at 20 V and (b) single-ion monitoring chromatogram of a solution of 0.75 mg l⁻¹ of ascorbic acid.

tages such as a smaller amount of matrix loaded in the column, thus reducing matrix effects of liquid samples, and the compensation for possible variability between samples of different origins or varieties.

8.3.2
Method Validation

The method had been validated previously for fruits and vegetables [41], so it was also considered necessary to validate it for fortified liquid and solid samples such as fruit juices, royal jelly, infant food, and biscuits.

Matrix effects were also studied to ensure bias-free analytical results. As no standard reference material was used and it was difficult to find blank samples,

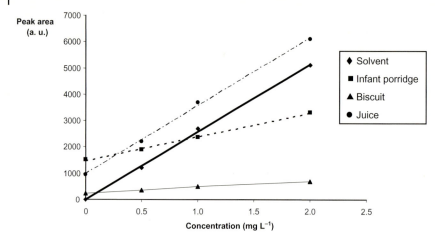

Figure 8.2 Evaluation of matrix effects in different fortified samples by recording calibration curves obtained in standard additions experiments.

the products analyzed, including biscuits, infant food, and breakfast cereals, were spiked with ascorbic acid at different concentrations (0.5–2.0 mg l^{-1}) and the slopes of the calibration curves were compared with the values obtained when pure solvent (methanol) was used for the calibration process. In all cases, a significant matrix effect was observed (Figure 8.2); the slopes were different from that of the solvent and they depended on the matrix used. In consequence, the standard additions methodology was essential in order to obtain reliable results. By contrast, when liquid samples were evaluated, it was observed that the calibration curves obtained using spiked samples (royal jelly and fruit juices) were not significantly different from that obtained by use of standard additions (Figure 8.2). Therefore, external calibration was used for quantification.

Peak area was used as the analytical response and linearity was studied by the standard additions method in the range 0.05–2.00 mg l^{-1}. Good linearity was obtained with correlation coefficients higher than 0.98.

Limits of detection (LODs) and limits of quantification (LOQs) were determined as the lowest ascorbic acid concentration injected that yielded a signal-to-noise (S/N) ratio of 3 and 10, respectively, when the quantification ion was monitored. Values of 0.015 and 0.05 mg l^{-1} were obtained for the LOD and LOQ, respectively.

Accuracy was evaluated by recovery studies. Before extraction, different aliquots of samples ($n = 6$) were spiked with ascorbic acid and were extracted with the developed method (S1). In addition, other aliquots of the same samples ($n = 6$) were extracted without spiking (S0), and recoveries were calculated as $R = 100 \times (S1 - S0)/C_{\text{spiked}}$. Good results were obtained and recoveries were in the range 92.8–103.5% for all the matrices analyzed.

The overall precision of the method was evaluated by performing repeatability and reproducibility experiments by carrying out five extractions during one day at a level of 50 mg kg^{-1} (repeatability), and two replicates on three different days (reproducibility). The relative standard deviations (RSDs) were calculated to be 4.5 and 9.8% for repeatability and reproducibility, respectively.

8.3.3
Application to the Analysis of Real Samples

The method was used for the determination of ascorbic acid in fortified samples of biscuit, infant food (instant porridge, infant cereals, and fruit desserts), two types of breakfast cereals (1 and 2), apple compote, apple juice, and three different royal jellies (1, 2, and 3).

In order to check the robustness of the system, internal quality control was applied to each batch of samples. This quality control was based on the evaluation of the recovery in one sample, as indicated previously, and on the evaluation of the linearity in the working concentration range.

The results obtained are given in Table 8.1. It can be observed that the experimental results are similar to the label values indicated on the commercial products. Furthermore, it can be noted that fortified royal jelly contains higher amounts of ascorbic acid than the other products analyzed, with the infant cereals having the lowest amount. Regarding precision, good repeatability was obtained during the analysis of real samples, with RSD values lower than 7%.

It can be highlighted that for the analysis of fortified liquid samples (apple juice and royal jelly), samples were injected directly into the HPLC–MS system after a dilution step, and an extraction process was avoided. No pretreatment process or cleanup step is necessary, so it allows the analysis of a large number of samples in a very short time. Figure 8.3 shows the chromatograms for two fortified samples, revealing that none of the extracts contained interfering substances.

Table 8.1 Ascorbic acid concentration in different fortified food products.

Sample	Found (mg kg^{-1})[a]	Labeled (mg kg^{-1})
Biscuit	334 (6.3)	350
Infant instant porridge	490 (2.0)	500
Infant cereals	232 (3.6)	250
Infant fruit dessert	340 (3.2)	350
Breakfast cereal 1	502 (4.2)	510
Breakfast cereal 2	480 (5.0)	510
Apple compote	377 (3.6)	400
Apple juice[b]	360 (10.1)	400
Royal jelly 1[b]	3716 (2.2)	4000
Royal jelly 2[b]	1837 (2.0)	2000
Royal jelly 3[b]	5448 (4.4)	6000

a) RSD (%) is given in parentheses ($n = 4$).
b) Concentration expressed as mg l^{-1}.

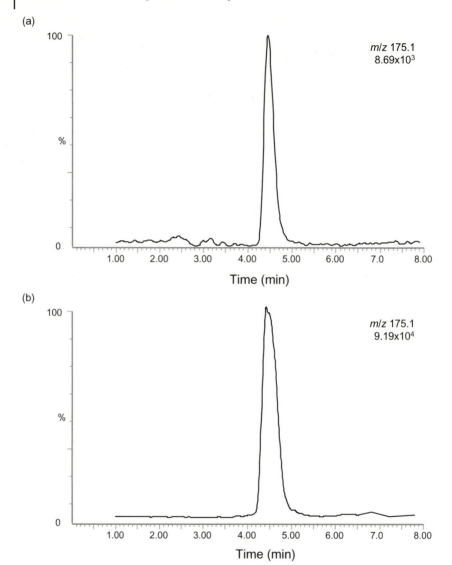

Figure 8.3 Single-ion monitoring chromatograms for (a) breakfast cereals and (b) royal jelly containing 480 mg kg^{-1} and 1837 mg l^{-1} of ascorbic acid, respectively.

Finally, it can be noted that the proposed method provides slightly lower amounts of ascorbic acid than the values indicated by the manufacturers. This could be due to the losses of compound as observed previously in other studies [42], where different percentages of ascorbic acid were lost during storage, depending on conditions such as oxygen and light.

In conclusion, our data indicate that HPLC–MS can be applied for the identification and determination of ascorbic acid in a wide range of fortified food products. The methodology is fast, with simple sample preparation, thus increasing sample throughput. This successful application of the method to several fortified food products indicate that it is highly attractive as a routine method in the food area.

References

1 Yin, C., Cao, Y., Ding, S., and Wang, Y. (2008) Rapid determination of water- and fat-soluble vitamins with microemulsion electrokinetic chromatography. *Journal of Chromatography A*, **1193**, 172–177.

2 Panel on Dietary Antioxidants and Related Compounds, Subcommittees on Upper Reference Levels of Nutrients and Interpretation and Uses of DRIs, Standing Committee on the Scientific Evaluation of Dietary Reference Intakes, Food and Nutrition Board, and Institute of Medicine (2000) *Dietary Reference Intakes for Vitamin C, Vitamin E, Selenium, and Carotenoids*, National Academiies Press, Washington, DC, p. 95.

3 Woodside, J., McCall, D., McGartland, C., and Young, I. (2005) Micronutrients: dietary intake v. supplement use. *Proceedings of the Nutrition Society*, **64**, 543–553.

4 Torres-Sequeiros, R.A., García-Falcón, M.S., and Simal-Gándara, J. (2001) Analysis of fluorescent vitamins riboflavin and pyridoxine in beverages with added vitamins. *Chromatographia*, **53**, S236–S239.

5 Teucher, B., Olivares, M., and Cori, H. (2004) Enhancers of iron absorption: ascorbic acid and other organic acids. *International Journal for Vitamin and Nutrition Research*, **74**, 403–419.

6 Fidler, M.C., Davidsson, L., Zeder, C., Walczyk, T., Marti, I., and Hurrell, R.F. (2004) Effect of ascorbic acid and particle size on iron absorption from ferric pyrophosphate in adult women. *International Journal for Vitamin and Nutrition Research*, **74**, 294–300.

7 Fawzi, W.W., Msamanga, G.I., Spiegelman, D., Wei, R., Kapiga, S., Villamor, E., Mwakagile, D., Mugusi, F., Hertzmark, E., Essex, M., and Junter, D.J. (2004) A randomised trial of multivitamin supplements and HIV disease progression and mortality. *New England Journal of Medicine*, **351**, 23–32.

8 Lanzillotti, J.S. and Tang, A.M. (2005) Micronutrients and HIV disease: a review pre- and post-HAART. *Nutrition and Clinical Care*, **8**, 16–23.

9 Flynn, A., Moreiras, O., Stehle, P., Fletcher, R.J., Müller, D.J.G., and Rolland, V. (2003) Vitamins and minerals: a model for safe addition to foods. *European Journal of Nutrition*, **42**, 118–130.

10 Medline Plus (2007) Medical Encyclopaedia: Vitamin C. http://www.nlm.nih.gov/medlineplus/ency/article/002404.htm (accessed 8 September 2008).

11 Codex Alimentarius Commission (1991) Codex Alimentarius A1. General Principles for the Addition of Essential Nutrients to Foods, vol 4. CAC/GL 09-1987 (amended 1989, 1991). Codex Alimentarius Commission, Rome.

12 Rosenberg, I.H. (2007) Further evidence that food fortification improves micronutrient status. *British Journal of Nutrition*, **97**, 1051–1052.

13 European Commission (2006) Commission Directive 2006/141/EC of 22 December 2006 on Infant Formulae and Follow-on Formulae and Amending Directive 1999/21/EC. *Official Journal of the European Union*, **L401**, 15.

14 Rahman Khan, M.M., Rahman, M.M., Islam, M.S., and Begun, S.A. (2007) A simple UV-spectrophotometric method for the determination of vitamin C content in various fruits and vegetables at Sylhet area in Bangladesh. *Journal of Biological Sciences*, **6**, 388–392.

15 Luque-Pérez, E., Ríos, A., and Valcárcel, M. (2000) Flow injection spectrophotometric determination of ascorbic acid in soft drinks and beer. *Fresenius' Journal of Analytical Chemistry*, **366**, 857–862.

16 Pérez-Ruiz, T., Martínez-Lozano, C., and Sanz, A. (1995) Flow-injection chemiluminometric determination of ascorbic acid based on its sensitized photooxidation. *Analytica Chimica Acta*, **308**, 299–307.

17 Dilgin, Y. and Nisli, G. (2005) Fluorimetric determination of ascorbic acid in vitamin C tablets using methylene blue. *Chemical and Pharmaceutical Bulletin*, **53**, 1251–1254.

18 Wu, X., Diao, Y., Sun, C., Yang, J., Yang, J., Wang, Y., and Sun, S. (2003) Fluorimetric determination of ascorbic acid with o-phenylenediamine. *Talanta*, **59**, 95–99.

19 Pérez-Ruiz, T., Martínez-Lozano, C., Tomas, V., and Fenol, J. (2001) Fluorimetric determination of total ascorbic acid by a stopped-flow mixing technique. *Analyst*, **126**, 1436–1439.

20 Sánchez-Mata, M.C., Cámara-Hurtado, M., Díez-Marqués, C., and Torija-Isasa, M.E. (2000) Comparison of high-performance liquid chromatography and spectrofluorimetry for vitamin C analysis of green beans (*Phaseolus vulgaris* L.). *European Food Research and Technology*, **210**, 220–225.

21 Peng, Y., Zhang, Y., and Ye, J. (2008) Determination of phenolic compounds and ascorbic acid in different fractions of tomato by capillary electrophoresis with electrochemical detection. *Journal of Agricultural and Food Chemistry*, **56**, 1838–1844.

22 Law, W.S., Kubán, P., Zhao, J.H., Li, S.F.Y., and Hauser, P.C. (2005) Determination of vitamin C and preservatives in beverages by conventional capillary electrophoresis and microchip electrophoresis with capacitively coupled contactless conductivity detection. *Electrophoresis*, **26**, 4648–4655.

23 Versari, A., Mattioli, A., Parpinello, G.P., and Galassi, S. (2004) Rapid analysis of ascorbic and isoascorbic acids in fruit juice by capillary electrophoresis. *Food Control*, **15**, 355–358.

24 Galiana-Balaguer, L., Roselló, S., Herrero-Martínez, J.M., Maquieira, A., and Nuez, F. (2001) Determination of L-ascorbic acid in *Lycopersicon* fruits by capillary zone electrophoresis. *Analytical Biochemistry*, **296**, 218–224.

25 Hernández, Y., Lobo, M.G., and González, M. (2006) Determination of vitamin C in tropical fruits: a comparative evaluation of methods. *Food Chemistry*, **96**, 654–664.

26 Fontannaz, P., Kilinc, T., and Heudi, O. (2006) HPLC–UV determination of total vitamin C in a wide range of fortified food products. *Food Chemistry*, **94**, 626–631.

27 Kabasakalis, V., Siopidou, D., and Moshatou, E. (2000) Ascorbic acid content of commercial fruit juices and its rate of loss upon storage. *Food Chemistry*, **70**, 325–328.

28 AOAC (1990) *Official Methods of Analysis of the Association of Official Analytical Chemists*, 15th edn., AOAC, Arlington, VA, pp. 1058–1059.

29 Pournaghi-Azar, M.H., Razmi-Nerbin, H., and Hafezi, B. (2002) Amperometric determination of ascorbic acid in real samples using an aluminum electrode, modified with nickel hexacyanoferrate films by a simple electroless dipping method. *Electroanalysis*, **14**, 206–212.

30 Verdini, R.A. and Lagier, C.M. (2000) Voltammetric iodometric titration of ascorbic acid with dead-stop end-point detection in fresh vegetables and fruit samples. *Journal of Agricultural and Food Chemistry*, **48**, 2812–2817.

31 Chávez-Servín, J.L., Castellote, A.I., Rivero, M., and López-Sábater, M.C. (2008) Analysis of vitamins A, E and C, iron and selenium contents in infant milk-based powdered formula during shelf-life. *Food Chemistry*, **107**, 1187–1197.

32 Meléndez-Martínez, A.J., Vicario, I.M., and Heredia, F.J. (2006) Provitamin A carotenoids and ascorbic acid contents of the different types of orange juices marketed in Spain. *Food Chemistry*, **101**, 177–184.

33 Blake, C.J. (2007) Analytical procedures for water-soluble vitamins in foods and dietary supplements: a review. *Analytical and Bioanalytical Chemistry*, **389**, 63–76.

34 Sánchez-Moreno, C., Plaza, L., De Ancos, B., and Cano, M.P. (2003) Vitamin C, provitamin A carotenoids, and other carotenoids in high-pressurized orange juice during storage. *Journal of Agricultural and Food Chemistry*, **51**, 647–653.

35 Gil, M.I., Tomás-Barberán, F.A., Hess-Pierce, B., and Kader, A.A. (2002) Antioxidant capacities, phenolic compounds, carotenoids, and vitamin C contents of nectarine, peach and plum cultivars from California. *Journal of Agricultural and Food Chemistry*, **50**, 4976–4982.

36 Burini, G. (2007) Development of a quantitative method for the analysis of total L-ascorbic acid in foods by high-performance liquid chromatography. *Journal of Chromatography A*, **1154**, 97–102.

37 Dodson, K.Y., Young, E.R., and Soliman, A.G.M. (1992) Determination of total vitamin C in various food matrices by liquid chromatography and fluorescence detection. *Journal of AOAC International*, **75**, 887–891.

38 Leubolt, R. and Klein, H. (1993) Determination of sulphite and ascorbic acid by high performance liquid chromatography with electrochemical detection. *Journal of Chromatography A*, **640**, 271–277.

39 Gentili, A., Caretti, F., D'Ascenzo, G., Marchese, S., Perret, D., Di Corcia, D., and Mainero Roca, L. (2008) Simultaneous determination of water-soluble vitamins in selected food matrices by liquid chromatography/electrospray ionization tandem mass spectrometry. *Rapid Communications in Mass Spectrometry*, **22**, 2020–2043.

40 Chen, Z., Chen, B., and Yao, S. (2006) High-performance liquid chromatography/electrospray ionization-mass spectrometry for simultaneous determination of taurine and 10 water-soluble vitamins in multivitamin tablets. *Analytica Chimica Acta*, **569**, 169–175.

41 Garrido Frenich, A., Hernández Torres, M.E., Belmonte Vega, A., Martínez Vidal, J.L., and Plaza Bolaños, P. (2005) Determination of ascorbic acid and carotenoids in food commodities by liquid chromatography with mass spectrometry detection. *Journal of Agricultural and Food Chemistry*, **53**, 7371–7376.

42 Johnston, C.S. and Bowling, D.L. (2002) Stability of ascorbic acid in commercially available orange juices. *Journal of the American Dietetic Association*, **102**, 525–529.

9

Quantitation of Pantothenic Acid in Fortified Foods by Stable Isotope Dilution Analysis and Method Comparison with a Microbiological Assay

Michael Rychlik

9.1
Introduction

(R)-Pantothenic acid (PA), also denoted vitamin B_5, is a component of conjugates such as coenzyme A (CoA) and acyl carrier protein (ACP), which are essential for the metabolism of steroids, fatty acids, and phosphatides. As PA occurs in many foods, reports on deficiencies in the Western world are rare. However, adolescents are considered to have an increased PA need [1] and may be at risk of inadequate intake. Therefore, PA is a common component of foods multiply fortified with vitamins.

The most commonly employed method of PA analysis is to measure turbidi-metrically the growth of *Lactobacillus plantarum* in a PA-deficient medium [2] However, as the microbiological assay (MA) is lengthy and lacks specificity, there are increasing efforts to replace it by more accurate methods. Chromatographic assays, however, have not been found suitable for several reasons. High-performance liquid chromatography, on the one hand, suffers from PA's weak UV absorption above 230 nm and with non-reproducible formation of fluorescent derivatives [3]. On the other hand, gas chromatography (GC) requires derivatiza-tion of PA and, therefore, the addition of an internal standard (IS) is essential. Banno *et al.* [4] used (R)-5-[(2,4-dihydroxy-3,3-dimethyl-1-oxobutyl)amino]penta-noic acid, a homolog of PA, as the IS, but the structural difference may cause discrimination between it and PA.

By contrast, enzyme-linked immunosorbent assays (ELISAs) have been devel-oped that gave similar results to those of MAs [5]. However, the major drawbacks of ELISAs are their lower sensitivity and the lack of commercial availability.

For about a decade, mass spectrometry (MS) coupled to liquid chromatography (LC) has appeared to be suitable for the selective, sensitive, and universal detection most water-soluble vitamins. However, as the ion yields in common atmospheric pressure ionization sources show wide differences within a single chromato-graphic run, the use of ISs is essential.

As we reported for the analysis of folates [6], stable isotope dilution assays (SIDAs) exhibit excellent sensitivity and reliability and are accurate alternatives to

Fortification of Foods with Vitamins, First Edition. Edited by Michael Rychlik.
© 2011 Wiley-VCH Verlag GmbH & Co. KGaA. Published 2011 by
Wiley-VCH Verlag GmbH & Co. KGaA.

other quantification methods. By using isotopologs of the analytes as the IS, losses during cleanup and derivatization and also ionization interferences can be exactly corrected.

The objectives of the present study were, therefore, (i) to develop SIDAs for PA based on the use of a suitable isotopolog as the IS and (ii) to perform a site-by-site comparison of identical samples analyzed by SIDA and MA as the current reference method.

9.2
Materials and Methods

9.2.1
Chemicals

The following chemicals were obtained commercially: [^{15}N,^{13}C$_3$]-β-alanine, N, O-bis(trimethylsilyl)trifluoroacetamide (BSTFA), (R)-(–)-pantolactone, and (R)-pantothenic acid hemicalcium salt from Aldrich (Steinheim, Germany), acetone, ammonium sulfate, dichloromethane, diethylamine, ethyl acetate, hydrochloric acid, KHCO$_3$, methanol, NaHCO$_3$, pyridine, sodium acetate, Na$_2$SO$_4$, and toluene from Merck (Darmstadt, Germany), and alkaline phosphatase and pigeon liver acetone powder from Sigma (Steinheim, Germany).

Calcium [^{15}N,^{13}C$_3$]-(R)-pantothenate was synthesized as reported recently [7].

9.2.2
Stable Isotope Dilution Assay for the Determination of Free Pantothenic Acid in Foods

Tablets and sweets were ground in a mortar and breakfast cereals in a grain mill (Bosch, Munich, Germany). The resulting powders or flours (0.5 g) were stirred for 1 h at 20 °C in extraction buffer containing calcium [^{15}N,^{13}C$_3$]-(R)-pantothenate (10 μg). To juices and whey products the labeled standards were added directly.

The extracts were filtered and, after passing through a syringe filter (0.4 μm) (Millipore, Bedford, MA, USA), were analyzed by liquid chromatography–tandem mass spectrometry (LC–MS/MS).

For gas chromatography–mass spectrometry (GC–MS), the extracts were alkalinized to pH 8 by addition of aqueous sodium hydroxide (1 mol l^{-1}) and washed with dichloromethane (2 × 10 ml). Then, the aqueous solutions were adjusted to pH 2–3 by addition of hydrochloric acid (1 ml, 18 mol l^{-1}), ammonium sulfate (3 g) was added, and the solution was extracted with ethyl acetate (2 × 10 ml). The extracts in ethyl acetate were dried over anhydrous Na$_2$SO$_4$ and evaporated to dryness in a stream of nitrogen. Then, BSTFA (100 μl) and pyridine (100 μl) were added and the mixture was heated for 60 min at 80 °C in a closed vial. After cooling to room temperature, the solution was evaporated to dryness in a stream of nitrogen and hexane (100 μl) was added.

9.2.3
Enzyme Hydrolysis for Quantification of Total Pantothenic Acid

Pigeon liver pantetheinase was prepared from pigeon liver acetone powder as reported recently [7]. For liberation of bound PA, the filtered extract was sterilized at 120 °C for 15 min and cooled to 37 °C. Then, solutions of pantetheinase (0.4 ml), alkaline phosphatase (0.8 ml, 2%), NaHCO$_3$ (1 ml, 0.85%) and an aqueous solution of calcium [^{15}N,^{13}C$_3$]-(R)-pantothenate (2 µg) were added to the sterilized extract (5 ml). The mixture was then alkalinized to pH 8 by addition of aqueous sodium hydroxide (1 mol l^{-1}) and incubated for 8 h at 37 °C. Subsequently, the hydrolyzate was acidified to pH 5.6 by addition of hydrochloric acid (1 ml, 18 mol l^{-1}) and subjected to LC–MS/MS.

9.2.4
Liquid Chromatography–Tandem Mass Spectrometry

LC–MS/MS was performed by means of an LCQ mass spectrometer (Finnigan MAT, Bremen, Germany) coupled to a Spectra Series high-performance liquid chromatograph (Thermo Separation Products, San Jose, CA, USA) equipped with an Aqua C$_{18}$ reversed-phase column (250 × 4.6 mm i.d.; 5 µm) (Phenomenex, Aschaffenburg, Germany). Volumes of 50 µl of the sample solutions were chromatographed using gradient elution with variable mixtures of aqueous formic acid (0.1%, solvent A) and acetonitrile (solvent B) at a flow-rate of 0.8 ml min^{-1}. After flushing the column for 9 min with 7% B, a 13 min linear gradient was programmed to 17% B followed by a further 3 min linear gradient to 25% B. Then, the concentration of B was immediately raised to 100%, maintained there for 5 min and subsequently brought back to the initial mixture for a further 5 min to allow for column equilibration. During the first 8 min of the gradient program, the column effluent was diverted to waste to ensure an adequate spray stability. For [^{15}N,^{13}C$_3$]PA the mass transitions (m/z precursor ion/m/z product ion) 224/206 and 224/188 and for unlabeled PA the mass transitions 220/202 and 220/184 were chosen.

9.2.5
Gas Chromatography–Mass Spectrometry

Mass chromatograms in the electron ionization (EI) mode were recorded by means of an MD 800 quadrupole mass spectrometer (Fisons Instruments, Manchester, UK) coupled to a Type 8000 gas chromatograph (ThermoQuest, Egelsbach, Germany) equipped with a DB-5 capillary column (30 m × 0.32 mm i.d. fused-silica capillary, film thickness of the stationary phase d_f = 0.25 µm) (Fisons Instruments, Mainz, Germany). The samples were applied by split injection at 230 °C and a splitting ratio of 1 : 20. After injecting the sample (2 µl), the temperature of the oven was raised from 100 to 250 °C at a rate of 20 °C min^{-1}, then to 280 °C at a rate of 5 °C min^{-1} and held at that temperature for 5 min. The flow-rate of the carrier gas helium was $2\,\text{ml}\,\text{min}^{-1}$. Trimethylsilyl (TMS)-[^{15}N,^{13}C$_3$]PA acid and TMS-PA were detected in the mass traces at m/z 295 and 291, respectively. The ionization energy in the EI mode was $70\,\text{eV}$.

9.2.6
Microbiological Assay

The extracts were assayed for pantothenic acid with *Lactobacillus plantarum* (ATTC 8014) in the medium of Strohecker and Henning ([8] (Difco Pantothenate Assay Medium, from Becton Dickinson, Sparks, MD, USA). *Lactobacillus plantarum* growth was measured spectrophotometrically with a Uvikon 933 spectrophotometer (Kontron Instruments, Eching, Germany) at 460 nm after incubation for 18 h.

9.2.7
Precision and Recovery

Intra-assay precision was evaluated by analyzing a multivitamin sweet as detailed earlier ($n = 5$).

Recovery was determined by adding $6\,\mu\text{g}$ of PA to edible corn starch (1 g) and performing SIDA in quadruplicate as detailed earlier.

9.3
Results and Discussion

9.3.1
Synthesis of a Stable Isotope-Labeled Internal Standard

To be used as isotope-labeled IS, fourfold-labeled PA was synthesized starting from commercially available [^{15}N,^{13}C$_3$]-β-alanine, which was converted into its hemicalcium salt and condensed with (R)-(–)-pantolactone in the presence of diethylamine to form the hemicalcium salt of [^{15}N,^{13}C$_3$]-(R)-PA, calcium [^{15}N,^{13}C$_3$]-(R)-pantothenate [7]. The purity and constitution of calcium [^{15}N,^{13}C$_3$]-(R)-pantothenate were confirmed by ^{1}H and ^{13}C NMR experiments.

9.3.2
Gas Chromatography–Mass Spectrometry of Derivatized Pantothenic Acid

Owing to its high separation performance, capillary GC was first chosen to determine PA. Since PA is not volatile enough for GC analysis, it has to be derivatized. In this study, the TMS derivative was generated by reacting PA with a mixture of BSTFA and pyridine following the method reported by Banno *et al.* [4]. The mass spectra in the chemical ionization mode and also in the EI mode (data reported in [7]) of labeled tris(TMS)PA revealed a mass shift of 4u compared with the unlabeled isotopolog. This finding is consistent with the incorporation of the labeled nitrogen and the three labeled carbon atoms into the PA molecule. As the $[^{15}N,^{13}C_3]$-β-alanine employed provides stability and completeness of labeling, the linearity of the response function depends solely on the contribution of the unlabeled analyte on the m/z value chosen for monitoring the labeled IS. Regarding the mass spectra of the TMS derivatives, in the EI mode the fragment ion pairs m/z 295 and 291 revealed a slope close to 1.0. This finding can undoubtedly be attributed to the elemental composition of the m/z 295/291 ions containing only one silicon atom in contrast to other possible ions. Therefore, the fragment ion $[M - 2TMS + 2H]^+$ of unlabeled tris(TMS)PA shows only negligible ion intensity from naturally occurring Si-30 isotopes falling at m/z 295 monitored for the respective fragment of labeled tris(TMS)PA.

For quantification of PA, foods were extracted into sodium acetate buffer (pH 5.65) according to AOAC Official Method 945.74 [9]. After addition of labeled PA, the analyte and IS were extracted into ethyl acetate, converted into the TMS derivatives and analyzed by GC–MS. Mass chromatography of the m/z 291 and 295 ions enables the ion abundance ratio, mass ratio of PA to the IS, and the PA content in the analyzed sample to be computed successively. The quantitation of a rice sample is displayed in Figure 9.1.

9.3.3
Liquid Chromatography–Tandem Mass Spectrometry of Pantothenic Acid

Derivatization can be omitted if PA is analyzed by LC. For MS detection, most suitable was the positive electrospray mode, giving as the base peak the protonated

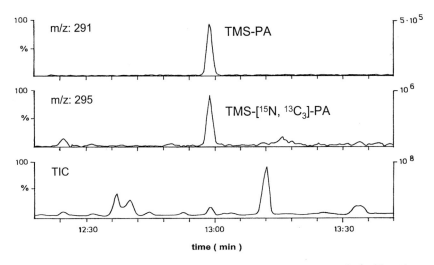

Figure 9.1 GC–MS traces for an unpolished rice sample containing 5.56 mg kg^{-1} of free PA. The IS TMS-$[^{15}N, ^{13}C_3]$PA) is detected in the m/z 295 trace and unlabeled TMS-PA in the m/z 291 trace. TIC, total ion current.

molecule at m/z 220 in addition to two minor signals at m/z 242 and 461, which can be assigned to $[M + Na]^+$ and $[2M + Na]^+$, respectively. Analogously, $[^{15}N,^{13}C_3]$ PA gave signals at m/z 224, 246, and 469.

As the recently developed SIDA of folates was based on LC–MS/MS due to matrix interferences [6], tandem MS was also applied to PA isotopologs if necessary for unequivocal quantification. By applying collision-induced dissociation (CID) to the $[M + H]^+$ ion of isotopologic PA, the spectrum shown in Figure 9.2 was obtained. Two signals at $[M - 18]^+$ and $[M - 36]^+$ are discernible, which are obviously due to consecutive losses of H_2O and which can be used for differentiation and quantification of the isotopologs. The calibration curves of different ratios of the isotopologs revealed linearity for $[M + H]^+$ in single-stage MS and for the product ions $[M - 18]^+$ and $[M - 36]^+$ from $[M + H]^+$ in MS/MS over two decades of isotope ratios.

9.3.4
Extraction and Analysis of Free and Total Pantothenic Acid by LC–MS/MS

Sample preparation for LC–MS of free PA proved to be very simple, as LC–MS/MS was specific enough to differentiate PA isotopologs from sample interferences. After extracting the powdered samples in an buffer containing definite amounts of $[^{15}N,^{13}C_3]$PA, the extracts only had to be filtered and passed through a syringe filter. LC–MS of fortified foods displayed in single-stage ESI-MS signals for the isotopologs clearly resolved from matrix peaks (Figure 9.3).

In those foods containing endogenously bound PA, the conjugates were hydrolyzed by treatment with pigeon liver pantetheinase and alkaline phosphatase after extraction [10].

9.3.5
Performance Criteria

To evaluate whether the sensitivity of LC–MS was sufficient for quantifying PA in foods, the detection limit (DL) was determined in edible starch according to the method of Hädrich and Vogelgesang [11].

The calculations resulted in a DL of 89 µg per 100 g and a quantification limit (QL) of 200 µg kg^{-1} in starch-containing foods. The recoveries was evaluated by adding 6 mg kg^{-1} of PA to edible starch and was found to be 103%.

For examining precision, a multivitamin sweet was analyzed and revealed intra- and inter-assay coefficients of variation of 3.2 and 3.1%, respectively.

9.3.6
Method Comparison of the SIDA Based on LC–MS/MS and GC–MS Detection and Microbiological Assay

For comparison of the chromatographic methods, the extracts of whole egg powder measured by LC–MS/MS were also quantified by GC–MS after extracting the PA

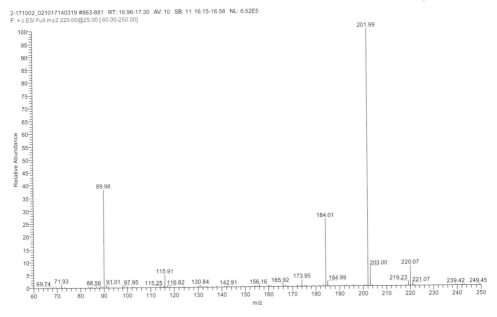

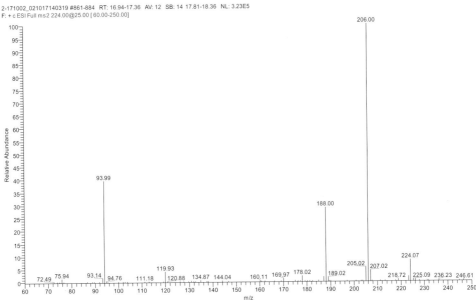

Figure 9.2 Tandem mass spectra of PA (above) and [^{15}N,^{13}C$_3$]PA (below) after CID of the protonated molecules in the positive electrospray ionization mode.

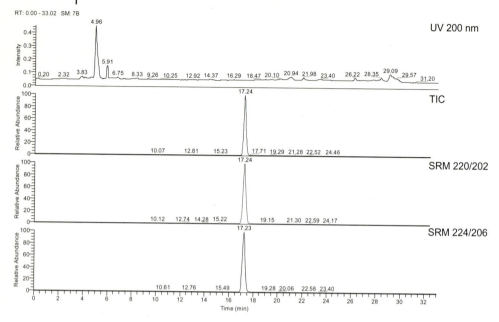

Figure 9.3 MS/MS chromatograms of fortified fruit juice in the positive electrospray ionization mode after CID of the protonated molecules. The IS [^{15}N,^{13}C$_3$]PA is detected in the *m/z* 224/206 trace and unlabeled PA in the *m/z* 220/202 trace. UV, UV absorption; TIC, total ion current.

isotopologs and subsequent trimethylsilylation. The GC–MS traces revealed une-quivocal signals of the isotopologs in the extracts, whereas in LC–MS the tandem MS/MS mode was required due to a matrix interference. As expected, comparison of the results gave no significant difference ($p > 0.05\%$) between the data obtained by GC–MS (8.4 mg per 100 g) and LC–MS/MS (8.7 mg per 100 g) according to the *t*-test. However, the former method takes 1 h longer due to the additional steps of sample cleanup, extraction, and derivatization of PA.

A further method comparison between the MA and the SIDA of total PA by LC–MS/MS was performed for a number of foods and feedstuffs. The results for plant-derived samples are given in Table 9.1.

For cereals and potatoes, the results of MA and SIDA of total PA were in fairly good agreement, with a maximum deviation of 21% for wheat. In the case of orange juice, which contains relatively small amounts of PA, the results of the two methods were also consistent. For cocoa, mushrooms, and soy beans, the two methods were also in good agreement. For hazelnuts, a significant difference of 25% between MA and SIDA was observed. The reasons for this discrepancy remain open, but the data imply that nuts contain substances that stimulate the growth of the microorganisms without being measured by SIDA, even after all enzyme treatments. These compounds might be yet unknown PA forms or sub-

Table 9.1 Comparison of results for PA obtained by SIDA and by MA.

Sample	PA concentration (mg per 100 g)		
	SIDA		MA
	Free PA	Total PA	
Millet	0.37	0.61	0.76
Wheat	0.64	0.84	1.07
Corn	0.22	0.65	0.65
Potato	0.25	0.25	0.28
Cocoa	0.67	1.39	1.36
Hazelnuts	0.85	1.22	1.64
Soy beans	1.45	1.76	1.91
Orange juice	0.08	0.08	0.18
Mushrooms	1.17	1.29	1.23
Whole-egg powder	7.78	8.09	9.53
Skim-milk powder	5.23	5.61	6.33
Porcine liver	4.35	5.23	6.59

stances that the microorganisms can utilize for their growth and that are structurally different from PA conjugates. This might also hold true for millet, wheat, and orange juice, for which the MA gave significantly higher PA contents than SIDA. In the case of cocoa, corn, and mushrooms, the slightly higher values given by SIDA compared with MA were not significantly different and obviously due to random dispersion.

For comparing the PA data in animal-derived products, whole-egg powder, skimmed milk powder, and porcine liver were analyzed by the two methods. As expected, these samples contained much more PA than the foods and feeds of plant origin and appear to be a much better source of dietary PA. In analogy with the latter products, the PA values given by SIDA were lower than those given by MA, but the differences were much more pronounced and amounted to 1.5 mg per 100 g for whole-egg powder. This observation corresponded to that of Walsh et al. [12], who performed a method comparison between radioimmunoassay and MA for PA analysis. They found, consistent with our results, good agreement of the data from plant-derived foods and significantly higher data from the MA of meat.

Regarding the contents of free and total PA, the part of bound PA for these products was below 10% of the total PA content, which is in agreement with our earlier reports [7, 10] and those of other authors [13, 14].

From this finding and from the results of treating the extract for MA with or without papain and diastase [15], it was concluded that MA is also able to measure a significant amount of bound PA.

9.3.7
Results for Analysis of Foods

Several foods fortified with various vitamins were surveyed to demonstrate the suitability of the new SIDA. Of liquid products, four fortified fruit juices and two whey products were quantified. In addition, two samples of breakfast cereals, two sweets, and one multivitamin pharmaceutical were analyzed. Finally, we quantified a meal for weight reduction and a breakfast drink. The results are presented in Table 9.2.

PA contents ranged from 0.39 per 100 g in the wheat flour to 125 mg per 100 g in the pharmaceutical. In nearly all products the PA label claim was exceeded by 10–50%, except for one fruit juice containing only 60%, one cereal containing 240%, and one whey containing 225% of the amounts declared on the label. As already reported by Romera *et al.* [16] for infant formulas, over-fortifications by up to 260% of the PA label claim are common, which was confirmed by our data. In contrast, the PA content of the pharmaceutical was exactly the labeled amount. Although an overdosage is thought to be reasonable to anticipate losses during manufacture and storage, these differences from the label values appear too high. According to a consensus of the Working Group on Vitamins in Germany [17], the content of PA should not differ by more than 30% from the label value. Therefore, manufacturers are called upon to adjust folate contents more accurately and official laboratories to monitor the contents more frequently.

Table 9.2 Analyzed and labeled contents of PA in foods.

Sample fortified	Concentration (mg per 100 g)		% of claim
	Analyzed	Label claim	
Fruit juices	3.58	2.50	143
	2.66	2.00	133
	3.79	3.00	126
	1.86	3.00	62
Whey products	2.03	0.90	225
	1.35	1.20	112
Breakfast cereals	7.94	5.10	156
	14.65	6.00	244
Sweets	23.31	18.00	130
	36.10	29.70	122
Pharmaceutical	115.1	115.00	100
Breakfast drink	0.91	0.60	152
Meal for weight reduction	0.84	0.65	130

References

1 Kathman, J.V. and Kies, C. (1984) Pantothenic acid status of free living adolescent and young adults. *Nutrition Research*, **4**, 245–250.

2 Tanner, J.T., Barnett, S.A., and Mountford, M.K. (1993) Analysis of milk-based infant formula. Phase V. Vitamins A and E, folc acid, and pantothenic acid: Food and Drug Administration–Infant Formula Council: collaborative study. *Journal of AOAC International*, **76**, 399–413.

3 Hudson, T.S., Subramanian, S., and Allen, R.J. (1984) Determination of pantothenic acid, biotin, and vitamin B_{12} in nutritional products. *Journal of AOAC International*, **67**, 994–998.

4 Banno, K., Matsuoka, M., Horimoto, S., and Kato, J. (1990) Simultaneous determination of pantothenic acid and hopantenic acid in biological samples and natural products by gas chromatography–mass fragmentography. *Journal of Chromatography*, **525**, 255–264.

5 Gonthier, A., Fayol, V., Viollet, J., and Hartmann, D.J. (1998) Determination of pantothenic acid in foods: influence of the extraction method. *Food Chemistry*, **63**, 287–294.

6 Freisleben, A., Schieberle, P., and Rychlik, M. (2003) Specific and sensitive quantification of folate vitamers in foods by stable isotope dilution assays using high-performance liquid chromatography–tandem mass spectrometry. *Analytical and Bioanalytical Chemistry*, **376**, 149–156.

7 Rychlik, M. (2000) Quantification of free and bound pantothenic acid in foods and blood plasma by a stable isotope dilution assay. *Journal of Agricultural and Food Chemistry*, **48**, 1175–1181.

8 Strohecker, R. and Henning, H.M. (1963) *Vitaminbestimmungen*, Verlag Chemie, Weinheim.

9 AOAC International (1997) *Official Methods of Analysis of AOAC International*, 16th edn., Method 945.74, AOAC International, Gaithersburg, MD.

10 Rychlik, M. (2003) Pantothenic acid quantification by a stable isotope dilution assay based on liquid chromatography–tandem mass spectrometry. *Analyst*, **128**, 832–837.

11 Hädrich, J. and Vogelgesang, J. (1999) Concept 2000 – a statistical approach for analytical practice. Part 1: limits of detection, identification and determination. *Deutsche Lebensmittel-Rundschau*, **95**, 428–436.

12 Walsh, J.H., Wyse, B.W., and Hansen, R.G. (1979) A comparison of microbiological and radioimmunoassay methods for the determination of pantothenic acid in foods. *Journal of Food Biochemistry*, **3**, 175.

13 Guilarte, T.R. (1989) A radiometric microbiological assay for pantothenic acid in biological fluids. *Analytical Biochemistry*, **178**, 63.

14 Song, W.O., Chan, G.M., Wyse, B.W., and Hansen, R.G. (1984) Effect of pantothenic acid status on the content of the vitamin in human milk. *American Journal of Clinical Nutrition*, **40**, 317.

15 Rychlik, M. and Roth-Maier, D.A. (2005) Pantothenic acid quantitation: method companion of a stable isotope dilution assay and a microbiological assay. *International Journal of Vitamin and Nutrition Research*, **75**, 218–223.

16 Romera, J.M., Ramirez, M., and Gil, A. (1996) Determination of pantothenic acid in infant milk formulas by high performance liquid chromatography. *Journal of Dairy Science*, **79**, 523–526.

17 Working Group on Nutritional Issues of the German Chemical Society GDCh (2009) Recommendations for tolerances for variations in the contents of nutrients and their labeling (in German). *Lebensmittelchemie*, **63**, 98.

10
Optimization of HPLC Methods for Analyzing Added Folic Acid in Fortified Foods

Padmanaban G. Krishnan, Sudheer R. Musukula, Michael Rychlik, David R. Nelson, Jonathan W. DeVries, and John L. MacDonald

10.1
Introduction

Vitamins of the folate group function as physiological methyl donors and cofactors involved in DNA synthesis. In this context, folate is expected to help prevent neural tube defects, Alzheimer's disease, and cardiovascular disease. Folic acid enrichment programs have been credited with reducing birth defects in the United States in recent years [1]. For this reason, fortification of staple foods has become mandatory in various countries.

The resulting increase in folic acid levels in the food supply has prompted the need for analytical methods for quantifying both the added synthetic form of folic acid and also the natural or endogenous forms. The official microbiological method, AACC (American Association of Cereal Chemists, now AACC International) Method 84-67, measures total folate content and expresses this as a single number [2]. In contrast, high-performance liquid chromatography (HPLC) methods are capable of measuring both natural and enriched forms of the vitamin and are popular analytical techniques. A multi-laboratory study of HPLC methods initiated by the AACC Vitamin Methods Committee involving seven laboratories in the United States, Singapore, Germany, and Australia yielded valuable data that showed enormous variability in folic acid values determined using enzymatic extraction, solid-phase cleanup, and isocratic HPLC quantification [3]. There was sufficient variation in procedures and techniques that an investigation of sources of variability appeared reasonable.

The objectives of the present study were, therefore, to determine optimal techniques for the determination of added folic acid [pteroylglutamic acid (PGA)] in food matrices [ready-to-eat (RTE) breakfast cereals, wheat flour products, infant formulas, baking mixes] and to delineate procedures that limit data variability. Robustness, accuracy, and precision were also evaluated.

Fortification of Foods with Vitamins, First Edition. Edited by Michael Rychlik.
© 2011 Wiley-VCH Verlag GmbH & Co. KGaA. Published 2011 by
Wiley-VCH Verlag GmbH & Co. KGaA.

10.2
Materials and Methods

10.2.1
Samples

A set of 21 cereal samples were purchased from a local grocery store. Each of the samples was individually mixed, ground using an A-10 mill, partitioned into sample sets, packaged, coded, and shipped to participating laboratories along with a detailed analytical protocol. Each sample set contained six blind-duplicates for determination of within-laboratory precision. Samples were selected to represent various cereal matrices (rice, oats, wheat, corn, barley, and products thereof) and a wide range of fortification levels (10, 25, 50, and 70% of the daily recommended intake for folic acid). A NIST infant formula sample was also provided by the US Food and Drug Administration, Office of Labeling, for inclusion in the sample set. This sample was previously determined to have a total folate concentration of $1.29\,\mu g\,g^{-1}$ in a collaborative study of the microbiological determination of folate [2].

10.2.2
Extraction at Different pH Values

A 2 g amount of each food sample was stirred for 1 h on a magnetic stir plate in a 125 ml conical flask with 25 ml of HEPES–CHES buffer, at pH 7.85 or pH 11 ($50\,mmol\,l^{-1}$ each HEPES and CHES), containing 2% (w/v) sodium ascorbate and 2-mercaptoethanol ($20\,mmol\,l^{-1}$). The sample extract was then centrifuged at $5000\,g$ for 10 min at 5 °C. The supernatant was collected and stored at 4 °C until extract purification and HPLC analysis.

10.2.3
Dual Enzyme Treatment

A dual enzyme extraction involving α-amylase and protease was employed, similar to that described by Pfeiffer *et al.* [4] but without conjugase treatment. A 2 g amount of each food sample was extracted with 20 ml of HEPES–CHES buffer, pH 7.85 ($50\,mmol\,l^{-1}$ each HEPES and CHES), containing 2% (w/v) sodium ascorbate and 2-mercaptoethanol ($10\,mmol\,l^{-1}$) in a 50 ml centrifuge tube. After mixing with a vortex mixer, all the tubes were placed in a boiling water-bath for 10 min, cooled rapidly in ice, and then homogenized using a tissue homogenizer (Biospec Products, Bartlesville, OK, USA) for 1 min. The contents of the tubes were then subjected to a dual enzyme treatment consisting of α-amylase (1 ml, $20\,mg\,ml^{-1}$) treatment for 4 h at 37 °C followed by protease (2 ml, $2\,mg\,ml^{-1}$) treatment for 1 h at 37 °C. After addition of each enzyme, the tubes were mixed on a vortex mixer. At the end of the enzyme treatments, the tubes were heated in a boiling water–bath for 5 min, cooled in ice, and centrifuged at $5000\,g$ for 10 min at 5 °C. The pellets were redissolved in 5 ml of extraction buffer and centrifuged at $5000\,g$ for 10 min at 5 °C. The supernatants were then combined, mixed well, and stored at 4 °C until extract purification and analysis.

10.2.4
Sample Purification Using Solid-Phase Extraction (SAX) Columns

A 6 ml strong ion-exchange solid-phase extraction (SPE) column (Phenomenex Strata SAX) was used to purify and concentrate the sample extracts using a Baker vacuum manifold. Hexane, methanol, and deionized water (10 ml each) were used in that order to activate each cartridge. A 5 ml aliquot of sample extract was applied slowly to the SAX column under vacuum. The column was then washed with 5 ml of deionized water and folic acid was eluted with 5 ml of sodium acetate solution (0.1 mol l^{-1}) containing 10% (w/v) sodium chloride and 1% (w/v) ascorbic acid. This final eluate was refrigerated at 4 °C until HPLC analysis. Prior to injection on to the column and HPLC analysis, all samples were filtered through a 0.45 μm filter (Millipore, Bedford, MA, USA).

10.2.5
Gradient HPLC Analysis

After equilibrating the HPLC system with mobile phase consisting of 95% phosphate buffer (33 mmol l^{-1}) and 5% acetonitrile, 100 μl of sample extract or working standard solutions were injected on to the column. A gradient program of phosphate buffer (33 mmol l^{-1}) and acetonitrile was employed, the acetonitrile composition being varied between 10 and 30% over a 26 min run. Chromatographic separation was achieved on a LiChrosorb C$_{18}$ reversed-phase column (250 cm × 4 mm i.d., 5 μm particle size) and the ultraviolet (UV) absorbance of the eluate was monitored at a wavelength of 285 nm.

10.2.6
Liquid Chromatography–Mass Spectrometry

Liquid chromatography–tandem mass spectrometry (LC–MS/MS) analysis was performed according to the procedure described by Rychlik [5]. In essence, the homogenized powders or flours (0.5 g) were stirred for 1 h at 20 °C in extraction buffer containing [^{2}H$_4$]folic acid (400 ng). The extracts were filtered and, after passing through a syringe filter (0.4 μm) (Millipore), were analyzed by LC–MS/MS. A typical LC–MS chromatogram is provided in Figure 10.1.

10.3
Results and Discussion

The gradient HPLC–UV method employed allowed the resolution of folic acid from other vitamin peaks found in enriched samples. Figure 10.2 shows typical

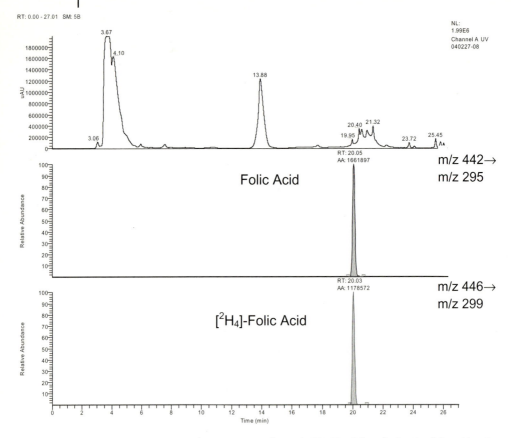

Figure 10.1 LC–MS/MS chromatograms of sample 732 (Total Cereal) showing folic acid and its coeluting isotopologic standard.

chromatograms for pure folic acid and a sample of breakfast cereal. The folic acid peak eluted at 15 min and was resolved from other vitamin peaks found in enriched samples. The method proved to be sensitive enough, as the lower limit of analysis was 0.1 μg ml^{-1} in the sample extract, corresponding to 30 ng g^{-1} for enriched commercial samples. The LiChrosorb C$_{18}$ column yielded reproducible chromatograms and appeared stable to 100 μl injections of standards and analytes. The folic acid peak appeared well before the time when the acetonitrile content was increased to 30% at 22.5 min. Therefore, folic acid was eluted from the column during an isocratic segment of the gradient program. The presence of numerous other vitamins in each enriched sample, however, required elevation to 30% acetonitrile and a further re-equilibration phase under the initial conditions.

For measuring the recovery of spiked flour samples, unfortified wheat flour spiked with folic acid at 1 mg g^{-1} and 10, 5, 2.5, and 1.25 μg g^{-1} served as an effective means of evaluating vitamin extraction and purification schemes. The recovery of spiked folic acid ranged from 89.6 to 97.8%.

(a)

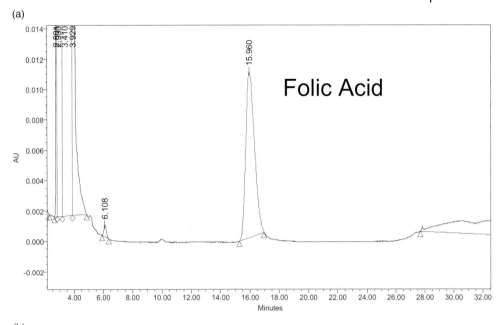

(b)

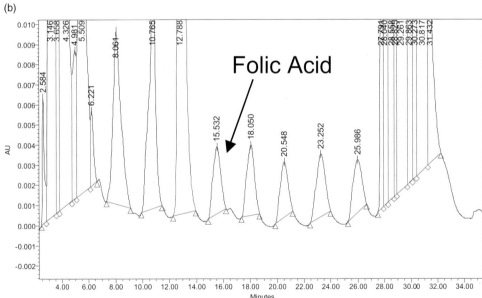

Figure 10.2 Typical chromatograms of 10% enriched white cornmeal (a) and 50% fortified Wheaties (b).

10.3.1
Comparison of Extraction and Purification Treatments

Table 10.1 provides statistical comparisons between the analytical treatments investigated in this study. LC–MS/MS determinations were based on replicate aqueous extractions, filtration, and direct tandem mass spectrometry with the aid

Table 10.1 Comparison of folic acid data obtained using SIDA, an HPLC–UV method in an interlaboratory study (IS) and with different extraction conditions.

Sample type	Sample	Folic acid ($\mu g\,g^{-1}$)					
		Label	SIDA	Mean HPLC IS value	HPLC with pH 11 extraction	HPLC with pH 7.85 extraction	HPLC with pH 7.85 extraction, dual treatment
Breakfast cereals	Rice Krispies	3.03	2.82[a]	3.32	2.32[b]	2.98[c]	4.07[d]
	Krispies	9.66	9.65[a),b)]	7.87	10.19[a]	9.13[b]	8.36[c]
	Krispies	8.48	9.84[a),b)]	4.80	10.13[a]	8.59[b]	7.62[c]
	Corn flakes	3.57	7.42[a]	5.92	7.46[a]	5.75[b]	9.97[c]
	Wheat rings	6.66	6.30[a]	5.88	5.43[a]	5.43[a]	8.34[b]
	Wheat pops	6.66	6.55[a]	5.61	10.45[b]	4.34[c]	6.70[a]
	Wheat bars	1.00	1.23[a]	1.32	1.26[a]	0.70[b]	4.70[c]
Noodles	Macaroni	1.79	1.62[a]	2.16	1.49[a]	2.98[b]	6.36[c]
Cookies	Cookies	1.33	0.81[a]	0.94	1.45[b]	1.42[b]	1.81[c]
	Whole wheat	13.3	12.39[a]	11.49	12.69[a]	8.30[b]	12.44[a]
	Crackers	2.67	1.35[a]	1.56	2.23[b]	1.63[a]	3.50[c]
Flours	Corn meal	1.48	1.93[a]	2.15	1.94[a]	1.43[b]	1.82[a]
	Wheat flour	1.33	1.63[a]	1.22	0.93[b]	1.03[b]	2.32[c]
Bread	Wheat bread	0.88	1.38[a]	1.62	1.45[a]	1.99[b]	1.93[b]
Infant formula	NIST Standard[e]	1.29*	1.18[a]	0.73	1.07[a]	1.64[a]	2.58[b]

a)–d) Means with the same superscript letter within rows are not statistically significantly different ($p \geq 0.05$).

e) The asterisk indicates total folate value as measured by microbiological analysis.

of stable isotopologs (Figure 10.1) as internal standards. Stable isotope dilution assay (SIDA) using LC–MS/MS analysis, in view of its selectivity, therefore reflects the most reliable measured data for our samples. In general, the SIDA data were in good agreement with the labeled contents. Label values are generally target enrichment levels intended by food manufacturers, but often include overages to insure that minimum levels are achieved over the whole shelf-life to meet regulatory standards. Table 10.1 shows that the folic acid values given by SIDA and HPLC at pH 11 did not differ for 10 of the 15 diverse samples.

Aqueous extraction at pH 7.85 yielded folic acid values that were similar to those for pH 11 extractions for only four samples (corn flakes, NIST infant formula, enriched flour, and cookies). Although alkaline pH favors dissolution of folic acid, it is not clear if the pH of the dissolution reagent alone can explain the differences in folic acid values yielded by the two methods, as extracts at pH 7.85 did not give lower values in all cases.

Enzymic extraction using amylase and protease enzymes is intended to liberate vitamins trapped in starch and protein matrices, particularly in cooked or processed foods. Enzyme hydrolysis and solid-phase extraction yielded values for only three samples that did not differ from the SIDA values (wheat rings, whole wheat, and white corn meal). As the highly specific LC–MS/MS method gave significantly lower values, the high signals in HPLC–UV analysis obviously are attributable to interfering compounds.

A folic acid standard ($0.893 \, \mu g \, g^{-1}$) taken through the same treatment at pH 7.85 and the same cleanup procedure on strong anion-exchange SPE cartridges yielded a recovery of 101.8%. The latter observation revealed that the pH of the extraction medium alone did not contribute to differences in folic acid values and points to enzymic treatments as a source of the variability.

10.4
Conclusion

Extraction and purification schemes have been identified as sources of variability in folic acid measurement by the HPLC–UV method. A simple aqueous extraction at pH 11 gave a good recovery of added folic acid in spiked flour and commercially enriched RTE breakfast cereals and cereal products. Commercially available solid-phase extraction devices employing a strong anion-exchange resin proved effective in extract cleanup and purification. A diverse set of RTE cereal samples analyzed using the refined method yielded accurate folic acid values, as judged by their closeness to mass spectrometric and label values. A SIDA using LC–MS/MS detection provided "true" folic acid values for the real samples. Blind duplicate samples within the sample set yielded between-laboratory repeatability and method precision. A gradient HPLC procedure employing 10–30% acetonitrile and phosphoric acid buffer yielded good separation of added folic acid from other enrichment ingredients. In order to evaluate interlaboratory precision, the simplified extraction protocol, solid-phase cleanup and gradient HPLC procedure will form the basis of an AACC collaborative study.

References

1 Yazdy, M.M., Honein, M.A., and Xing, J.
(2007) Reduction in orafacial clefts
following folic acid fortification of the
U.S. grain supply. *Birth Defects Research A.
Clinical and Molecular Teratology,* **79** (10),
16–23.

2 DeVries, J.W., Keagy, P.M., Hudson, C.A.,
and Rader, J.I. (2001) AACC collaborative
study of a method for determining total
folate in cereal products – microbiological
assay using trienzyme extraction (AACC
Method 86-47). *Cereal Foods World,* **46** (5),
216–219.

3 Krishnan, P., Reddy, S., Nelson, D.,
Rychlik, M., DeVries, J., and MacDonald,
J. (2005) Optimization of HPLC methods
for folic acid determination of RTE cereal
foods. Abstracts Book of 2005 AACC
International Annual Meeting, Orlando,
FL, p. 135.

4 Pfeiffer, C.M., Rogers, L.M., and Gregory,
J.F. (1997) Determination of folate in
cereal grain food products using
trienzyme extraction and combined
affinity and reverse phase liquid
chromatography. *Journal of Agricultural
and Food Chemistry,* **45**, 407–413.

5 Rychlik, M. (2003) Simultaneous analysis
of folic acid and pantothenic acid in foods
supplemented with vitamins by stable
isotope dilution assays. *Analytica Chimica
Acta,* **495**, 133–141.

11
Studies on New Folates in Fortified Foods and Assessment of Their Bioavailability and Bioactivity

Michael Rychlik and Dora Roth-Meier

11.1
Introduction

Fortification of wheat flour with folic acid (FA) is mandatory in various countries, especially on the American continent. The merits are obvious, as the average daily folate intake in the USA was increased by $200\,\mu g$ and the occurrence of neural tube defects declined by 3.8 cases per 1000 pregnancies [1]. Moreover, further possible benefits include decreases in the incidence of coronary heart disease, colon cancer, and Alzheimer's disease.

However, FA consumption masking vitamin B_{12} deficiency and possible adverse effects of unmetabolized FA in plasma have prevented many countries from implementing mandatory fortification with FA so far. Another potential risk may arise from thermal treatment of FA and its reactions with other ingredients in foods.

In this context, a nonenzymatic glycation involving carbohydrates and FA acting as an amino compound was reported in 2002 [2]. The product, N^2-[1-(1-carboxy) ethyl]folic acid (CEF) was structurally characterized and its formation was followed in models involving different carbohydrates and their degradation products. As sugar products such as corn syrup and high-fructose corn syrup (HFCS) are common sweeteners, the question arises of the occurrence of CEF and further glycation products in fortified foods.

As we reported for the analysis of endogenous folates in foods [3], the use of stable isotopologs of folate vitamers enables one to correct for losses during extraction, cleanup, high-performance liquid chromatography (HPLC) and mass spectrometry (MS) detection. The purpose of the present study was, therefore, first to elucidate whether CEF is present also in foods and other glycation products can arise from FA. Second, a sensitive and accurate method for the quantification of CEF and further possible glycation products was to be developed and products fortified with FA were to be screened for their contents.

Fortification of Foods with Vitamins, First Edition. Edited by Michael Rychlik.
© 2011 Wiley-VCH Verlag GmbH & Co. KGaA. Published 2011 by
Wiley-VCH Verlag GmbH & Co. KGaA.

11.2
Materials and Methods

11.2.1
Chemicals

The following chemicals were obtained commercially: acetonitrile, dihydroxyace-tone, FA, formic acid, fructose, glucose, glyoxal, 2-mercaptoethanol, lactose, meth-anol, $NaHCO_3$, KH_2PO_3, Na_2HPO_3, sodium acetate, sodium chloride, and sucrose from Merck (Darmstadt, Germany) and 2-(cyclohexylamino)ethanesulfonic acid (CHES), 4-(2-hydroxyethyl)-1-piperazineethanesulfonic acid (HEPES), and sodium ascorbate from Sigma (Deisenhofen, Germany).

11.2.2
Extraction and Incubation Buffer

For storage and extractions of folates, the following buffer system according to Wilson and Horne [4] was used, henceforth referred to as "extraction buffer": aqueous HEPES ($50\,mmol\,l^{-1}$) and aqueous CHES ($50\,mmol\,l^{-1}$) at pH 7.85 contain-ing sodium ascorbate (2%) and 2-mercaptoethanol (20 mM).

11.2.3
Synthesis of Carboxyethylfolic Acid and Carboxymethylfolic Acid

CEF and [$^2H_4$]CEF were synthesized as reported [5], and N^2-[1-(1-carboxy)methyl] folic acid (CMF was prepared analogously with the use of glyoxal instead of dihy-droxyacetone as the reactant:

Folic acid (315.7 mg; 0.72 mmol) in phosphate buffer (10 ml, $1\,mol\,l^{-1}$, pH 7.4) was reacted with glyoxal (40%, 0.5 ml, 2 mmol) at 100 °C for 24 h. Then, aliquots (50 µl) of the mixture were subjected to preparative HPLC using a Nucleosil column (5 µm, 250 × 3 mm i.d.) (Macherey-Nagel, Düren, Germany) coupled to an HPLC system (Biotek, Eching, Germany) and eluted with a linear gradient starting from a mixture of methanol in aqueous formic acid (0.1%; 12 + 88, v/v) to 55% methanol within 40 min and then raised to 100% methanol within 1 min at a flow-rate of $0.8\,ml\,min^{-1}$. CMF was eluted as the highest peak at 280 nm and was col-lected from 50 runs. The pooled fraction finally was lyophilized.

The mass and tandem mass spectra of the protonated molecule in the positive electrospray ionization mode revealed the protonated molecular ion at m/z 500 and the main fragment at m/z 353.

11.2.4
Commercial Food Samples

Two multivitamin juices and four baby foods were obtained from local retail stores in Munich, Germany. The baby foods were infant formulas from different producers. Six different types of commercial cookies were purchased at supermarkets in Washington, DC, USA.

11.2.5
Extraction and Quantification of CEF and CMF in Foods

Solid samples were frozen in liquid nitrogen and minced in a blender (Privileg, Quelle, Fürth, Germany). The resulting powder or liquid samples (1 g) were mixed with extraction buffer (10 ml) containing $[^2H_4]FA$ (1 µg) and $[^2H_4]CEF$ (0.1 µg). The mixtures were then stirred for 1 h and subsequently centrifuged for 10 min at 13 000 rpm. The supernatant was passed through a syringe filter (0.4 µm) (Millipore, Bedford, MA, USA) or subjected to solid-phase extraction (SPE) prior to liquid chromatography–tandem mass spectrometry (LC–MS/MS).

11.2.6
Quantitation of CEF in Blood Plasma

Plasma samples (1 ml) were spiked with $[^2H_4]$-5-methyltetrahydrofolate (20 ng) and equilibrated for 30 min at room temperature. The solution was then centrifuged (10 min; 2000 g) and diluted with extraction buffer (2 ml) before being subjected to SPE as described below.

As plasma samples contained 5-methyltetrahydrofolate as the only vitamer and showed lower matrix interferences than the food extracts, gradient elution was shortened. Elution started at 10% acetonitrile and was raised to 20% acetonitrile within 6 min. Then, the acetonitrile concentration was sharply raised to 80% within 1 min and to 100% within a further 1 min. This concentration was maintained for 2 min before lowering the acetonitrile concentration back to the initial value and allowing the column to equilibrate for a further 3 min. Each plasma extract was analyzed in triplicate. Amounts of 5-methyltetrahydrofolate and CEF were calculated from the peak areas in the respective mass traces using the calibration functions as detailed below.

11.2.7
Sample Cleanup by Solid-Phase Extraction (SPE)

Extracts were purified by SPE according to the method described by Gounelle et al. [6], using a 12-port vacuum manifold (Alltech, Bad Segeberg, Germany) equipped with Bakerbond strong anion-exchange (SAX) cartridges (quaternary amine, 500 mg, No. 7091-3) (Baker, Gross-Gerau, Germany). The cartridges were successively activated with two volumes of hexane, methanol, and water, and then conditioned with 7–8 volumes of phosphate buffer (pH 7.5, 0.01 mol l^{-1}, containing 0.2% mercaptoethanol).

After applying the sample extracts (6 ml), the columns were washed with six volumes of conditioning buffer and the folates were eluted with 3 ml of aqueous sodium chloride (5%, containing 1% sodium ascorbate and 0.1 mol l^{-1} sodium acetate). 2-Mercaptoethanol (100 μl) was added to each eluate and the purified extracts were subjected to LC–MS/MS.

11.2.8
LC–MS/MS

The samples (20 μl) were injected into a high-performance liquid chromatograph equipped with a Nucleosil C_{18} reversed phase column (250 × 3 mm i.d.; 5 μm) (Macherey-Nagel) that was connected to a Surveyor diode-array detector and a TSQ triple-quadrupole mass spectrometer (Thermo Science, Bremen, Germany).

The mobile phase consisted of variable mixtures of aqueous formic acid (0.1%) (eluent A) and acetonitrile (eluent B) at a flow-rate of 0.4 ml min^{-1}. Gradient elution started at 10% B for 6 min, followed by raising the concentration of B linearly to 100% within 19 min. Subsequently, the mobile phase was maintained at 100% B for 5 min before equilibrating the column for 5 min at the initial mixture.

During the first 10 min of the gradient program, the column effluent was diverted to waste. The spectrometer was operated in the positive electrospray ionization mode using selected reaction monitoring (SRM). The spray voltage was set to 5.5 kV, the capillary temperature to 200 °C, and the capillary voltage to 24.3 V. Collision-induced dissociation (CID) was performed using a source CID collision energy of 10% and the further conditions detailed in [5].

11.2.9
Determination of Response Factors for LC–MS/MS

Solutions of CEF, CMF and [^{2}H$_4$]CEF in extraction buffer (10 ml) were mixed in five mass ratios ranging from 0.2 to 5 to give a total folate content of 2 μg. Subsequently, the CEF–CMF mixtures were subjected to LC–MS/MS as outlined earlier and response factors (R_f) were calculated.

11.2.10
Recovery

Recovery was determined by adding 160 ng of CEF and CMF to edible corn starch (1 g) and performing SIDA as detailed earlier in quadruplicate.

11.2.11
Microbiological Assay (MA)

In the microbiological assay (MA), the glycation products CEF and CMF were assayed for growth response to folates with *Lactobacillus casei* (NCIMB 10463) in Micro Inoculum Broth (Difco 0320-17-4; Becton Dickinson, Sparks, MD, USA) according to Strohecker and Henning [7]. *L. casei* growth was measured spectrophotometrically at 660 nm after an incubation time of 24 h at 37 °C (Uvicon 933; Kontron Instruments, Eching, Germany). FA was used as the reference standard for calibration in amounts ranging between 0.5 and 5 ng in aqueous ascorbic acid (2%). Calibration curves were recorded at an optical density below 1.5. Each calibrant of standards and samples was analyzed in triplicate.

11.2.12
Human Study

The bioavailability study was designed as a short-time protocol including the consumption of one test meal consisting of 250 g of commercial cookies containing 14.5 μg of CEF and 105 μg of FA by one non-smoking male volunteer (42 years old with a body mass index of 21.6 kg m^{-2}). Two weeks before consumption of the test meal, tissue from the volunteer was saturated by supplementation with FA (800 μg per day). Two days before the test, the gavage of FA was stopped to allow for stabilization of the plasma level.

The test meal was consumed by the volunteer at 8 a.m. and consumption was followed by blood sampling each hour for 3 h and then every 2 h for a further 8 h. The plasma obtained was stored at −60 °C until analysis.

11.3
Results and Discussion

As Schneider *et al.* [2] found a glycation product of FA, CEF, in models containing sugars and FA, CEF is a conceivable compound in fortified foods that are similarly composed and undergo a similar treatment to that in the models used by Schneider *et al.* Possible examples are multivitamin juices containing carbohydrates that are pasteurized to make them shelf-stable. Furthermore, cookies produced from folate-fortified flour and glucose or fructose syrup may contain CEF.

When considering the glycation of amino compounds, the possible generation of further Maillard products appeared interesting. As carboxymethyllysine (CML) is an important glycation product of lysine [8], the question arose of whether the respective derivative of FA, CMF (Figure 11.1), might be generated. For unambiguous identification and quantification of this hypothetical product, it was necessary to synthesize the pure reference compound.

11.3.1
Synthesis of N^2-[1-(1-Carboxy)ethyl]folic Acid and N^2-[1-(1-Carboxy)methyl]folic Acid

As CEF is not commercially available, initial experiments were performed to synthesize it as a reference compound. The most successful way was that described by Schneider *et al.* [2] by reacting FA with dihydroxyacetone (DHA) and purifying CEF from unreacted FA by preparative HPLC [5]. NMR experiments revealed similar data to those reported by Schneider *et al.*

Analogously to the proposed generation of CML [8] and the synthesis of CEF, we reacted FA with glyoxal and obtained two peaks upon HPLC. As peak 2 revealed

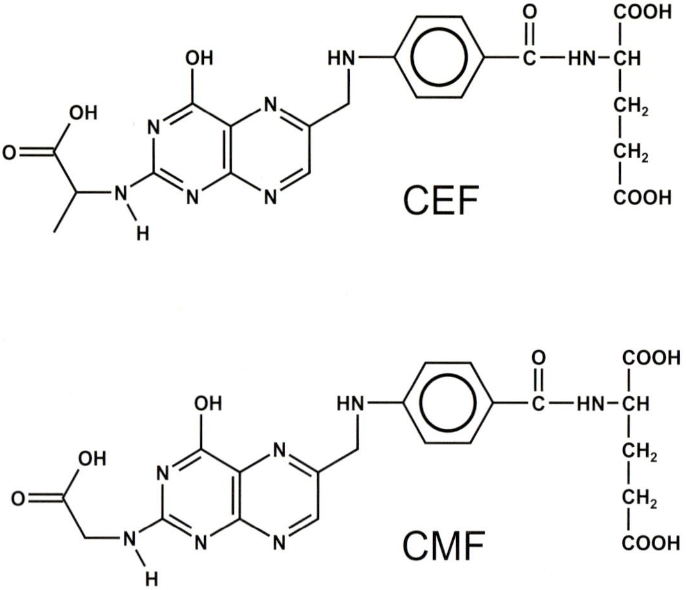

Figure 11.1 Structures of N^2-[1-(1-carboxyethyl)]folic acid (CEF) and N^2-[1-(1-carboxymethyl)] folic acid (CMF).

in LC–MS an intense signal at m/z 500 corresponding to the protonated molecular mass of CML, it was purified by preparative HPLC. Further one- and two-dimensional NMR experiments were performed to confirm the structure of the compound. ^{1}H NMR analysis showed, apart from the typical signals of FA, a singlet at δ 4.19 ppm in comparison with the spectrum of CEF, which showed at δ 4.3 ppm a multiplet originating from the methine group at C-2′ coupling with the methyl hydrogen at C-3′. From this result, it can be concluded that the singlet at δ 4.19 ppm corresponded to C-2′ and no further adjacent CH bond was present. The assignment of C-2′ was substantiated by heteronuclear multiple bond coherence (HMBC) and heteronuclear multiple quantum coherence (HMQC) NMR experiments.

Similarly to its homolog CEF, CMF revealed in positive electrospray ionization an intense signal of its protonated molecule and upon CID in MS/MS the loss of the glutamate moiety, typical of folates.

Our earlier studies showed that LC–MS/MS in combination with a SIDA is a sensitive and accurate tool for the quantitation of folates [3]. Therefore, the development of a SIDA based on labeled CEF as the internal standard (IS) appeared straightforward. For this purpose, [^{2}H$_4$]CEF was prepared from [^{2}H$_4$]FA by reaction with dihydroxyacetone and subsequent HPLC purification, in analogy with the synthesis of unlabeled CEF. Due to its structural similarity, [^{2}H$_4$]CEF was also chosen as the IS for CMF.

For detection of CMF in demanding matrices such as cookies, we adapted our existing method for folates and CEF. After extraction with Wilson–Horne buffer, folates were separated from interfering compounds on a SAX cartridges. Upon final LC–MS/MS, CMF was eluted 1 min earlier than CEF due to its lower hydrophobicity.

11.3.2
Development of a Stable Isotope Dilution Assay

For relating the measured ion intensities to the mass ratios of labeled CEF and unlabeled CEF and also CMF, two graphs were calculated from calibration mixtures of known mass ratios and the corresponding peak area ratios in LC–MS/MS. Good response linearity was demonstrated in both cases for mass ratios ranging from 0.2 to 5.

Anion exchange chromatography (AEC) provided purified extracts with less background noise than without sample cleanup. Interestingly, the intensity ratio between CMF and CEF after AEC was very similar to that without using this cleanup, thus indicating that neither CEF nor CMF was discriminated when binding to the anion-exchange material. This behavior resulted also in similar recoveries for CEF and CMF, which were found to be 98.5 and 89%, respectively.

When applying the newly developed SIDA to multivitamin juices, multivitamin sweets, and fortified milk products such as baby foods, we could not detect either CEF or CMF. However, cookies purchased in the USA were found to contain considerable amounts of CEF. In contrasts, CMF was not detectable (Figure 11.2).

In particular, cookies made from fortified flour and glucose, glucose syrup or fructose syrup revealed CEF concentrations ranging from 5.1 to 9.2 µg per 100 g. In contrasts, CEF was not measurable in cookies either produced from non-fortified flour or containing artificial sweeteners instead of sugars (Table 11.1).

The absence of CMF was surprising as the potential precursors of CEF and CMF, 2-oxopropanal and glyoxal, respectively, generally occur simultaneously when reducing carbohydrates are degraded thermally [9]. The same result was obtained for cookies made from different carbohydrates [5]. As we did not quantify the precursors 2-oxopropanal and glyoxal, it has to be speculated that either glyoxal is formed to a smaller extent or reacts less effectively with FA than 2-oxopropanal in these kind of products.

11.3.3
Microbiological Assay (MA)

To gain a first insight into folate activity or folate antagonism of the new glycation products, a bioassay using the microorganism L. casei was applied.

As generally for bioassays, the growth of an organism for which the analyte to be determined is essential is measured in a specifically deficient nutrient solution. The medium is then supplemented with a sample or a vitamin standard, and the

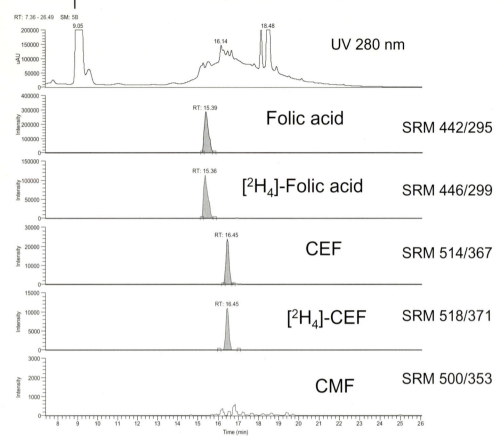

Figure 11.2 LC–MS/MS chromatograms of a cookie extract in the positive electrospray ionization mode after CID of the protonated molecules. Unlabeled FA, unlabeled CEF, unlabeled CMF, and the ISs [²H₄]FA and [²H₄] CEF are detected in the traces SRM 442/295, 514/367, 500/353, 446/299, and 518/371, respectively. UV, UV absorption. SRM traces: m/z precursor ion/m/z product ion.

Table 11.1 Contents of FA and CEF in different foods fortified with FA; CMF was not detectable.

Sample	No. of samples	Folic acid (µg per 100 g)	CEF (µg per 100 g)[a]
Multivitamin juices	2	51–77	n.d.
Baby food	4	48–137	n.d.
Multivitamin sweets	2	670–700	n.d.
Cookies	3	32–103	5.1–7.1
Cookies, sugar-free	2	55–68	n.d.
Cookies made from non-fortified flour	1	0.4	n.d.

a) n.d., not detectable.

Table 11.2 Responses of the glycation products CEF and CMF in the microbiological assay.

Sample	Response (%)
Folic acid ($1\,ng\,ml^{-1}$)	100
CEF ($1\,ng\,ml^{-1}$)	<5
CMF ($1\,ng\,ml^{-1}$)	<5
Folic acid + CEF ($1\,ng\,ml^{-1}$ each)	98
Folic acid + CMF ($1\,ng\,ml^{-1}$ each)	96

resulting growth is measured at intervals, for example turbidimetrically or neph-elometrically. Most commonly, growth is reflected by the turbidity of the assay, which can be conveniently measured in a UV–visible spectrophotometer.

The first test was to evaluate the growth response of *L. casei* to the pure compounds CEF and CMF. In contrast to FA, the response to which was set to 100%, both CEF and CMF showed no response above the detection limit, which is 5% of the response to FA (Table 11.2). In complementary trials, CEF and CMF were added in equimolar amounts to FA and the response of the MA was recorded, the results being 98 and 96% for CEF and CMF, respectively, of the response to that of pure FA, which is not significantly different from the response of the pure compound in a *t*-test ($p < 0.05$). From these tests, it can be concluded that there is neither a folate activity nor a folate antagonism of CEF or CMF.

11.3.4
Human Study to Assess the Bioavailability of CEF

For a preliminary risk evaluation of the new glycation product CEF, we conducted a pilot human study. As the test meal, 250 g of commercial cookies containing 14.5 µg of CEF and 105 µg of FA were chosen. The volunteer's folate stores were saturated by predose gavage of FA to simulate the situation in a country where fortification with FA is mandatory. After consumption of the test meal, seven plasma samples were drawn within 11 h. The samples were analyzed for the naturally occurring vitamer 5-methyltetrahydrofolate and for CEF. Whereas the 5-methyltetrahydrofolate level was found to follow a common plasma response curve after dosage of FA, no CEF was detected in any of the samples. An exemplary LC–MS/MS chromatogram of a sample is shown in Figure 11.3. However, as the detection limit of CEF is 10-fold higher than that of nonglycated folates, the dose of 14.5 µg of CEF would have been close to its detection limit, even when assuming passive diffusion and distribution solely in the volunteer's plasma volume of 3000 ml [10]. Nevertheless, CEF obviously has no critical effect on resorption and metabolism of FA. Therefore, from the results of this first bioactivity trial, the new glycation products of FA do not appear to be an acute threat to the consumer. However, as this preliminary study lacks representativeness, further investigations including a larger number of volunteers need to be performed.

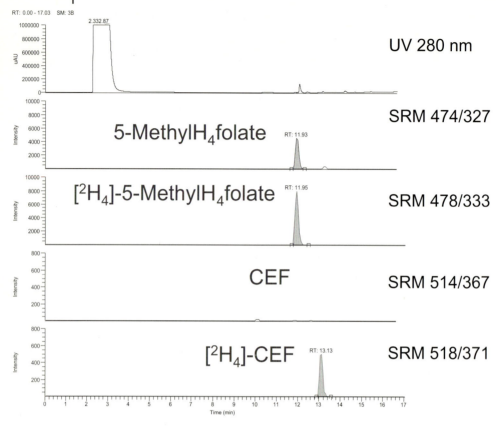

Figure 11.3 LC–MS/MS chromatograms of a plasma sample taken from a volunteer 2 h after consumption of cookies containing 14 μg of CEF. Unlabeled 5-methyltetrahydrofolate, unlabeled CEF, and the ISs [^{2}H$_4$]-5-methyltetrahydrofolate and [^{2}H$_4$]CEF are detected in the traces SRM 474/327, 514/367, 478/333, and 518/371, respectively. UV, UV absorption. SRM traces: *m/z* precursor ion/*m/z* product ion.

11.3.5
Glycation Products of Endogenous Folates

Following the identification of CMF and CEF as glycation products of FA, it appeared straightforward to extend these experiments also to naturally occurring folates in foods. A first step to achieve this task was described recently by Verlinde *et al.*, who reported the generation and identification of the analogous compound N^2-[1-(carboxy)ethyl]-5-methyl-5,6,7,8-tetrahydrofolic acid (Figure 11.4) from reaction of 5-methyltetrahydrofolic acid with reducing carbohydrates and their degradation product dihydroxyacetone [11]. According to kinetic studies, the reaction of dihydroxyacetone with the oxidation product of 5-methyltetrahydrofolic acid, namely 5-methyl-7,8-dihydrofolic acid, was postulated as the ultimate step in the

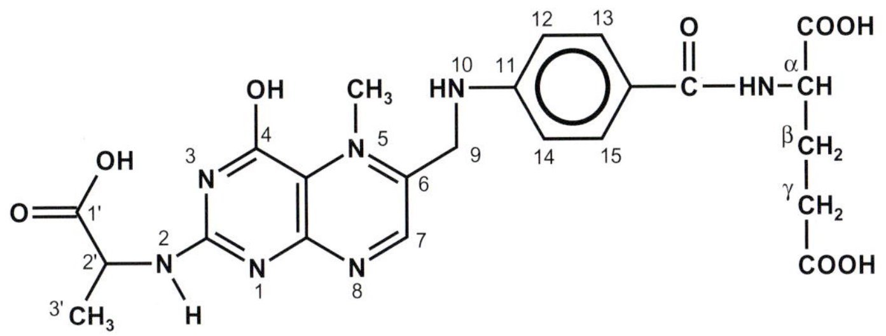

Figure 11.4 Structure of N^2-[1-(carboxy)ethyl]-5-methyl-5,6,7,8-tetrahydrofolic acid, a new glycation product of 5-methyl-5,6,7,8-tetrahydrofolic acid [12].

formation of the new glycation product. However, the occurrence of carboxyethylated 5-methyltetrahydrofolic acid in foods has to be confirmed.

References

1 De Wals, P., Tairou, F., Van-Allen, M.I., Uh, S.-H., Lowry, R.B., Sibbald, B., Evans, J.A., Van den Hof, M.C., Zimmer, P., Crowley, M., Fernandez, B., Lee, N.S., and Niyonsenga, T. (2007) Reduction in neural-tube defects after folic acid fortification in Canada. *New England Journal of Medicine*, **357**, 135–142.

2 Schneider, M., Klotzsche, M., Werzinger, C., Hegele, J., Waibel, R., and Pischetsrieder, M. (2002) Reaction of folic acid with reducing sugars and sugar degradation products. *Journal of Agricultural and Food Chemistry*, **50**, 1647–1651.

3 Freisleben, A., Schieberle, P., and Rychlik, M. (2003) Comparison of folate quantification in foods by high-performance liquid chromatography–fluorescence detection to that by stable isotope dilution assays using high-performance liquid chromatography–tandem mass spectrometry. *Analytical Biochemistry*, **315**, 247–255.

4 Wilson, S.D. and Horne, D.W. (1984) High-performance liquid chromatographic determination of the distribution of naturally occurring folic acid derivatives in rat liver. *Analytical Biochemistry*, **142**, 529–535.

5 Rychlik, M. and Mayr, A. (2005) Quantitation of N^2-[1-(1-carboxy)ethyl]

folic acid, a nonenzymatic glycation product of folic acid, in fortified foods and model cookies by a stable isotope dilution assay. *Journal of Agricultural and Food Chemistry*, **53**, 5116–5124.

6 Gounelle, J.-C., Ladjimi, H., and Prognon, P. (1989) A rapid and specific extraction procedure for folates–determination in rat liver and analysis by high-performance liquid chromatography with fluorimetric detection. *Analytical Biochemistry*, **176**, 406–411.

7 Strohecker, R. and Henning, H. (1963) *Vitamin-Bestimmungen*, Verlag Chemie, Weinheim.

8 Ahmed, M.U., Thorpe, S.R., and Baynes, J.W. (1986) Identification of N^ε-carboxymethyllysine as a degradation product of fructoselysine in glycated protein. *Journal of Biological Chemistry*, **261**, 4889–4894.

9 Hofmann, T. (1999) Quantitative studies on the role of browning precursors in the Maillard reaction of pentoses and hexoses with L-alanine. *Zeitschrift für Lebensmittel-Untersuchung und -Forschung*, **209**, 113–121.

10 International Commission of Radiological Protection (ICRP) (2002) Basic Anatomical and Physiological Data for Use in Radiological Protection:

reference values. ICRP Publication 89. Annals of the ICRP, **32**, 1–277.

11 Verlinde, H.C.J., Oey, I., Lemmens, L., Deborggraeve, W.M., Hendrickx, M.E., and Von Loey, A.M. (2010) Influence of reducing carbohydrates on (6S)-5-methyltetrahydrofolic acid degradation during thermal treatments. *Journal of Agricultural and Food Chemistry*, **58**, 6190–6199.

12
Analysis of Vitamin B$_{12}$ by HPLC

Bo Chen and Da-jin Yang

12.1
Introduction

As chromatographic methods are very common analytical techniques, assays based on these techniques are potential candidates for official methods for the analysis of vitamin B$_{12}$ in food products. High-performance liquid chromatography (HPLC), coupled with appropriate detection procedures, constitutes an efficient measurement system for vitamin B$_{12}$, and has been improved by advances in reversed-phase column technology, HPLC detectors, and sample pretreatment technology. In general, the vitamin B$_{12}$ concentration in food products is very low (in the range 3–50 ng g^{-1}). The cyanocobalamin form of B$_{12}$ is the most widespread in the food industry. Although the concentration of vitamin B$_{12}$ in fortified food products is always higher than that in natural food products, the sensitivity of analytical methods is the most critical index for a successful analysis of vitamin B$_{12}$. A number of HPLC methods coupled with ultraviolet (UV) or visible (Vis) photometric detection have been proposed for the determination of vitamin B$_{12}$. However, cyanocobalamin is a tetrapyrrole complex which contains an atom of cobalt in its molecule, and lacks an adequate chromophore. Because of the low sensitivity and poor specificity of UV–Vis detection, various other on-line detection techniques, for example, inductively coupled plasma mass spectrometry (ICP–MS) and electrospray ionization mass spectrometry (ESI-MS), have been suggested.

This chapter is focused on the developments in sample pretreatment, separation, and detection for the HPLC analysis of vitamin B$_{12}$ by various methods.

12.2
Sample Preparation

12.2.1
Solid-Phase Extraction (SPE)

Solid-phase extraction (SPE) is an extraction method that uses a solid phase and a liquid phase to isolate one, or one type, of analyte from a solution. It is usually used to clean up a sample before using a chromatographic or another analytical method to determine the amount of analyte(s) in the sample. For the determination of trace amounts of vitamin B$_{12}$, HPLC with UV–Vis detection after SPE is a useful strategy. Among the packed materials for SPE, reversed-phase (RP) and immunoaffinity (IA) sorbents are the most widely used.

12.2.1.1 Reversed-Phase Solid-Phase Extraction (RP-SPE)
Although RP-SPE has been widely used in HPLC analyses of vitamin B$_{12}$, the SPE conditions must be optimized for the retention and elution of trace amounts of vitamin B$_{12}$ in sample solution. Different SPE conditions have been reported (Table 12.1).

12.2.1.2 Immunoaffinity Solid-Phase Extraction (IA-SPE)
The IA-SPE technique is based on a molecular recognition mechanism. The high affinity and high selectivity of antigen–antibody interactions allow the specific extraction and concentration of the analytes of interest in one step. In the pharmaceutical and biological fields, where most often matrices are complex and analytes are present at trace levels, this approach constitutes a unique tool for fast and solvent-free sample preparation.

In 2005, Pakin *et al.* [4] proposed a novel IA-SPE method for the purification of vitamin B$_{12}$ from sample solution. A 6 mm × 8 mm i.d. prototype immunoaffinity column (R-Biopharm, Saint-Didier au Mont d'Or, France) was used to purify

Table 12.1 RP-SPE conditions reported for vitamin B$_{12}$ analysis.

SPE cartridge	Elution	Detection	Matrices	Ref.
Bond-Elut C$_{18}$ (500 mg)	50% acetonitrile in water	Vis 550 nm	Nutrient	[1]
C$_{18}$ AR (30 mg)	Deionized water and methanol–water (6:4, v/v)	UV 362 nm	Multivitamin formulation	[2]
Supelclean LC-18 (500 mg)	Methanol–water (85:15, v/v)	UV 230 nm	Multivitamin formulation, serum, urine	[3]

vitamin B_{12}. After loading the sample solution on the immunoaffinity column, the latter was washed successively with 10 ml of phosphate buffer (100 mmol l^{-1}, pH 7) and 5 ml of distilled water and then air dried by passing 10 ml of air through the column with a syringe. Vitamin B_{12} was then eluted with 3 ml of methanol. The recovery of the suggested method as determined for pig liver was very satisfactory (98%). Moreover, the purification enabled the foodstuff extract to be concentrated by a factor of up to six for samples with very low vitamin B_{12} contents. Hence the minimum quantifiable concentration of vitamin B_{12} in foodstuffs was about 3 ng g^{-1}.

Heudi et al. [5] developed a method which was slightly different from Pakin et al.'s method for the determination of vitamin B_{12} in food products and in premixes by IA-SPE, HPLC, and using a similar immunoaffinity column for the purification of vitamin B_{12}. Cyanocobalamin was extracted from samples with sodium acetate buffer at pH 4.0, followed by a purification step on the immunoaffinity column before HPLC analysis. They also obtained a recovery of vitamin B_{12} of about 100%. The recovery was similar to that obtained under the protocol recommended by the supplier of the immunoaffinity column, that is, phosphate buffer as washing solution and pH 7 for loading the sample on to the column.

Nowadays, immunoaffinity columns for the purification of vitamin B_{12} are commercially available, for example, EASI-EXTRACT® Vitamin B_{12} (R-Biopharm Rhône, Glasgow, UK). According to the working principle, the immunoaffinity binding takes place between the antibody and the vitamin on the this column. Its suitable working pH range is 4.5–7.0, so, the pH of sample solutions must be kept within this range. Different samples have different pH backgrounds. Therefore, the sample solutions should be adjusted to a suitable pH range.

Although few methods for immunoaffinity purification for vitamin B_{12} have been reported in the literature, and perhaps there is a disadvantage of a lack of specificity to all of the cobalamins, active forms, and inactive analogs, present in foodstuffs, it can be an appropriate choice for the selective extraction and purification of vitamin B_{12} from food products, especially from fortified foods, as it showed a higher specificity than RP-SPE.

12.3
Separation

So far, separation methods for the analysis of vitamin B_{12} have been either conventional HPLC or microbore HPLC. In general, reversed-phase conditions using either C_8 or C_{18} columns have been used with acetonitrile or methanol as the organic solvent components of the mobile phase.

For the HPLC analysis of vitamin B_{12}, different mobile phases have been proposed. The typical separation conditions reported in the literature are listed in Table 12.2, from which it can be seen that C_{18} columns were mostly used for the separation of vitamin B_{12}.

Table 12.2 Chromatographic conditions reported in the literature.

Column	Mobile phase	Gradient or isocratic elution	Detection	t_r (min)	Ref.
C_{18}	0.085 M H_3PO_4 (pH 3.1, adjusted with NH_4OH) + acetonitrile 20% (v/v)	Gradient	365 nm	18	[6]
C_{18}	Methanol solution containing 1% (v/v) acetic acid	Isocratic	361 nm	8.5	[7]
C_{18}	20% (v/v) methanol solution containing 1% (v/v) acetic acid + methanol containing 1% (v/v) acetic acid	Gradient	278 nm	19	[8]
C_{18}	20 mM KH_2PO_4 (pH 2.1, adjusted with H_3PO_4)–acetonitrile (90:10, v/v)	Isocratic	550 nm	12	[1]
C_{18}	0.05 M NH_4OAc + methanol	Gradient	362 nm	13	[2]
C_{18}	0.025% TFA (pH 2.6) + acetonitrile	Gradient	210 nm	15	[9]
C_{18}	0.1 M NaH_2PO_4–methanol (63:27, v/v)	Isocratic	361 nm	16.5	[10]
C_{18}	Acetonitrile–water (pH 3.45) (40:60, v/v)	Isocratic	240 nm	1.5	[11]
C_{18}-CN			ICP-MS 254 nm		[12]
C_{18}	75% 0.25 M NaH_2PO_4 (pH 3.5) + 25% methanol	Isocratic	371 nm	7.1	[13]
C_{18}	MeOH–water (30:70)	Isocratic	Fluorescence		[14]
Phenylpropanol–amine bonded silica	30 mM phosphate buffer (pH 3.00) containing 6% (v/v) acetonitrile	Isocratic	361 nm	3.5	[15]
C_{18}	Water–acetonitrile	Gradient	ESI-MS	9.8	[16]
C_{18}	Methanol–phosphate buffer (10:90) and trimethylamine (pH 3.55)	Isocratic	Electrochemical	16.4	[17]
C_{18}	0.025% TFA (pH 2.6)–acetonitrile	Gradient	361 nm	17.2	[5]
C_{18}	0.05 M phosphate buffer–10% methanol and 0.018 M trimethylamine (pH 3.55)	Isocratic	Coulochemical	16.2	[18]

12.4
Detection

12.4.1
Ultraviolet–Visible Detection

The most common method used for the quantification of vitamin B_{12} was HPLC with UV–Vis detection. The wavelengths for the vitamin B_{12} reported in the recent literature are listed in Table 9.2; the sensitivity of detection at different wavelengths showed no significant difference.

12.4.2
Fluorescence Detection

Vitamin B_{12} is rather difficult to determine by HPLC owing to its very low concentration in foods, and HPLC–UV methods revealed a low sensitivity because of the lack of a suitable chromophore in the vitamin B_{12} structure. To improve the sensitivity of detection of vitamin B_{12}, different detection methods coupled with HPLC were investigated.

In general, fluorescence detection has a higher specificity and/or sensitivity than UV–Vis detection. Li *et al.* developed an HPLC method with direct fluorescence detection for the analysis of vitamin B_{12} in multivitamin tablets and fermentation [14, 19]. The column eluate was monitored with a fluorescence detector at 305 nm with excitation at 275 nm. The linear range was from 1 to 100 ng ml^{-1} and the limit of detection (LOD) was 0.1 ng ml^{-1}. No interference from different substances such as K^+, Na^+, NH_4^+, NO_3^-, Cl^-, SO_4^{2-}, HCO_3^-, Ca^{2+}, Mg^{2+}, Cu^{2+}, Zn^{2+}, Co^{2+}, Fe^{3+}, Al^{3+}, vitamin B_1, vitamin B_2, vitamin B_6, and vitamin C was observed. This method was simple and sensitive for the determination of vitamin B_{12} in fortified food products. In 2005, a novel fluorescence detection method for the HPLC analysis of vitamin B_{12} was proposed [4]. Vitamin B_{12} was converted in a precolumn into the fluorescent α-ribazole after treatment with sodium hydroxide (2.5 mol l^{-1} at 100 °C for 15 min) and alkaline phosphatase (at 37 °C and pH 8 for 16 h). Fluorescence detection was performed at an excitation wavelength of 250 nm and an emission wavelength of 312 nm. Although the method was sensitive and can be applied to all foodstuff samples, it was first necessary to separate vitamin B_{12} from α-ribazole pre-existing in the sample.

12.4.3
Electrochemical Detection

HPLC coupled with electrochemical detection (ED) is a sensitive and selective method for the determination of redox-active compounds. It is one of the new analytical methods that has been recommended for the determination of B vitamins in pharmaceutical formulations and biological matrices. A number of papers on the electrochemistry of cyanocobalamin, including polarographic studies and cyclic voltammetry, have been published in recent years. Some results provide evidence that vitamin B$_{12}$ has a redox relation to the cobalt atom. Vitamin B$_{12}$ containing Co(III) can be reduced to vitamin B$_{12}$ containing Co(II), and subsequently to vitamin B$_{12}$ containing Co(I), all in aqueous media [20, 21]. For HPLC–ED analysis of vitamin B$_{12}$, in 2005 Marszałł *et al.* developed a method involving HPLC coupled with coulometric ED [18]. The method was applied to the analysis of pharmaceutical formulations. A coulometric electrochemical detector equipped with a dual analytical cell and guard cell was employed. The guard cell was connected in-line before the injection port and was used to remove oxidizable impurities in the mobile phase, in order to eliminate disturbances of baseline stability. The working electrode was a porous graphite electrode. The linear range was 5–27.7 ng ml^{-1}, the LOD was 0.08 ng ml^{-1}, and the limit of quantification (LOQ) was 0.28 ng ml^{-1}. Although the method showed high sensitivity, its linear range was narrow. In addition, the experimental conditions were stringent, that is, water for the experiment must be of the highest purity and the mobile phase must be continually degassed, which restricted conventional application of the method. Nevertheless, this was the first method to combine UV and ED for the simultaneous determination of vitamins B$_1$, B$_6$, and B$_{12}$. In 2007, the same group used the same method to analyze animal and plant food samples [17]. They confirmed that hydroxy-, methyl- and cyanocobalamin had same retention time under the separation conditions and the same response on the coulometric electrochemical detector. Hence prior conversion of cobalamins to cyanocobalamin can be omitted and cyanocobalamin can be used as an external standard to quantify total cobalamins. This advantage rendered the method more powerful than other detection techniques in the HPLC determination of vitamin B$_{12}$ in natural foodstuff sample.

12.4.4
Inductively Coupled Plasma Mass Spectrometry (ICP-MS) Detection

ICP-MS was developed in the late 1980s to combine the easy sample introduction and rapid analysis of ICP technology with the accurate and low detection limits of a mass spectrometer. HPLC combined with inductively coupled argon plasma mass spectrometry is one of the most commonly used hyphenated techniques for the analysis of metallo-organic compounds. HPLC allows a wide choice of columns showing various separation principles for different samples. On the other hand, ICP-MS is an element-specific detection technique, and allows multi-elemental, highly sensitive, and multi-isotopic analyses. It can be combined directly with HPLC, and HPLC–ICP-MS is a powerful tool for pharmaceutical and biomedical studies.

When applying ICP-MS detection, one must be aware of the effects of organic solvents, which can lead to changes in analyte sensitivity and also an increase in the number of carbon-containing polyatomic ions and deposition of carbon on the cones. Concerns related to the use of organic solvents with ICP-MS have limited their use in the development of hyphenated techniques. Microbore HPLC was used to interface with ICP-MS because of the reduced flow-rates that it provides. Capillary-based (<100 μm i.d.) microseparations would be even more appropriate because only nanoliter volumes of organic solvents reach the plasma, thus minimizing detrimental effects.

In fact, good detection limits (10–50 ng ml^{-1}) for the on-line detection of vitamin B_{12} have been reported for the combination of microbore HPLC and ICP-MS [12, 22]. Columns containing C_{18} packing material of 2 μm particle size were applied. Employing the gradient elution program 15–45% B, 13 min; 45–100% B, 7 min; 100% B, 6 min; 100–50% B, 10 min; 15% B, 4 min (A = 25 mM ammonium acetate in water; B = 25 mM ammonium acetate in 50% acetonitrile), four cobalamin species (cyanocobalamin, hydroxocobalamin, methylcobalamin, and 5′-deoxyadenosylcobalamin) were separated and detected.

12.4.5
Electrospray Ionization Mass Spectrometry (ESI-MS) Detection

The introduction of the atmospheric pressure ionization technique greatly expanded the number of compounds and matrices that can be analyzed by HPLC–MS. However, the application of LC–ESI-MS to the analysis of large molecules, such as vitamin B_{12} (molecular mass 1355 Da), is relatively limited due to their low ionization efficiency. To overcome this difficulty, Luo et al. [16] tested ESI techniques and found the application of a high cone voltage to be effective. In ESI-MS, the cone voltage can affect the detection sensitivity and fragmentation. It was necessary to optimize the cone voltage for each of the ions of interest. For ESI of vitamin B_{12}, there were mainly low abundances of pseudomolecular ions such as $[M + 2H]^{2+}$, $[M + H + Na]^{2+}$, $[M + H + K]^{2+}$, $[M + 2Na]^{2+}$, $[M + H]^{+}$, and $[M + Na]^{+}$ at low cone voltages. On increasing the cone voltage, the full-scan spectrum displays high abundances of different fragment ions, which can be explained by an initial dissociation of the axial side-chain groups. The m/z 1015.9 ion is found, representing $[M + H - base - sugar - PO_3]^{+}$ fragment ions. The m/z 989.8 ion corresponds to $[M + H - base - sugar - PO_3 - CN]^{+}$ and m/z 930.8 correspond to $[M + H - base - sugar - PO_3 - CN - Co]^{+}$ fragment ions. The results indicate that the optimal peak area of m/z 930.8 takes place at a cone voltage of 180 V. The peak area of m/z 930.8 is about double that of m/z 1015.9 at 130 V and that of m/z 690.0 at 30 V, and more than 15 times higher than that of m/z 1356.1 at 70 V and m/z 1394.1 at 100 V. These results imply that ESI-MS is a highly sensitive detection technique. Figure 12.1 shows the fragmentation of vitamin B_{12} at a high cone voltage in the ESI mode.

Vitamin B_{12} was quantified with ginsenoside Re as internal standard (IS) after their separation on a C_{18} column with a gradient of water and acetonitrile. MS in the selected ion monitoring (SIM) mode at m/z 930.8 was used for quantification of vitamin B_{12}. Under the optimum analysis conditions, linearity was studied over the concentration range 6–150 ng ml^{-1} for vitamin B_{12} in the presence of 0.1 μg ml^{-1} IS. The LOD was about 2 ng g^{-1} for vitamin B_{12} in milk powder.

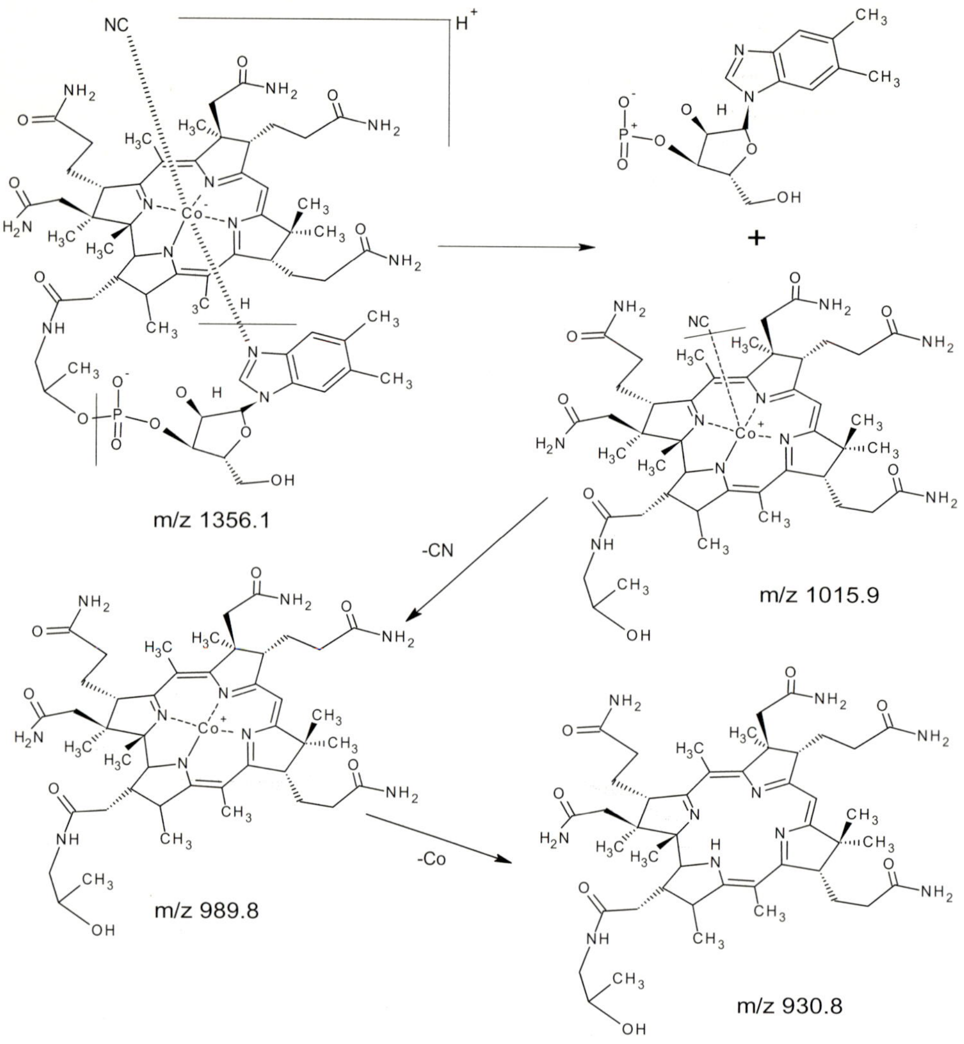

m/z 1356.1

m/z 1015.9

m/z 989.8

m/z 930.8

-CN

-Co

Figure 12.1 Fragmentation of vitamin B₁₂ at high cone voltage in the ESI mode.

12.5
Reference Materials for Vitamin B₁₂ Analysis

In 1989, the Community Bureau of Reference (BCR) commenced a program to improve the quality of vitamin analysis in food. The preparation of the certified reference materials (CRMs) was a key section of the program. For the analysis of vitamin B₁₂, pig liver and milk power were lyophilized and used as CRMs [23].

The total cyanocobalamin content was analyzed in pig liver and milk powder and the data sets from eight and seven laboratories, respectively, were evaluated.

The analytical methods included an affinity binding method (carried out in two laboratories) and a microbiological method (used by the other laboratories). The microbiological assay seemed to produce higher values than the affinity binding method for milk power, whereas statistical agreement between the two methods was found for lyophilized pig liver. The within-laboratory and between-laboratories variations (repeatability relative standard deviation and reproducibility relative standard deviation) were good, 4.2–4.6 and 9.1–13.1%, respectively. Greater variation between laboratories was reported earlier in an interlaboratory comparison of serum vitamin B_{12} determination; the mean relative standard deviation was >30% in plasma samples with a normal vitamin B_{12} content. Differences in sample matrix and the concentration levels can produce different variations. Variations in the matrices of seemingly similar materials but actually different matrices can require revalidation of the analytical method.

Various analytical methods are now available for the determination of vitamin B_{12}. CRMs provide a basis for further improvements in quality control programs in vitamin B_{12} analysis. The preparation and study of CRMs should be continued by harmonizing the analytical procedures internationally.

References

1 Hiroshi, I. and Ichiro, O. (1997) Determination of cyanocobalamin in foods by high-performance liquid chromatography with visible detection after solid-phase extraction and membrane filtration for the precolumn separation of lipophilic species. *Journal of Chromatography A*, **771**, 127–134.

2 Moreno, P. and Salvado, V. (2000) Determination of eight water- and fat-soluble vitamins in multi-vitamin pharmaceutical formulations by high-performance liquid chromatography. *Journal of Chromatography A*, **870**, 207–215.

3 Chatzimichalakis, P.F., Samanidou, V.F., Verpoorte, R., and Papadoyannis, I.N. (2004) Development of a validated HPLC method for the determination of B-complex vitamins in pharmaceuticals and biological fluids after solid phase extraction. *Journal of Separation Science*, **27**, 1181–1188.

4 Pakin, C., Bergaentzle, M., Aoude-Werner, D., and Hasselmann, C. (2005) α-Ribazole, a fluorescent marker for the liquid chromatographic determination of vitamin B_{12} in foodstuffs. *Journal of Chromatography A*, **1081**, 182–189.

5 Heudi, O., Kilinc, T., Fontannaz, P., and Marley, E. (2006) Determination of Vitamin B_{12} in food products and in premixes by reversed-phase high performance liquid chromatography and immunoaffinity extraction. *Journal of Chromatography A*, **1101**, 63–68.

6 Kelly, R.J., Gruner, T.M., and Sykes, A.R. (2005) Development of a method for the separation of corrinoids in ovine tissues by HPLC. *Biomedical Chromatography*, **19**, 329–333.

7 Miyamoto, E., Watanabe, F., Ebara, S., Takenaka, S., Takenaka, H., Yamaguchi, Y., Tanaka, N., Inui, H., and Nakano, Y. (2001) Characterization of a vitamin B_{12} compound from unicellular coccolithophorid alga (*Pleurochrysis carterae*). *Journal of Agricultural and Food Chemistry*, **49**, 3486–3489.

8 Watanabe, F., Katsura, H., Takenaka, S., Fujita, T., Abe, K., Tamura, Y., Nakatsuka, T., and Nakano, Y. (1999) Pseudovitamin B_{12} is the predominant cobamide of an algal health food, spirulina tablets. *Journal of Agricultural and Food Chemistry*, **47**, 4736–4741.

9 Heudi, O., Kilinc, T., and Fontannaz, P. (2005) Separation of water-soluble

vitamins by reversed-phase high performance liquid chromatography with ultra-violet detection: application to polyvitaminated premixes. *Journal of Chromatography A*, **1070**, 49–56.

10 Astier, A. and Baud, F.J. (1995) Simultaneous determination of hydroxocobalamin and its cyanide complex cyanocobalamin in human plasma by high performance liquid chromatography: application to pharmacokinetic studies after high-dose hydroxocobalamin as an antidote for severe cyanide poisoning. *Journal of Chromatography B*, **667**, 129–135.

11 González, L., Yuln, G., and Volonte, M.G. (1999) Determination of cyanocobalamin, betamethasone, and diclofenac sodium in pharmaceutical formulations, by high performance liquid chromatography. *Journal of Pharmaceutical and Biomedical Analysis*, **20**, 487–492.

12 Yanes, E.G. and Miller-Ihli, N.J. (2004) Cobalamin speciation using reversed-phase micro-high-performance liquid chromatography interfaced to inductively coupled plasma mass spectrometry. *Spectrochimica Acta B*, **59**, 891–899.

13 Quesada-Chanto, A., Schmid-Meyer, A.C., Schroeder, A.G., Fuchter, A., Carvalho-Jonas, M.F., Koehntopp, P.I., and Jonas, R. (1998) Comparison of methods for determination of vitamin B_{12} in microbial material. *Biotechnology Techniques*, **12**, 75–77.

14 Li, H.-B., Chen, F., and Jiang, Y. (2000) Determination of vitamin B_{12} in multivitamin tablets and fermentation medium by high-performance liquid chromatography with fluorescence detection. *Journal of Chromatography A*, **891**, 243–247.

15 Wongyai, S. (2000) Determination of vitamin B_{12} in multivitamin tablets by multimode high-performance liquid chromatography. *Journal of Chromatography A*, **870**, 217–220.

16 Luo, X., Chen, B., Ding, L., Tang, F., and Yao, S. (2006) HPLC–ESI-MS analysis of vitamin B_{12} in food products and in multivitamins–multimineral tablets. *Analytica Chimica Acta*, **562**, 185–189.

17 Lebiedzi _UNDEFINEDska, A., Marszałł, M.L., Kuta, J., and Szefer, P. (2007) Reversed-phase high-performance liquid chromatography method with coulometric electrochemical and ultraviolet detection for the quantification of vitamins B_1 (thiamine), B_6 (pyridoxamine, pyridoxal and pyridoxine) and B_{12} in animal and plant foods. *Journal of Chromatography A*, **1173**, 71–80.

18 Marszałł, M.L., Lebiedzińska, A., Czarnowski, W., and Szefer, P. (2005) High-performance liquid chromatography method for the simultaneous determination of thiamine hydrochloride, pyridoxine hydrochloride and cyanocobalamin in pharmaceutical formulations using coulometric electrochemical and ultraviolet detection. *Journal of Chromatography A*, **1094**, 91–98.

19 Wolf, C., Rösick, U., and Brätter, P. (2000) Quantification of the metal distribution in metallothioneins of the human liver by HPLC coupled with ICP-AES. *Fresenius' Journal of Analytical Chemistry*, **368**, 839–843.

20 Zagal, J.H., Aguirre, M.J., and Páez, M.A. (1997) O_2 reduction kinetics on a graphite electrode modified with adsorbed vitamin B_{12}. *Journal of Electroanalytical Chemistry*, **437**, 45–52.

21 Zheng, D. and Lu, T. (1997) Electrochemical reactions of cyanocobalamin in acidic media. *Journal of Electroanalytical Chemistry*, **429**, 61–65.

22 Chassaigne, H. and Lobinski, R. (1998) Determination of cobalamins and cobinamides by microbore reversed-phase HPLC with spectrophotometric, ion-spray ionization MS and inductively coupled plasma MS detection. *Analytica Chimica Acta*, **359**, 227–235.

23 Ollilainen, V., Finglas, P.M., van den Berg, H., and de Froidmont-Görtz, I. (2001) Certification of B-group vitamins (B_1, B_2, B_6, and B_{12}) in four food reference materials. *Journal of Agricultural and Food Chemistry*, **49**, 315–321.

13
Microbiological Detection of Vitamin B₁₂ and Other Vitamins

Fumio Watanabe and Yukinori Yabuta

13.1
Vitamin B₁₂

Vitamin B_{12} (B_{12}) is the largest (molecular mass 1355.4) and most complex of all the vitamins. Although the scientific use of the term "B_{12}" is usually restricted to cyanocobalamin, in this chapter B_{12} represents all potentially biologically active cobalamins. Cobalamin is the term used to refer to a group of cobalt-containing compounds (corrinoids) that have a lower axial ligand that contains the cobalt-coordinated nucleotide (5,6-dimethylbenzimidazole as a base) (Figure 13.1). Cyanocobalamin, which is used in most supplements, is readily converted to the coenzyme forms of cobalamin, namely methylcobalamin and 5′-deoxyadenosylcobalamin, in the human body [1]. B_{12} is synthesized only in certain bacteria [2]. The B_{12} synthesized by bacteria is concentrated mainly in the bodies of higher predatory organisms in the natural food chain system. Although various corrinoids carrying a base other than 5,6-dimethylbenzimidazole also occur in Nature, they are biologically inactive for humans [3].

The major signs of B_{12} deficiency are megaloblastic anemia and neuropathy [4]. Strict vegetarians have a greater risk of developing B_{12} deficiency relative to non-vegetarians [5] and must consume B_{12}-fortified foods or B_{12}-containing dietary supplements to prevent B_{12} deficiency. A considerable proportion of elderly subjects having low serum B_{12} levels without pernicious anemia have been reported to have malabsorption of protein-bound B_{12} (food-bound B_{12} malabsorption) [6]. Food-bound B_{12} malabsorption is found in patients with certain gastric dysfunctions, such as atrophic gastritis with decreased stomach acid secretion [7]. Because the bioavailability of crystalline B_{12} is not altered in patients with atrophic gastritis, the Institute of Medicine recommended that adults aged 51 years and older should obtain the majority of the recommended dietary allowance of B_{12} (2.4 µg per day in the US) through the consumption of foods fortified with crystalline B_{12} or B_{12}-containing supplements [4].

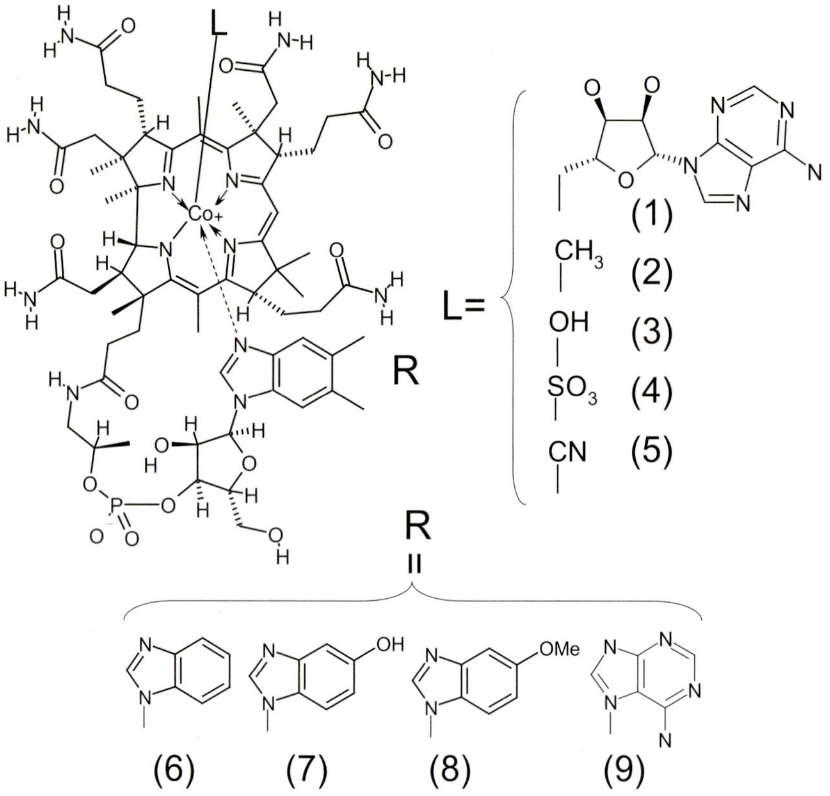

Figure 13.1 Structural formula of vitamin B₁₂ and partial structures of vitamin B₁₂-related compounds: **1**, 5′-deoxyadenosylcobalamin; **2**, methylcobalamin; **3**, hydroxocobalamin; **4**, sulfitocobalamin; **5**, cyanocobalamin (vitamin B₁₂); **6**, benzimidazolylcyanocobamide; **7**, 5-hydroxybenzimidazolylcyanocobamide; **8**, 5-methoxybenzimidazolylcyanocobamide; **9**, 7-adeninylcyanocobamide.

13.1.1
Analysis of Vitamin B₁₂ in Foods

Historically, the B₁₂ content of foods has been determined by bioassay with certain B₁₂-requiring microorganisms such as *Lactobacillus delbrueckii* subsp. *lactis* ATCC7830 (formerly *Lactobacillus leichmannii*) [8]. Radioisotope (RI) dilution assay with radiolabeled B₁₂ and hog intrinsic factor (IF) (the most specific B₁₂-binding protein) has also been used for the determination of the B₁₂ content of foods [9]. Now, various types of non-RI B₁₂ analyzers are being manufactured and used clinically for the routine assay of human serum B₁₂ worldwide. We evaluated the applicability of the IF-based chemiluminescent B₁₂ analyzer in food analysis, and found an excellent correlation coefficient between both methods in most foods tested, although in some specific foods the values determined by the microbiological method were several times higher than those determined using the machine [10]. This difference may be due to the fact that *L. delbrueckii* used for the microbiological assay of food B$_{12}$ utilizes corrinoid compounds inactive for humans. Furthermore, it is known that both deoxyribosides and deoxynucleotides (known as the alkali-resistant factor) can substitute B$_{12}$ in this lactic bacterium [11].

If IF-based clinical assay kits or analyzers are used for measuring food B$_{12}$ contents, the results may not represent only B$_{12}$ because of the possibility that the binding of B$_{12}$ to IF suffers slight interference from certain food ingredients or inactive corrinoid compounds. The difficulty of evaluating whether certain foods contain B$_{12}$ or inactive corrinoids should be easily resolved by the use of a simple technique, namely bioautography with B$_{12}$-dependent *Escherichia coli* mutant after separation of the sample by silica gel 60 thin-layer chromatography (TLC) [12]. This bioautography has great advantages with regard to simplicity, flexibility, speed, and relative cheapness for the analysis of B$_{12}$ compounds in foods.

13.2
Analysis of Food Vitamin B$_{12}$ by Bioautography

13.2.1
Vitamin B$_{12}$-Fortified Foods

Ready-to-eat cereals fortified with B$_{12}$ contribute a great proportion of dietary B$_{12}$ intake in the US [4]. Several groups of investigators have suggested that feeding a breakfast cereal fortified with folic acid, vitamin B$_{12}$, and vitamin B$_6$ increases blood concentrations of these vitamins and decreases plasma total homocysteine concentrations in elderly populations [13]. Fortified breakfast cereals have become a particularly valuable source of B$_{12}$ for vegetarians and elderly people.

Figure 13.2a shows a typical bioautogram of the extracts of certain cereal and multivitamin supplement. By using bioautography, loss of B$_{12}$ would be readily evaluated during storage of B$_{12}$-fortifed foods or supplements.

13.2.2
Edible Cyanobacteria

Some species of the cyanobacteria, including *Spirulina, Aphanizomenon,* and *Nostoc,* are produced at annual rates of 500–3000 tons for the food and pharmaceutical industries worldwide [14]. Tablets containing *Spirulina* spp. are sold as a health food fad, since it is known to contain a large amount of B$_{12}$ [15]. We found that commercially available spirulina tablets contained 127–244 µg of B$_{12}$ per 100 g [16]. When two corrinoid compounds were purified and characterized from the spirulina tablets, the major (83%) and minor (17%) compounds were identified as pseudovitamin B$_{12}$ (compound **9** shown in Figure 13.1) and vitamin B$_{12}$, respectively [16]. Other edible cyanobacteria often contain a large amount of pseudovitamin B$_{12}$ [17–19], which is known to be biologically inactive in humans [3]. As

(a)   (b)

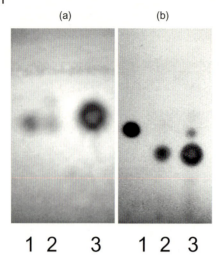

1 2   3   1 2 3

Figure 13.2 Bioautogram of the vitamin B$_{12}$ extracts of a vitamin B$_{12}$-fortified food, multivitamin supplement, and edible cyanobacterium. (a) 1, Authentic B$_{12}$; 2, ready-to-eat cereal; 3, multivitamin supplement. (b) 1, Authentic B$_{12}$; 2, authentic pseudovitamin B$_{12}$; 3, edible cyanobacterium Suizenji-nori.

shown in Figure 13.2b, *E. coli* 215 bioautography of an extract of *Aphanothece sacrum* (Suizenji-nori) indicates that pseudovitamin B$_{12}$ is the predominant corrinoid of the bacterial cells. By using bioautography, it can be determined whether certain foods contain true B$_{12}$ or not, without purification of any corrinoid compounds.

Therefore, these cyanobacteria are not suitable for use as a source of B$_{12}$ for the prevention of B$_{12}$ deficiency among high-risk populations such as vegans and elderly subjects.

13.2.3
Analysis of Vitamin B$_{12}$ Degradation Products During Storage of Multivitamin–Mineral Supplements

Appreciable loss of biologically active B$_{12}$ occurs in multivitamin–mineral supplements containing B$_{12}$ [20], since B$_{12}$ is converted to inactive B$_{12}$ compounds by the addition of substantial amounts of vitamin C in the presence of copper [21]. Some of the B$_{12}$ degradation products have been reported to block B$_{12}$ metabolism in mammalian cells [20].

As shown in Figure 13.3a, B$_{12}$ was destroyed significantly in the presence of vitamin C and copper; some B$_{12}$ degradation products were separated from the concentrated solution of the treated B$_{12}$ by silica gel 60 TLC. The two B$_{12}$ degradation products could be detected by bioautography even when the treated B$_{12}$ sample was diluted to $10\text{–}20\,\mu g\,l^{-1}$ (Figure 13.3a). No information is available on chemical structures of the degradation products.

(a)   (b)

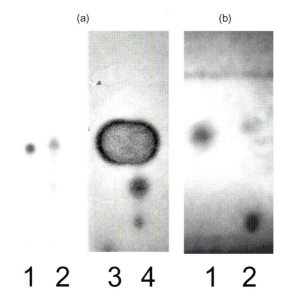

1 2   3 4   1 2

Figure 13.3 Bioautogram of vitamin B_{12} degradation products during cooking and food processing. (a) Concentrated solutions of authentic B_{12} (1) and B_{12} treated with vitamin C and copper (2) were analyzed by silica gel 60 TLC. Each solution was diluted significantly and then analyzed by bioautography (3, authentic B_{12}; 4, the treated B_{12}). (b) B_{12} extracts of milk with (2) and without (1) microwave treatment.

13.2.4
Analysis of Vitamin B_{12} Degradation Products During Cooking or Processing of Foods

Although the B_{12} content of various types of milk is not high (0.3–0.4 μg per 100 g) [22], milk and dairy products are significant contributors of B_{12} intakes, since the intake of dairy products is high in the general population [4]. Appreciable losses of B_{12} have been reported during the processing of milk; boiling for 2–5 and 30 min resulted in losses of 30 and 50%, respectively [1]. Microwave cooking for 5 min led to a 50% loss and a 5–10% loss on pasteurization [1, 23]. As shown in Figure 13.3b, a B_{12} degradation product can be detected by bioautography of an extract of the 5 min microwave-treated milk. One of the B_{12} degradation products after microwave heating was isolated and its chemical structure was analyzed [23].

13.3
Other Vitamins Determined by Microbiological Analysis

The US Food and Drug Administration has mandated the fortification of cereal-grain products enriched with folic acid because of its relationship to diseases such as neural tube defects. Therefore, it is important to determine accurately folic acid concentrations in food to ensure its proper dietary intake. As the trienzyme extraction (α-amylase, protease, and folate conjugase) method allows for a more complete extraction of folate trapped in carbohydrate or protein matrices in food than the traditional extraction method [24], trienzyme extraction in combination with microbiological assay is recommended in food folate analysis [25].

Vitamin B$_1$ (thiamin), vitamin B$_2$ (riboflavin), vitamin B$_6$ (pyridoxine), niacin, pantothenic acid, and biotin were assayed microbiologically in food [26]. Radiometric–microbiological assays were developed to eliminate some of the technical problems associated with the conventional microbiological method [27]. The radiometric–microbiological method is based on the measurement of $^{14}CO_2$ from the metabolism of a ^{14}C-labeled substrate (L-[1-^{14}C]valine) by the test organism (*Kloeckera brevis*) in the presence of the specific vitamin (niacin, biotin, thiamine, pantothenic acid, and vitamin B$_6$) to be analyzed.

References

1 Ball, G.F.M. (1998) *Vitamin B$_{12}$ Bioavailability and Analysis of Vitamins in Foods*, Chapman & Hall, London, pp. 497–515.

2 Scheider, Z. and Stroiñski, A. (1987) Biosynthesis of vitamin B$_{12}$, in *Comprehensive B$_{12}$: Chemistry, Biochemistry, Nutrition, Ecology, Medicine* (eds Z. Schneider and A. Stroiński), Walter de Gruyter, Berlin, pp. 93–110.

3 Brandt, L.J., Goldberg, L., Bernstein, L.H., and Greenberg, G. (1979) The effect of bacterially produced vitamin B-12 analogues (cobamides) on the *in vitro* absorption of cyanocobalamin. *American Journal of Clinical Nutrition*, **32**, 1832–1836.

4 Institute of Medicine (1998) Vitamin B$_{12}$, in *Dietary Reference Intakes for Thiamin, Riboflavin, Niacin, Vitamin B$_6$, Folate, Vitamin B$_{12}$, Pantothenic Acid, Biotin, and Choline*, National Academy Press, Washington, DC, pp. 306–356.

5 Millet, P., Guilland, J.C., Fuchs, F., and Klepping, J. (1989) Nutrient intake and vitamin status of healthy French vegetarians and nonvegetarians. *American Journal of Clinical Nutrition*, **50**, 718–727.

6 Baik, H.W. and Russell, R.M. (1999) Vitamin B$_{12}$ deficiency in the elderly. *Annual Review of Nutrition*, **19**, 357–377.

7 Park, S. and Johson, M.A. (2006) What is an adequate dose of oral vitamin B-12 in older people with poor vitamin B-12 status? *Nutrition Reviews*, **64**, 373–378.

8 Schneider, Z. (1987) Purification and estimation of vitamin B$_{12}$, in *Comprehensive B$_{12}$: Chemistry, Biochemistry, Nutrition, Ecology, Medicine* (eds Z. Schneider and A. Stroiński), Walter de Gruyter, Berlin, pp. 111–155.

9 Casey, P.J., Speckman, K.R., Ebert, F.J., and Hobbs, W.E. (1982) Radioisotope dilution technique for determination of vitamin B$_{12}$ in foods. *Journal of AOAC International*, **65**, 85–88.

10 Watanabe, F., Takenaka, S., Abe, K., Tamura, Y., and Nakano, Y. (1998) Comparison of a microbiological assay and a fully automated chemiluminescent system for the determination of vitamin B$_{12}$ in food. *Journal of Agricultural and Food Chemistry*, **46**, 1433–1436.

11 Resources Council, Science and Technology Agency (1995) *Standard Tables of Food Composition in Japan – Vitamin K, B$_6$ and B$_{12}$*, Resources Council, Science and Technology Agency, Tokyo, pp. 6–56.

12 Tanioka, Y., Yabuta, Y., Miyamoto, E., Inui, H., and Watanabe, F. (2008) Analysis of vitamin B$_{12}$ in food by silica gel 60 TLC and bioautography with

vitamin B$_{12}$-dependent *Escherichia coli* 215. *Journal of Liquid Chromatography and Related Technologies*, **31**, 1977–1985.

13 Tucker, K.L., Olson, B., Bakun, P., Dallal, G.E., Selhub, J., and Rosenberg, I.H. (2004) Breakfast cereal fortified with folic acid, vitamin B-6, and vitamin B-12 increases vitamin concentrations and reduces homocysteine concentrations: a randomized trial. *American Journal of Clinical Nutrition*, **79**, 805–811.

14 Pulz, O. and Gross, W. (2004) Valuable products from biotechnology of microalgae. *Applied Microbiology and Biotechnology*, **65**, 635–648.

15 van den Berg, H., Dagnelie, P.C., and van Staveren, W.A. (1988) Vitamin B$_{12}$ and seaweed. *Lancet*, **i**, 242–243.

16 Watanabe, F., Katsura, H., Takenaka, S., Fujita, T., Abe, K., Tamura, Y., Nakatsuka, T., and Nakano, Y. (1999) Pseudovitamin B$_{12}$ is the predominate cobamide of an algal health food, spirulina tablets. *Journal of Agricultural and Food Chemistry*, **47**, 4736–4741.

17 Miyamoto, E., Tanioka, Y., Nakao, T., Barla, F., Inui, H., Fujita, T., Watanabe, F., and Nakano, Y. (2006) Purification and characterization of a corrinoid compound in an edible cyanobacterium *Aphanizomenon flos-aquae* as a nutritional supplementary food. *Journal of Agricultural and Food Chemistry*, **54**, 9604–9607.

18 Watanabe, F., Miyamoto, E., Fujita, T., Tanioka, Y., and Nakano, Y. (2006) Characterization of a corrinod compound in the edible (blue–green) algae, suizenji-nori. *Bioscience, Biotechnology, and Biochemistry*, **70**, 3066–3068.

19 Watanabe, F., Tanioka, Y., Miyamoto, E., Fujita, T., Takenaka, H., and Nakano, Y. (2007) Purification and characterization of corrinoid compounds from the dried powder of an edible cyanobacterium, *Nostoc commune* (Ishikurage). *Journal of Nutrition Science and Vitaminology*, **53**, 183–186.

20 Kondo, H., Binder, M.J., Kolhouse, J.F., Smythe, W.R., Podell, E.R., and Allen, R.H. (1982) Presence and formation of cobalamin analogues in multivitamin-mineral pills. *Journal of Clinical Investigation*, **70**, 889–898.

21 Takenaka, S., Sugiyama, S., Watanabe, F., Abe, K., Tamura, Y., and Nakano, Y. (1997) Effects of carnosine and anserine on the destruction of vitamin B$_{12}$ with vitamin C in the presence of copper. *Bioscience, Biotechnology, and Biochemistry*, **61**, 2137–2139.

22 USDA (2007) National Nutrient Database for Standard Reference, Release 18. Vitamin B-12 (mg) Content of Selected Foods per Common Measure, Sorted by Nutrient Content. US Department of Agriculture, Washington, DC.

23 Watanabe, F., Abe, K., Fujita, T., Goto, M., Hiemori, M., and Nakano, Y. (1998) Effects of microwave heating on the loss of vitamin B$_{12}$ in foods. *Journal of Agricultural and Food Chemistry*, **46**, 206–210.

24 Hyun, T.H. and Tamura, T. (2005) Trienzyme extraction in combination with microbiologic assay in food folate analysis: an updated review. *Experimental Biology and Medicine*, **230**, 444–454.

25 DeVries, J.W., Rader, J.I., Keagy, P.M., Hudson, C.A., Angyal, G., Arcot, J., Castelli, M., Doreanu, N., Hudson, C., Lawrence, P., Martin, J., Peace, R., Rosner, L., Strandler, H.S., Szpylka, J., van den Berg, H., Wo, C., and Wurz, C. (2005) Microbiological assay–trienzyme procedure for total folates in cereals and cereal foods: collaborative study. *Journal of AOAC International*, **88**, 5–15.

26 Angyal, G. (1996) *Methods for the Microbiological Analysis of Selected Nutrients*. AOAC International, Gaithersburg, MD.

27 Guilarte, T.R. (1991) Radiometric microbiological assay of B vitamins. Part 1: assay procedure. *Journal of Nutritional Biochemistry*, **2**, 334–338.

14
Multimethod for Water-Soluble Vitamins in Foods by Using LC–MS

Alessandra Gentili and Fulvia Caretti

14.1
Introduction

The water-soluble vitamins are a heterogeneous group of compounds with different structures and physico-chemical properties [1]. Moreover, a single vitamin consists of several biologically active forms, known as vitamers, which introduce a further element of heterogeneity due to the subtle differences in their chemical structures [2]. For these reasons, it is very difficult to find experimental conditions suitable for their simultaneous determination, and this is further complicated by the labile nature of some vitamins and by the complexity of food matrices. In the case of naturally occurring vitamins, other factors that may cause additional analytical problems include their possible linkage with macromolecular components and their low concentrations in foods. When considering fortified foodstuffs, the natural vitamin content should be determined accurately in order to make additions in conformity with the terms established by legislation. Within the European Union, Regulation 1925/2006/EC [3] states the forms and the total amount of a specific vitamin (natural occurring and supplemented) which may be added to a food. This amount should not exceed the maximum tolerance that the Commission would submit by January 2009. On these bases, the most suitable analytical approach seems to be that addressed to the determination of the complete vitamin bioavailability.

As a consequence of all the above-described difficulties, nowadays it is preferred to analyze each vitamin singly [4], but also in this case the problems concerning the simultaneous determination of all its possible forms, those used for fortification and the endogenous ones, can be difficult to overcome. An often adopted solution is to apply extraction conditions which permit the determination of only the free form, expression of the total vitamin content, and freeing all bound forms either by acidic hydrolysis or by an enzymatic digestion (e.g., with takadiastase, protease) for labile vitamins [4, 5].

Fortification of Foods with Vitamins, First Edition. Edited by Michael Rychlik.
© 2011 Wiley-VCH Verlag GmbH & Co. KGaA. Published 2011 by
Wiley-VCH Verlag GmbH & Co. KGaA.

14.2
Extraction and Cleanup with Conventional Methods of Analysis

In the ambit of an analytical procedure, aimed at the analysis of water-soluble vitamins in food products, the extraction and cleanup represent the most delicate, complex, and time-consuming steps [5]. Each vitamin requires unique conditions for its extraction, depending on a series of factors that have to be minutely examined: (i) type of food; (ii) form in which the vitamin occurs naturally and/or added to food; (iii) its bond to the food matrix; (iv) effects of pH, temperature, and other experimental parameters (such as exposure to light, oxygen partial pressure, ionic strength, and occurrence of heavy metal ions) on its stability and on that of its vitamers, if their individual determination is required; (v) nature and amount of potentially interfering compounds; (vi) sensitivity and selectivity of the separation and detection method.

In order to define a procedure for a multivitamin extraction, knowledge of the procedures currently used for each vitamin is fundamental.

Naturally occurring forms of vitamin B_1 (B_1)are thiamin and its phosphorylated esters (thiamin monophosphate, thiamin pyrophosphate, and thiamin triphosphate), whereas thiamin hydrochloride and thiamin mononitrate are the forms that may be added to foodstuffs, according to Regulation 1925/2006/EC. For the determination of the total content of B_1, it is used to free the protein-bound and phosphorylated forms by an acid extraction (HCl–H_2SO_4, $0.1\,mol\,l^{-1}$, at 100 °C or in an autoclave at 121 °C), followed by enzymatic hydrolysis with diastatic and phosphorolytic enzymes [6].

Vitamin B_2 (B_2) exists naturally in three principal forms: riboflavin (RF), flavin mononucleotide (FMN), and flavin adenine dinucleotide (FAD). RF hydrochloride and RF 5′-phosphate are the different forms of B_2 that may be utilized to fortify food products. Generally, the method of extraction is similar to that applied for estimating the total content of B_1 [5], involving mineral acid/enzymatic hydrolysis to convert all forms to RF. When the individual vitamers have to be quantified, the extraction procedure is conducted between pH 5 and 7 and under subdued light, as a consequence of their instability towards alkaline and acidic pH [7].

Vitamin B_3 (B_3; niacin or vitamin PP) occurs mainly as the free acid (nicotinic acid) and as its corresponding amide (nicotinamide). Nicotinamide is the form of niacin typically used in nutritional supplements and in food fortification, but Regulation 1925/2006/EC also indicates nicotinic acid for the same uses. In cereals, nicotinic acid is chemically bound to polysaccharides, peptides, and glyco-peptides; these macromolecular compounds, known as niacinogen or niacytin, are not hydrolyzed by intestinal enzymes, so they are metabolically unavailable [2, 4, 5]. It is generally accepted that the measurement of free niacin in foods provides an accurate evaluation of its biologically available content. In this case, either acidic hydrolysis (mineral acid, $0.1\,mol\,l^{-1}$) or a simple extraction with water or ethanol, followed by cleanup and/or filtration, can be applied [5]. The determination of total niacin requires more extreme conditions: extraction by autoclaving with alkali or mineral acid ($0.5–1\,mol\,l^{-1}$) [4].

Regarding vitamin B_5 (B_5), three biologically active forms can be found in Nature: pantothenic acid, coenzyme A (CoA), and acyl carrier protein (ACP). Calcium or sodium pantothenate is generally used for the supplementation of infant formulas. For the total quantification of B_5, alkaline or acidic media cannot be employed to free its bound forms, since the hydrolysis of its amidic bond would also be catalyzed at these pH values. Currently, pantothenic acid release from CoA is achieved by enzymatic digestion; however, this treatment is not able to achieve the same result from ACP [8, 9].

Naturally active forms of vitamin B_6 (B_6) are pyridoxine, pyridoxal, pyridoxamine, and their 5'-phosphate esters. Its total content is measured as the sum of all these vitamers. If pyridoxine is the form used for the quantification, the whole procedure [5] includes acidic hydrolysis, enzymatic dephosphorylation, and conversion of pyridoxamine into pyridoxal (with glyoxylic acid in the presence of Fe^{2+}) and of pyridoxal into pyridoxine (by sodium borohydride in alkaline medium). Nevertheless, if the aim is the individual determination of all the vitamers, milder conditions have to be applied, but in this case their chromatographic separation is not easy achievable [10]. Regulation 1925/2006/EC cites pyridoxine hydrochloride, pyridoxine 5'-phosphate, and pyridoxine dipalmitate as the utilizable forms to enrich foods.

Vitamin B_8 (B_8) consists of two biologically active forms: D-(+)-biotin, being the sole active stereoisomer among the eight possible, and D-biocytin (ϵ-N-biotinyl-L-lysine) [2, 11]. These vitamers are present in foods in their free form and covalently bound to proteins in variable proportions. Extraction of total biotin is accomplished by acidic hydrolysis (sulfuric acid, 1–3 mol l^{-1}, by autoclaving at 121 °C) that breaks bonds to proteins and totally converts D-biocytin into D-biotin [11, 12]. For the extraction of available biotin, enzymatic hydrolysis with papain can be utilized. Biotin is also the form used for food enrichment [12].

Various natural forms of folates occur in food products [13], all at low levels [2, 4]. The folacin complex includes folic acid (pteroylglutamic acid), dihydrofolic acid, tetrahydrofolic acid (H_4folic acid), 5-formyl-H_4folic acid, 5-methyl-H_4folic acid, and their derivatives characterized by chains containing 2–8 glutamic acid residues [1, 14]. The determination of the total B_9 content usually requires the release of folates from the food matrix, enzymatic deconjugation (folic acid conjugase) of the poly-glutamates to the monoglutamate forms, and a final cleanup step [4, 5]. Some precautions have to be taken during their extraction in order to avoid the pH-dependent interconversion of some species, oxidative losses (by adding antioxidants such as ascorbate and 2-mercaptoethanol), and thermal degradation [13]. In cereals, the endogenous folate concentration is low and folic acid is the form utilized for their fortification; for these foods, the use of folate conjugase may be omitted ([4] see Chapter 10).

Vitamin B_{12} (B_{12}) is the generic term adopted to refer to a group of cobalt-containing organometallic compounds, known as cobalamins, that exhibit anti-pernicious anemic activity. The naturally predominant vitamers are hydro-xocobalamin, 5'-deoxyadenosylcobalamin (coenzyme B_{12}), and methylcobalamin [15]. During the extraction procedure, the conversion of all cobalamins into cyanocobalamin is forced by dissolving the food sample in a buffered solution (pH 4.5) in the presence of KCN [16]. In fact, this form is fairly stable at pH 4–6, even at high temperatures, but it is degraded in alkaline media or in the presence of reducing agents such as ascorbic acid [1]. Since B_{12} vitamers occur in foodstuffs in free and protein-bound forms, their release is achieved either by autoclaving the sample at 121 °C or by using enzymatic hydrolysis with pepsin [17]. Due to its stability, cyanocobalamin is the form used in fortified foods, food supplements, and pharmaceutical formulations [3]. Regulation 1925/2006/EC also cites hydroxocobalamin.

L-Ascorbic acid (AA) and L-dehydroascorbic acid (DHAA) are the two biologically active forms of vitamin C. Considering that DHAA is easily converted into AA in the human body, both of them should be determined for estimating the total content of vitamin C [5]. D-Isoascorbic acid (D-IAA), also known as erythorbic acid or D-araboascorbic acid, has about 5% of the vitamin activity of L-AA; this isomer does not exist naturally in foods and it is approved within the European Union (EU) as a food antioxidant additive due to its reducing properties [18]. In this case, the quantification of the two ascorbic acid isomers and their primary oxidation products is desirable [19]. AA is very sensitive to light, heat, air, and the presence of halogens and metal ions such as Cu and Fe, all factors which stimulate its oxidation [1]. For these reasons, extraction of vitamin C is a critical step and some precautions have to be taken, particularly in the choice of the stabilizing agent and pH. Acidic stabilizers (metaphosphoric acid, perchloric acid, and orthophosphoric acid) are efficient in preventing oxidation of AA, the degradation of which is minimal at pH 2, while EDTA is used to chelate the minerals occurring in foods, which catalyze its oxidation [19]. Cysteine or dithioerythritol can be also employed to convert dehydroascorbic acid into its reduced form [20]. The possible forms to be added to a food are ascorbic acid, sodium, calcium, or potassium L-ascorbate and L-ascorbyl 6-palmitate.

14.3
Liquid Chromatographic Methods for the Analysis of Water-Soluble Vitamins

The great advantage of chromatographic techniques lies in the possibility of performing quantitative multianalyte determinations. The sensitivity of performing trace analyses depends on the coupled detection system. Gas chromatography (GC), using capillary columns, is characterized by a high resolving power, but for the analysis of polar, non-volatile, and heat-sensitive compounds, such as water-soluble vitamins, liquid chromatography (LC) is the method of choice. Moreover, LC has fewer limitations regarding the molecular weight of the molecule that has to be analyzed; this could be particularly advantageous for the direct analysis of high molecular weight vitamers, such as cobalamins, polyglutamate forms, FAD, and CoA.

LC is now often proposed for the individual determination of thiamin, riboflavin, B_6 and total vitamin C [4]. Its ability in separating and quantifying the several biologically active forms of a specific vitamin has been utilized for the direct determination of vitamers B_3 [21] and B_6 [22] in different types of foods such as infant formulas, cereals, honey, and fruit products. Furthermore, there have been several studies of the simultaneous analysis of water-soluble vitamins by LC coupled with conventional detection systems [ultraviolet (UV) and fluorescence detection] [23–26]. A future trend is to develop fast LC methods based on the use of new technologies such as ultra-performance liquid chromatography (UPLC) and mass spectrometry (MS) for detection.

The most common mode of high-performance liquid chromatography (HPLC) for analyzing vitamins is the reversed-phase (RP) mode, and ion exchange is occasionally adopted [4, 5]. RP chromatography is also the most suitable mode for developing multivitamin methods.

LC analysis of a single vitamin is not always simple to achieve. For instance, the main difficulties in the case of thiamin are due to its basic nature and its polarity. In fact, broadening and tailing phenomena of its chromatographic peak can occur as a consequence of the interactions of its basic site with the acidic silanol groups of a C_{18} stationary phase. The use of highly deactivated columns eliminates this constraint. Another problem is the short retention time due to its polarity and low molecular weight. Working with high percentages of water in the mobile phase and a suitable ion pair agent allows the thiamin retention time on a C_{18} column to be increased [5, 6]. For the same reasons, other vitamins also suffer from a retention time that is too close to the void volume, which may lead to errors in quantitation, especially when a conventional detection system is employed. Ion-pair RP chromatography is a solution often adopted for the individual analysis of nicotinic acid, pyridoxine, folic acid, and ascorbic acid [4, 5], but it is more difficult to adapt to a multivitamin analysis than ion suppression RP chromatography.

The choice of the detector to be coupled with HPLC has an influence on the sample preparation, on the conditions applied for LC separation, and on the method sensitivity and selectivity. At present, UV absorbance is the most common detection method owing to its simplicity, but fluorescence and electrochemical detection are preferred when higher sensitivity and selectivity are required (e.g., when determining the low amounts of vitamins naturally occurring in foods) and when the physico-chemical properties of the analyte permit it.

As thiamin is characterized by a low molar absorptivity, the use of UV detection (254 nm) is indicated for the analysis of fortified foods where high concentrations of the vitamin are present [5]. The sensitivity is greatly improved by the employment of fluorescence detection, after the pre- or postcolumn oxidation of the thiochrome group [4]. Riboflavin is a naturally fluorescent compound, but in some cases the less sensitive UV detection is also utilized [5]. For estimating total niacin, a UV detector set at 260 nm is used for the determination of nicotinic acid. Another approach employs derivatization and fluorescence detection to distinguish nicotinamide and nicotinic acid (i.e., bioavailable niacin). The two vitamers are converted into fluorescent compounds by UV irradiation of the postcolumn effluent in the presence of H_2O_2 and Cu^{2+} in the mobile phase; however, a substantial drawback is the long time required for chromatographic separation (1 h) [4]. Pantothenic acid shows a very weak absorbance in the low-UV range ($\lambda < 205$ nm). Therefore, UV detection can only be applied for vitamin determination in supplemented food-stuffs due to the low sensitivity and specificity of this detection mode. However, this LC approach is not suitable for infant formulas owing to interfering peaks from peptides. A postcolumn derivatization of pantothenic acid to a fluorescent compound (formation of β-alanine by hot alkaline hydrolysis of pantothenic acid and reaction with o-phthaldialdehyde) has been proposed for its determination in a large variety of food products, thus overcoming sensitivity and selectivity prob-lems [4]. B_6 is commonly quantified as pyridoxine by fluorimetric detection, also employed for the analysis of the phosphate esters after a postcolumn derivatization with sodium hydrogensulfite, in order to enhance their fluorescence [4, 5].

The low concentration of D-biotin and the absence of a strong chromophore in the molecule are the main difficulties for its HPLC analysis, which, nevertheless, has the advantage of differentiating between biotin and its analogs, for example, dethiobiotin, biotin sulfoxides, and biotin sulfone [11, 12]. Biotin detection can be achieved without any previous derivatization (UV) or after proper derivatization (UV, fluorescence) in an attempt to improve the detection limit of the assay [4]. UV detection is applied for the analysis of foods with a normal content of folates but, for low concentrations, fluorimetric detection becomes indispensable [13, 27]. The determination of cyanocobalamin by LC–UV detection is rather difficult to perform in non-supplemented foodstuffs owing to the low sensitivity of the detec-tion system employed. The limit of detection (LOD) can be improved by fluori-metric detection but, as this molecule is not naturally fluorescent, a chemical and/ or enzymatic hydrolysis is necessary to release a characteristic fragment of B_{12}, α-ribazole, which can be utilized as a fluorescent marker. Since this fragment is involved in the metabolism of B_{12}, it may pre-exist in foods; therefore, it is essential to perform an extensive purification of the extracts before carrying out the hydroly-sis [28]. UV detection at a single wavelength is the most common mode used for the HPLC analysis of vitamin C. Fluorimetric detection was sometimes used after pre- or postcolumn derivatization [4, 5].

Several authors have used both UV–diode array and fluorescence detection to develop multivitamin LC methods, but some problems remain open. In fact, fluorimetric detection is hardly suitable for performing simultaneous analyses whereas, by using a UV detector, some vitamins (thiamin, pantothenic acid, and D-biotin) weakly absorb in the low-UV region, where the selectivity is poor as a consequence of absorption of interfering compounds.

Albalá-Hurtado *et al.* [26] presented an ion-pair LC method for the simultaneous analysis of eight water-soluble vitamins (B_1, B_2, nicotinamide, pyridoxine, pyri-doxal, pyridoxamine, B_9 and B_{12}) in liquid and powered infant milk, using a variable-wavelength UV detector; in this way, each vitamin was detected at its optimal wavelength of absorption. Sample preparation was rapid, involving acidic deproteination, centrifugation, and gravity filtration.

Ndaw *et al.* [7] proposed an LC–fluorimetric detection method for the analysis of B_1, B_2, and B_6 in various food samples (cereals, vegetables, milk, meats, fish, yeast) that were not supplemented with vitamins. They tested a treatment with a mixture of enzymes (α-amylase, papain, and acid phosphatase) to release the different forms (phosphorylated and protein-bound) in a single step. The analyzed extracts showed that the vitamin contents were at least as high as those found when the enzymatic treatment was performed with diastase. The protease present in the mixture of enzymes made acidic hydrolysis superfluous.

Viñas *et al.* [23] developed an LC method for the determination of nine water-soluble vitamins (B_1, B_2, two B_3 vitamers, pyridoxine, pyridoxal, B_9, B_{12}, and inosine) in different baby foods such as infant formulas, cereals and fruit products. Separation was carried out on an RP-amide C_{16} column and detection was performed with a photodiode-array detector. Analytes were extracted from various foods by a combination of acidic and enzymatic digestion to release protein-bound and phosphorylated vitamins.

Zafra-Gómez *et al.* [25] devised a fast method for the determination of eight water-soluble vitamins (B_1, B_2, B_3, B_5, B_6, B_9, B_{12}, and C) in supplemented milk, infant nutrition products, and a milk powder used as a certified reference material. The main advantage was the simplicity of the extraction procedure, based on precipitation and sample centrifugation, that allowed quantitative recovery of analytes. The conditions for HPLC separation were based on those described by Albalá-Hurtado *et al.* [26] with some modifications; the authors confirmed pH to be a critical factor; in particular, a pH lower than 3.0 was fundamental to resolve the pairs B_3–B_6 and B_2–B_{12}. Analytes were detected at different wavelengths by either fluorescence (B_1 and B_2) or UV–visible detection (all others).

Heudi *et al.* [24] analyzed nine water-soluble vitamins (B_1, B_2, B_3, B_5, B_6, B_8, B_9, B_{12}, and C) in premixes used for fortification of infant nutrition products by ion suppression RP chromatography coupled with diode–array detection at two wavelengths (275 and 210 nm). A disadvantage of this method is that the sample composition has to be known in advance. According to European legislation, for example, foods might be fortified with riboflavin phosphate or thiamin phosphate, that is, vitamers that are not included in the simultaneous separations described.

Although refractive index measurement, as a bulk property detection method, is suitable for multianalyte determination, it is seldom used in vitamin analysis due to its poor sensitivity and selectivity. A detector of promise in developing multivitamin LC methods is the mass spectrometer, but its performance depends on the interface employed for the analyte ionization. The first applications go back to the mid-1980s [29] and early 1990s [30, 31], when the electrospray ionization (ESI) technique had just been invented and the ionization techniques available at that time (particle beam and thermospray) were not able to achieve low detection limits.

14.4
LC–MS in Multianalyte Confirmation Analyses

When LC is coupled with common detection systems (UV, diode-array and fluorescence detection), parameters for the identification of each analyte are the characteristic retention time, the wavelength corresponding to the absorption maximum (UV), the UV–visible spectrum (diode-array), and the excitation/emission wavelengths (fluorescence). The resolving power of the LC technique is less than that of GC and the separation of homologous compounds, such as vitamers of a specific vitamin, may be only partial due to their structural analogies. Quantitative analysis on not fully resolved chromatographic peaks lacks precision and accuracy. In these cases, the use of a mass spectrometer as a chromatographic detector may be the method of choice. In fact, each analyte is identified by the characteristic mass/charge ratios (m/z) of its (pseudo)molecular ion (if stable enough) and of its fragmentation products. This means that in the absence of chromatographic separation of two compounds, whose pseudomolecular ions have different m/z values, the extraction of their ion currents allows an accurate quantitative analysis to be performed. In addition to the very high level of selectivity, this hyphenated technique is also very sensitive and amounts of analyte less than 1 pg are sometimes sufficient for its analysis. Among the different types of interfaces developed for LC–MS coupling, atmospheric pressure ionization (API) sources have had the most success and commercial adoption, due to the well-known advantages that result when operating at atmospheric pressure. The technological problems in interfacing, the principles, the instrumentation, and the applications of LC–MS have been extensively explained in several books [32–34] and reviews [35–37].

Currently, three API interfaces are commercially available: ESI, atmospheric pressure chemical ionization (APCI), and atmospheric pressure photoionization (APPI) [38], which complement each other well with regard to polarity of analytes. All of them are soft ionization techniques, that is, techniques which ionize the analyte avoiding or limiting its fragmentation. The ESI interface is the source of choice to promote the ionization of polar substances such as water-soluble vitamins, which, provided with acidic and/or basic groups, can be deprotonated (giving pseudomolecular anions $[M - H]^-$) or protonated (giving pseudomolecular ions $[M + H]^+$). In fact, the sensitivity of ESI is about 2–5 times higher for all vitamins [38], but some authors have found APCI to be the source that produces the best response for niacin vitamers [39]. In contrast to by other API sources, ESI is able to ionize high molecular weight molecules and to produce multicharged pseudomolecular ions such as $[M + nH]^{n+}$ or $[M - nH]^{n-}$. This peculiarity is valuable in the detection of cobalamins that give intense double-charged pseudomolecular ions. In recent years, high flow-rate ESI sources have been developed in order to decrease the instrumental LOD. In these sources, the application of a hot gas flow (drying gas) – in concurrent flow (Z-spray interface) or perpendicular to the spray (turboionspray interface) – enhances the rate of charged droplet desolvation, improves the efficiency of the ion evaporation process, and generally increases the chromatographic signal-to-noise ratio (S/N). The turboionspray source was employed to ionize water-soluble vitamins, testing the different temperatures (200, 250, 300, 350, 400 °C) used to heat the drying gas. The best S/N was achieved at 300 °C. In fact, temperatures exceeding 300 °C decreased the signal intensity of B_1 and, to a lesser extent, of vitamins B_{12} and C [38]. The use of a drying gas is particularly useful also when the mobile phase composition is characterized by a high percentage of water in order to separate very polar analytes on C_{18} columns, such as water-soluble vitamins. In effect, the high superficial tension of water is responsible for the production of a poor aerosol (large sized droplets), and hence the action of the temperature, in addition to that of the shear forces of the drying gas and curtain gas, assists the ion evaporation process.

Although the advantage of mild ionization conditions of the ESI source guarantees the direct determination of the analyte molecular weight, the absent or reduced fragmentation of the pseudomolecular ion does not give structural information, useful for accurate analyte identification. A solution to this problem is the fragmentation activated in the ion transport region [up front or in-source collision-induced dissociation (CID)] or the use of tandem mass spectrometry (MS/MS) [34, 40]. In-source CID may be poorly selective, since all ions (of the analyte and the possible interfering compounds) are dissociated at the same time. It is reliable when preceded by an efficient cleanup and chromatographic separation. The selectivity is greatly increased by means of MS/MS, since the pseudomolecular ion is selected by the first analyzer (e.g., first quadrupole) and fragmented in the collision cell (where a focusing quadrupole is located). The third analyzer [quadrupole or time-of-flight (TOF)] examines the fragmentation products. In this way, the origin of the observed fragment ions is sure, the potential interferences from the sample matrix and from the mobile phase are eliminated, there is a drastic reduction of the chemical noise, and increased instrumental sensitivity is obtained.

Triple quadrupole (QqQ) instruments are the most widespread mass spectrometers, due to their performance, low cost, and reduced hindrance (benchtop detectors). Four are the following MS/MS acquisition modes: (i) product ion scan (PIS); (ii) multireaction monitoring (MRM); (iii) neutral loss scan (NLS); and (iv) precursor ion scan (PrIS). In the PIS mode, Q1 selects the pseudomolecular ion and Q3 sweeps a given mass range in order to analyze the precursor ion (pseudomolecular ion) plus its fragment ions. This mode is used for the fragmentation study of a compound or for a full-scan LC–MS/MS analysis. In the MRM mode, Q1 is fixed on the pseudomolecular ion and Q3 on the fragment ion of interest. In this acquisition mode, QqQ is characterized by a high duty cycle and performs reliable quantitative analyses. The last two modes are often utilized for screening purposes owing to their capability to identify homologous compounds which, under CID conditions, may show a common fragment ion or the same neutral loss.

Recently, within the EU, Commission Decision 657/2002/CE [41] established a series of confirmation criteria for the unequivocal identification of chemical residues in foodstuffs by LC–MS (see Table 14.1). These criteria should also be extended in other analysis fields, since they allow the avoidance of false positives, also possible in the MRM mode, when complex matrices have to be analyzed. They can be summarized as follows: (i) use of an identification point system; (ii) the relative intensities of the diagnostic fragment ions must match those of the standard within a maximum permitted tolerance; and (iii) the analyte in the matrix should elute at the retention time of the standard (±2.5%).

Table 14.1 Commission Decision 2002/657/EC: performance criteria and other requirements for mass spectrometric detection.

Maximum permitted tolerances for relative ion intensities

Relative intensity (%)[a]	Tolerance in LC–MS, LC–MS^n (%)
>50	±20
>20 to 50	±25
>10 to 20	±30
≤10	±50

Identification point system			Examples of identification points earned		
MS technique	IPs	Technique	Number of ions		IPs
Low-resolution (LR) mass spectrometry	Precursor ion 1.0	LC–MS (LR)	N precursor		N
	Product ion 1.5	LC–MS^n (LR)	1 precursor and 2 product ions (for example selecting two MRM transitions)		4
High-resolution mass spectrometry (HRMS)	Precursor ion 2.0	LC–HRMS	N		$2N$
	Product ion 2.5	LC–HRMS	1 precursor and 2 product ions		7

a)   % of base peak.

Another approach to increase selectivity is the use of time-of-flight mass spectrometry (TOF-MS), singly or combined with a quadrupole (QqTOF-MS) [40]. The high-resolution TOF method allows (i) a higher number of identification points (IPs) number to be achieved than low-resolution MS (e.g., quadrupole and ion trap), being equal to the number of the selected fragment ions and (ii) confirmation of those analytes which, subjected to the CID process, generate one product ion only (4.5 points overall). Another advantage is the high sensitivity in the full-scan mode and the associated high identification power. Today, the use of TOF instruments is still limited, but their spread, being strictly linked to their cost, is increasing in both food and environmental analysis.

Some studies on the LC–MS analysis of water-soluble vitamins have recently been published. In the early 2000s, the ESI source was applied for the first time to pantothenic acid detection [42] and to distinguish the several vitamers of folacin complex in both the positive and negative ionization modes [43, 44]. In particular, stable isotope dilution LC–MS/MS assay was devised as a reference procedure for validating LC methods using UV or fluorescence detection ([14, 45]) (see also Chapter 1). ESI, successfully applied to the analysis of B$_{12}$ in some fortified foods [46], is a promising LC detection mode for the total determination of this vitamin and for the direct characterization of its vitamers naturally occurring in foods.

Recently, negative ESI was employed for vitamin C determination in several food commodities ([47]) (see also Chapter 8).

Despite the potential of LC–ESI-MS, only three papers have been published on multivitamin analysis in dietary supplements [48] and in food samples [38, 39], but before describing these methods it is useful to explain the ESI and CID fragmentation of water-soluble vitamins.

14.5
Electrospray Ionization and Collision-Induced Dissociation

Most water-soluble vitamins respond to ESI in both ionization modes due to the presence of acidic and basic functional groups within their molecules.

Positive ESI product ion scan mass spectra and the postulated fragmentation patterns of several water-soluble vitamins are shown in Figures 14.1–14.3.

Vitamin B_1 is directly observed as $[M]^+$, which, under CID conditions, generates two product ions at m/z 122 (pyrimidinic ring) and 144 (thiazolic ring).

Vitamin B_2 may be detected in positive and negative ionization modes. Protonation of one of the basic sites produces an intense ion at m/z 377, and deprotonation of the –NH– group in the 2-position gives an abundant pseudomolecolar anion at m/z 375, which is stable due to its resonance structure. In both cases, CID fragmentation involves the ribitol chain.

Pyridinic nitrogen protonation of nicotinamide and nicotinic acid leads to an abundant $[M + H]^+$ ion that, upon loss of urea and carbon dioxide, respectively, generates a fragment ion at m/z 80. Nicotinic acid may be also detected in the negative ion mode.

Vitamin B_5 responds to positive and negative ionization. The successive loss of water from $[M + H]^+$ results in two product ions at m/z 202 and 184. The cleavage of the $C\alpha$–CO bond of pantoic acid produces the acyl ion at m/z 116, whereas cleavage of the amidic bond gives protonated β-alanine at m/z 90. CID fragmentation of $[M - H]^-$ does not involve water loss and induces the production of a few fragment ions, at m/z 146, at m/z 88 (deprotonated β-alanine), and at m/z 71.

Phosphorylated forms of B_6 may be detected in the positive and negative ion modes, whereas free forms are only visible in positive ionization. Under CID conditions, $[M + H]^+$ fragments to give the following product ions:

- m/z 232, 151, and 134, upon loss of ammonia and/or phosphoric acid from pyridoxamine 5′-phosphate
- m/z 152 and 134, upon loss of ammonia and water from pyridoxamine
- m/z 150 and 122, upon loss of phosphoric acid and subsequent loss of formaldehyde from pyridoxal 5′-phosphate
- m/z 150, upon loss of water from pyridoxal
- m/z 152 and 134, upon subsequent losses of water from pyridoxine.

D-Biotin may be observed in both ionization modes. Fragmentation of $[M + H]^+$ leads to an intense ion at m/z 227 (−18 Da) and to other less intense fragments at

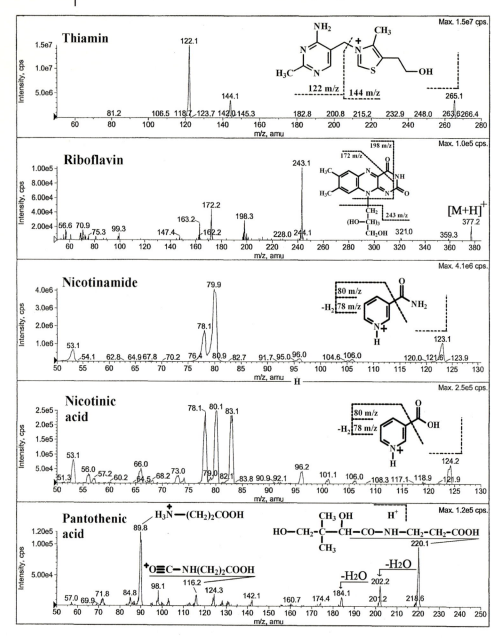

Figure 14.1 Product ion scan mass spectra and corresponding fragmentation schemes for thiamin, riboflavin, nicotinamide, nicotinic acid, and pantothenic acid.

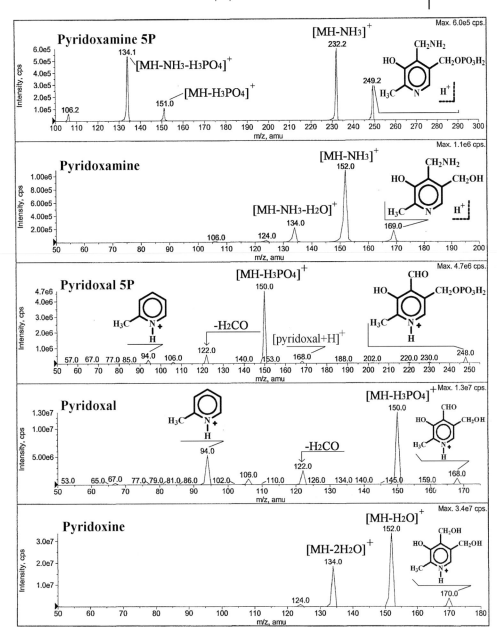

Figure 14.2 Product ion scan mass spectra and corresponding fragmentation schemes for B₆ vitamers: pyridoxamine 5′-phosphate, pyridoxamine, pyridoxal 5′-phosphate, pyridoxal, and pyridoxine.

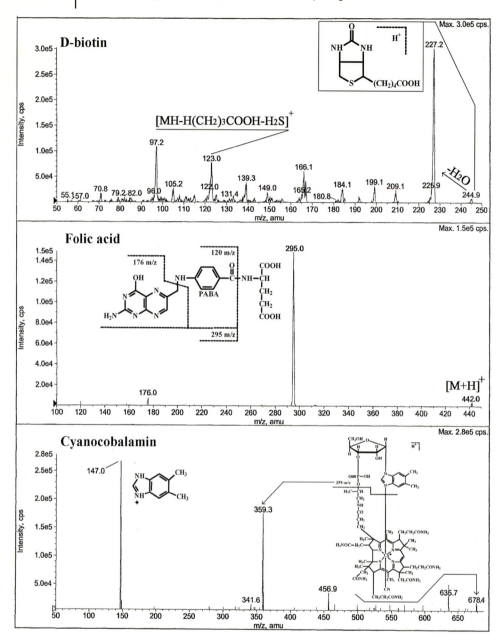

Figure 14.3 Product ion scan mass spectra and corresponding fragmentation schemes for D-biotin, folic acid, and cyanocobalamin.

m/z 209, 199, 166, 123, and 97. Fragmentation of $[M - H]^-$ is less extensive and produces three abundant ions at m/z 200, 166, and 122.

Vitamin B_9 may be studied in positive and negative ionization. Decomposition of $[M + H]^+$ reveals several product ions, all visible at different collision energies. At low voltages, the pseudomolecolar ion at m/z 442 loses glutamic acid to generate an abundant ion at m/z 295 (pteroic acid) that is cleaved into the pteridine portion (m/z 176) and into *p*-aminobenzoic acid (m/z 120) by suitably increasing the collision energy. In negative ionization, the pseudomolecolar anion requires a higher collision energy to start fragmenting: the CO_2 loss leads to the product ion at m/z 396, that of pteroic acid produces the fragment at m/z 311, while the combined loss of pteroic and *p*-aminobenzoic acid gives the ion at m/z 175.

Vitamin B_{12} responds to positive and negative ionization and, in both modes, it is possible to observe the presence of singly and doubly charged species. The product ion scan mass spectrum of the abundant ion $[M + 2H]^{2+}$ at m/z 678 is shown in Figure 14.3.

The high selectivity of a triple-quadrupole mass spectrometer was insufficient when the fragmentation of vitamin C was studied by infusing its standard solution. The Q1 scan spectrum showed a pseudomolecular ion $[M + H]^+$ at m/z 177, but the same spectrum was acquired when methanol, acetonitrile, or water was infused singly into the ESI source. The occurrence of an interfering isomeric compound complicated the fragmentation study of this vitamin. Its possible product ions were located at m/z 55, 57, 67, 69, 73, 83, 85, 95, 97, 111, 113, 123, 129, and 141, after subtracting the fragmentation spectrum of water from that of the vitamin C (each sum of 10 scans), for every collision energy used. All the above-mentioned ions were confirmed as fragments of vitamin C by LC–MS/MS analysis as they produced a peak at the retention time of ascorbic acid (see Figure 14.4). Vitamin C can be also detected in the negative ion mode as a consequence of the acidic character of its C-2 hydroxyl group ($pK_a = 4.04$).

14.6
LC–ESI-MS for Simultaneous Analysis of Water-Soluble Vitamins

In LC, the RP mode is more versatile and hence more appropriate for developing a multivitamin method. In order to avoid broadening phenomena of chromatographic peaks due to the acidic properties of some vitamins, ion-pair or ion-suppression conditions should be applied. Because of their volatility, ammonium acetate and formate are ion-pair agents suitable for MS analysis but their effect on increasing the retention times of polar and low molecular weight vitamins is similar to that obtained under ion-suppression conditions; moreover, the signal could be dispersed between $[M + H]^+$ and $[M + NH_4]^+$ adducts for those vitamins which form both of them. Ion suppression is easily performed by adding a volatile acid, such as formic or acetic acid, to the mobile phase and, in this way, ESI is also assisted when working in the positive ion mode [38].

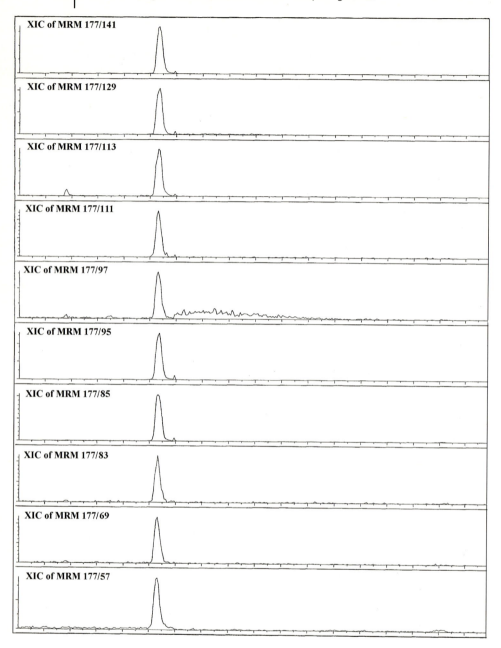

Figure 14.4 LC–MS analysis for verifying the actual fragment ions of ascorbic acid. The extract ion currents of the less intense fragment ions at *m/z* 123, 73, 67, and 55 are not shown (see text).

Most vitamins show an ESI response higher in positive than in negative ionization except for folic acid, vitamin B_{12}, pyridoxal 5'-phosphate, and pantothenic acid. The modern triple quadrupoles can combine positive and negative ionization in the same chromatographic run; nevertheless, the switch of polarity requires a greater pause time between the scans, which results in a decrease in sensitivity compared with that recorded when a single ionization mode is selected. Moreover, the ion-suppression conditions, applied for the chromatographic separation, inhibit the negative ESI response. For these reasons, ion-suppression RP chromatography followed by positive ESI detection appears to be a good compromise for the simultaneous LC–MS analysis of water-soluble vitamins. Figure 14.5 shows a representative LC–ESI(+)-MRM chromatogram of 14 compounds with vitamin activity. All analytes, including B_3 and B_6 vitamers, are identifiable on the basis of the experimental parameters shown in Table 14.2. This is an important advantage offered by MS detection in the case of poor chromatographic resolution; in fact, the difficulty in separating B_6 vitamers, due to the structural analogies of pyridoxine, pyridoxamine, and pyridoxal and to the acidity of their phosphorylated forms, can easily be overcome.

14.7
Simultaneous Extraction Procedures

The key step in the development of a multivitamin method is to establish an extraction procedure which is able to extract several vitamins concurrently and quantitatively from a particular food matrix. The simplest way to proceed, often reported in the literature as being suitable for product compliance monitoring, is the extraction of fortified levels of water-soluble vitamins. Nevertheless, European law [3] suggests the determination of the total vitamin content (fortified and endogenous forms) as the most suitable approach. Moreover, as labeling regulations are becoming stricter, it is fundamental that the sampling is representative and that the extraction is quantitative (maximum release from the matrix with the minimum loss of analyte).

The application of an acid treatment, in order to hydrolyze the bound forms, could be used for simultaneous extraction of B_1, B_2, B_3, B_6, B_8, B_{12}, and C vitamins, but certainly not for B_5 and B_9, which are sensitive to low pH. Leporati *et al.* [39] developed and validated an LC–ESI(+)-MS/MS method for the determination of several water-soluble vitamins in typical Italian pasta samples and in fortified pasta samples produced for the US market. Vitamins B_1, B_2, B_6 (pyridoxine, pyridoxal, and pyridoxamine), and B_3 (nicotinamide and nicotinic acid) were analyzed concurrently since they were extracted by acid hydrolysis from pasta samples, after being ground to homogeneity. Vitamins B_5 and B_9 needed an individual hydrolysis–extraction procedure and were analyzed in separate LC runs. Therefore, the actual simultaneous analysis was restricted to B_1, B_2, and B_6 vitamers with unquestionable advantages in the simplification of the sample pretreatment; in particular, the complex procedure to convert free forms of B_6 to pyridoxine, applied when LC

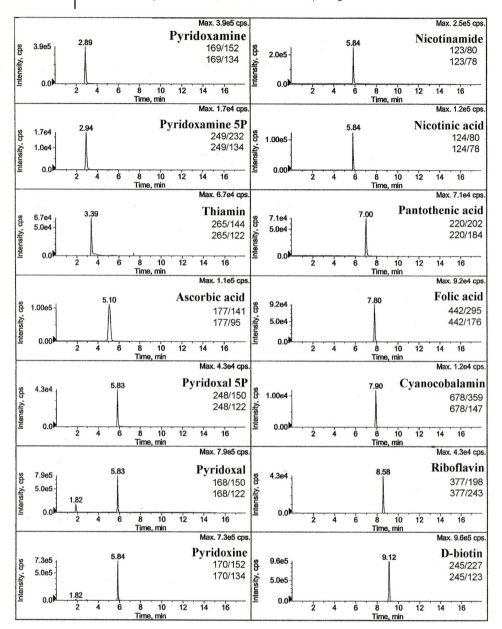

Figure 14.5 LC–ESI(+)-MRM chromatogram of the 14 water-soluble vitamins (5 ng injected for thiamin, nicotinamide, nicotinic acid, and B$_6$ vitamers; 500 ng injected for ascorbic acid; 20 ng injected for all others). For each analyte, the sum of the two extracted ion currents is shown.

Table 14.2 LC–MS/MS parameters for the identification of water-soluble vitamins selected in this study.

Analyte	Retention time (min)	Qualifier and quantifier MRM transitions[a]	Relative abundance (mean ± SD)[b]
Pyridoxamine	2.89	169/134 169/152	0.42 ± 0.02
Pyridoxamine 5′-phosphate	2.94	249/134 249/232	0.86 ± 0.02
Thiamin	3.39	265/144 265/122	0.151 ± 0.006
Ascorbic acid	5.10	177/95 177/141	0.59 ± 0.05
Pyridoxal 5′-phosphate	5.83	248/122 248/150	0.202 ± 0.005
Pyridoxal	5.83	168/122 168/150	0.103 ± 0.008
Pyridoxine	5.84	170/134 170/152	0.69 ± 0.02
Nicotinamide	5.84	123/78 123/80	0.24 ± 0.05
Nicotinic acid	5.84	124/78 124/80	0.90 ± 0.03
Pantothenic acid	7.00	220/184 220/202	0.65 ± 0.02
Folic acid	7.80	442/176 442/295	0.257 ± 0.008
Cyanocobalamin	7.90	678/359 678/147	0.28 ± 0.01
Riboflavin	8.58	377/198 377/243	0.213 ± 0.005
D-Biotin	9.12	245/123 245/227	0.164 ± 0.008

a) The first line gives the least intense MRM transition (qualifier) and the second line the most intense one (quantifier).
b) The relative abundance is calculated as ratio of qualifier intensity/quantifier intensity; the results are reported as the arithmetic average of six replicates plus the corresponding standard deviation.

with fluorimetric detection is used, was eliminated. The authors verified extraction yields ranging between 90 and 95%, except for vitamin B_2 (52%).

Gentili *et al.* [38] developed an LC–ESI(+)-MS/MS method for the simultaneous determination of vitamins B_1 (thiamin), B_2 (riboflavin), B_3 (nicotinamide and nicotinic acid), B_5, B_6 (pyridoxine, pyridoxamine, pyridoxamine 5′-phosphate, pyridoxal hydrochloride, pyridoxal 5′-phosphate monohydrate), B_8 (D-biotin), B_9 (folic acid), B_{12} (cyanocobalamin). and C (L-ascorbic acid) in unfortified foods (green kiwi, golden kiwi, maize flour, and tomato pulp). For the simultaneous extraction of selected analytes, the authors tried to apply the matrix solid-phase dispersion (MSPD), an extraction technique introduced in 1989 [49, 50]. By means of this technique, the complete disintegration of the matrix in a dispersion medium (diatomaceous earth, C_18, etc.) is obtained by blending an aliquot of sample and of solid support (in a ratio of 1:2 or 1:3, w/w) with a glass pestle until the whole mixture shows a uniform color and consistency. Using a C_{18} sorbent as dispersion medium and methanol as extractant, the authors obtained recoveries exceeding 77%, but severe losses were observed for pyridoxal 5′-phosphate, pyridoxamine 5′-phosphate, and vitamins B_9 and C. Degradation of ascorbic acid was even complete on extracting it from maize flour under these conditions. The preparation of a control sample confirmed the negative effect of heat, developed as a consequence both of the pestle friction during the matrix dispersion and of exposure to air, for the time necessary to perform the extraction. This technique, mainly chosen to support the release of forms bound to food proteins, appeared too aggressive and was replaced by a milder solid–liquid extraction. The extraction/purification cartridge was prepared by introducing first a layer of the C_{18} sorbent (0.5 g) and then a layer of the food sample (2 g of maize flour, tomato pulp, or homogenized kiwi), mixed with butylated hydroxytoluene (15 mg) as stabilizer. In the case of tomato pulp and kiwi, diatomaceous earth (1 g) was also blended to give a solid consistency to the two matrices described above. The C_{18} material served to retain proteins and to avoid the formation of foam. Analytes were extracted with an ethanol–water (50:50, v/v) solution and a large volume of eluate was directly injected on to the LC column without creating problems of peak broadening. An evaporation step, which is necessary for concentrating the extract, was eliminated as a consequence of the LC–MS/MS technique's high sensitivity. The analyte recovery depended on the type of matrix. In particular, the recovery of the analytes from maize flour exceeded 70%, with the exception of vitamin C (~20%), pyridoxal-5′-phosphate and vitamin B_9 (~40%); for tomato pulp, the recovery exceeded 64%, except for vitamin C (41%); for kiwi, the recovery exceeded 73%, except for nicotinamide (~30%).

The greatest difficulties were found during the extraction of vitamin C from maize flour, owing to its degradation, probably catalyzed by the presence of minerals such as iron in maize. The addition of EDTA to the extractant was very effective in preventing ascorbic acid oxidation. Vitamin C was recovered in a range between 80% (with 1.5 mM EDTA) and 70% (with 0.1 mM EDTA); nevertheless, its presence in the eluate had two negative effects: (i) it interfered with the chromatographic analysis of some other vitamins (B_1, pyridoxal-5′-phosphate, pyridoxine, pantothenic acid, B_3 vitamers) also at low concentration, and (ii) it caused ion suppression for most vitamins, which was very severe for B_2, B_9, pyridoxamine and D-biotin. These effects are shown in Figure 14.6.

In summary, the first results in developing a multivitamin method were obtained by this work. The use of rapid and simple extraction procedures (short exposure to light and air) and direct injection of the extract (avoiding any exposure to heat) were the two important outcomes achieved as a consequence of the high sensitivity and selectivity of LC–MS/MS. The mild extraction conditions prevented labile analytes from degradation, allowed the estimation of each vitamer singly and

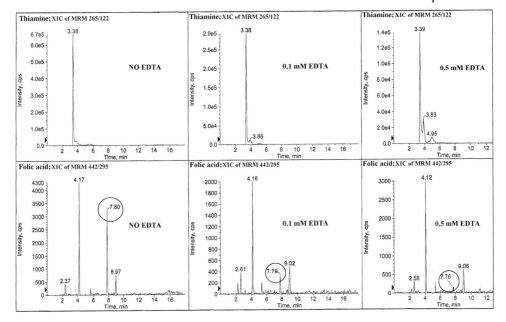

Figure 14.6 Examples of EDTA effects on LC–ESI-MS analysis of water-soluble vitamins in maize. The presence of this chelating agent interferes with the chromatography of thiamin (double peak) and suppresses the ESI response of folic acid. Extracted ion chromatograms for the most intense transition are shown. The less intense transition is already suppressed at 0.1 mM EDTA.

avoided artifact creation. In fact, cyanocobalamin and folic acid were directly identified and determined in kiwi and tomato pulp. Two questions remained unsolved: (i) the development of a mild procedure to free the forms bound to food proteins and (ii) the use of a proper stabilizer to obtain a good recovery of vitamin C.

14.8
Conclusions and Future Developments

Due to the chemical and structural diversity of water-soluble vitamins, suitable extraction and chromatographic conditions for their simultaneous analysis were not easy to find.

The difficulties concerning their LC separation can be overcome by using MS as the detection system, which allows multivitamin determination to be carried out in a single chromatographic run.

The development of a common extraction procedure is more complex to realize, but the use of LC–MS as an analytical technique is fundamental to achieve this aim. A major advantage is the high sensitivity, which permits the extract to be injected directly, eliminating the concentration step and exposure to heat, and the extraction procedures to be simplified, reducing the analysis time and minimizing exposure of the analyte to air and light.

If the objective is the control of the fortification levels of water-soluble vitamins added to a food, the development of a simultaneous method of analysis presents less complications. In fact, the number of analytes to be extracted is limited to the supplemented forms, and also LC methods based on the use of conventional detection systems are able to perform these determinations, as described in this chapter. Although the use of LC–MS simplified the extraction procedure, problems due to interfering compounds from the matrix still have to be solved.

If the target is the determination of the total level of each single vitamin, added to and naturally occurring in the food, the use of LC–MS becomes indispensable for developing a multi-analyte method. For developing such a method, two approaches are conceivable.

The first involves the adoption of a conventional extraction method such as the application of an acidic hydrolysis in order to free all bound forms and to analyze a single form for each vitamin. This choice precludes the concurrent analysis of pantothenic acid and folic acid, which are sensitive to low pH. On this basis, Leporati *et al.* [39] limited their analysis to thiamin, riboflavin, pyridoxine, pyridoxal, pyridoxamine, nicotinamide, and nicotinic acid.

The second is more difficult to realize and requires additional efforts. As detailed in this chapter, LC–MS is an analytical technique suitable to perform multi-analyte analyses. This ability allows an important analytical simplification to be obtained, that is, the direct analysis of all possible vitamers of each vitamin, avoiding the procedures for their conversion to a single form. LC–MS can analyze a large number of compounds simultaneously, maintaining good sensitivity; in the literature, studies of the application of LC–MS to the analysis of about 100 analytes in a single chromatographic run which involved two or more time segments have been described [51]. An additional advantage of this approach is that new and more detailed information about the vitamin profile of a specific food can be acquired. The application of a mild extraction procedure preserves all vitamers and avoids the creation of artifacts. The conditions applied by Gentili *et al.* [38] allowed the accomplishment of these objectives and an important result was the identification and direct determination of cyanocobalamin and folic acid in kiwi and tomato pulp. The main obstacle is the development of a mild procedure to free the forms bound to proteins and polysaccharides of a food. Combined enzymatic hydrolysis (with enzymes such as protease and amylase) could represent a solution, without using buffers at extreme pH values. Some authors have been working in this direction [52]. Another possible approach, still to be verified for vitamins, is the use of an extraction solvent which, by lowering the dielectric constant of the medium, induces deproteination and, perhaps, favors the probable release of those forms linked by tight electrostatic interactions. This might be particularly realizable when the selected food is milk. The accuracy of the data obtained may be verified by the use of standard additions method or stable isotope dilution assays, provided that suitable labeled analogs are available.

Further achievements in this analytical field will come from the continuous technological progress. For example, with the advent of the UPLC system [53, 54], it is now possible to operate at high pressures (1000 bar versus 400 bar in HPLC) and to use 1 mm i.d. columns with packings in the 1–2 μm size range at high flow-rates. In this way, the performance of LC–MS techniques can be further improved: the times for multi-residue analysis are shortened (around 10 min or less), and the resolution and the sensitivity are increased due to the very narrow chromatographic peaks. The best results are obtained when the UPLC system is coupled to a mass analyzer characterized by a high duty cycle, such as TOF-MS or triple-quadrupole instruments operating in the full-scan and the MRM mode, respectively.

References

1 Belitz, H.-D., Grosch, W., and Schieberle, P. (2004) Vitamins, in *Food Chemistry*, 3rd edn., Springer, Berlin, p. 409.

2 Ball, G.F.M. (1998) *Bioavailability and Analysis of Vitamins in Foods*, Chapman and Hall, London.

3 European Commission (2006) Regulation 1925/2006/EC of the European Parliament and of the Council of 20 December 2006 on the Addition of Vitamins and Minerals and of Certain Other Substances to Foods. *Official Journal of the European Union* **L404**, 26–38.

4 Blake, C.J. (2007) Analytical procedures for water-soluble vitamins in foods and dietary supplements: a review. *Analytical and Bioanalytical Chemistry*, **389**, 63–76.

5 Rizzolo, A. and Polesello, S. (1998) Chromatographic determination of vitamins in foods. *Journal of Chromatography*, **624**, 103–152.

6 Lynch, P.L.M. and Young, I.S. (2000) Determination of thiamine by high-performance liquid chromatography. *Journal of Chromatography A*, **881**, 267–284.

7 Ndaw, S., Bergaentzlé, M., Aoudé-Werner, D., and Hasselmann, C. (2000) Extraction procedures for the liquid chromatographic determination of thiamine, riboflavin and vitamin B_6 in foodstuffs. *Food Chemistry*, **71**, 129–138.

8 Gonthier, A., Fayol, V., Viollet, J., and Hartmann, D.J. (1998) Determination of pantothenic acid in foods: influence of the extraction method. *Food Chemistry*, **63**, 287–294.

9 Pakin, C., Bergaentzlé, M., Hubscher, V., Aoudé-Werner, D., and Hasselmann, C. (2004) Fluorimetric determination of pantothenic acid in foods by liquid chromatography with postcolumn derivatization. *Journal of Chromatography A*, **1035**, 87–95.

10 Argoudelis, C.J. (1997) Simple high-performance liquid chromatographic method for the determination of all seven vitamin B_6-related compounds. *Journal of Chromatography A*, **790**, 83–91.

11 Livaniou, E., Costopoulou, D., Vassiliadou, I., Leondiadis, L., Nyalala, J.O., Ithakissios, D.S., and Evangelatos, G.P. (2000) Analytical techniques for determining biotin. *Journal of Chromatography A*, **881**, 331–343.

12 Lahély, S., Ndaw, S., Arella, F., and Hasselmann, C. (1999) Determination of biotin in foods by high-performance liquid chromatography with postcolumn derivatization and fluorimetric detection. *Food Chemistry*, **65**, 253–258.

13 Quinlivan, E.P., Hanson, A.D., and Gregory, J.F. (2006) The analysis of folate and its metabolic precursors in biological samples. *Analytical Biochemistry*, **348**, 163–184.

14 Rychlik, M. and Freisleben, A. (2002) Quantification of pantothenic acid and folates by stable isotope dilution assays.

Journal of Food Composition and Analysis, **15**, 399–409.

15 Velísek, J. and Cejpek, K. (2007) Biosynthesis of food constituents: vitamins. 2. Water-soluble vitamins: Part 2–a review. *Czech Journal of Food Sciences,* **25**, 101–118.

16 Lebiedzińska, A., Marszall, M.L., Kuta, J., and Szefer, P. (2007) Reversed-phase high-performance liquid chromatography method with coulorimetric electrochemical and ultraviolet detection for the quantification of vitamins B_1 (thiamine), B_6 (pyridoxamine, pyridoxal and pyridoxine) and B_{12} in animal and plant foods. *Journal of Chromatography A,* **1173**, 71–80.

17 Heudi, O., Kilinç, T., Fontannaz, P., and Marley, E. (2006) Determination of vitamin B_{12} in food products and premixes by reversed-phase high performance liquid chromatography and immunoaffinity extraction. *Journal of Chromatography A,* **1101**, 63–68.

18 Versari, A., Mattioli, A., Parpinello, G.P., and Galassi, S. (2004) Rapid analysis of ascorbic and isoascorbic acids in fruit juice by capillary electrophoresis. *Food Control,* **15**, 355–358.

19 Kall, M.A. and Andersen, C. (1999) Improved method for simultaneous determination of ascorbic acid and dehydroascorbic acid, isoacsorbic acid and dehydroisoascorbic acid in food and biological samples. *Journal of Chromatography B,* **730**, 101–111.

20 Brause, A.R., Woollard, D.C., and Indyk, H. (2003) Determination of total vitamin C in fruit juices and related products by liquid chromatography: interlaboratory study. *Journal of AOAC International,* **86**, 367–374.

21 Lahély, S., Bergaentzlé, M., and Hasselmann, C. (1999) Fluorimetric determination of niacin in foods by high-performance liquid chromatography with postcolumn derivatization. *Food Chemistry,* **65**, 129–133.

22 Viñas, P., Balsalobre, N., López-Erroz, C., and Hernández-Córdoba, M. (2004) Determination of vitamin B_6 compounds in foods using liquid chromatography with postcolumn derivatization

fluorescence detection. *Chromatographia,* **59**, 381–386.

23 Viñas, P., López-Erroz, C., Balsalobre, N., and Hernández-Córdoba, M. (2003) Reversed-phase liquid chromatography on an amide stationary phase for the determination of the B group vitamins in baby foods. *Journal of Chromatography A,* **1007**, 77–84.

24 Heudi, O., Kilinç, T., and Fontannaz, P. (2005) Separation of water-soluble vitamins by reversed-phase high performance liquid chromatography with ultra-violet detection: application to polyvitaminated premixes. *Journal of Chromatography A,* **1070**, 49–56.

25 Zafra-Gómez, A., Garballo, A., Morales, J.C., and García-Ayuso, L.E. (2006) Simultaneous determination of eight water-soluble vitamins in supplemented foods by liquid chromatography. *Journal of Agricultural and Food Chemistry,* **54**, 4531–4536.

26 Albalá-Hurtado, S., Veciana-Nogués, M.T., Izquierdo-Pulido, M., and Mariné-Font, A. (1997) Determination of water-soluble vitamins in infant milk by high-performance liquid chromatography. *Journal of Chromatography A,* **778**, 247–253.

27 Arcot, J. and Shrestha, A. (2005) Folate: methods of analysis. *Trends in Food Science and Technology,* **16**, 253–266.

28 Pakin, C., Bergaentzlé, M., Aoudé-Werner, D., and Hasselmann, C. (2005) α-Ribazole, a fluorescent marker for the liquid chromatographic determination of vitamin B_{12} in foodstuffs. *Journal of Chromatography A,* **1081**, 182–189.

29 Azoulay, M., Desbene, P.-L., Frappier, F., and Georges, Y. (1984) Use of liquid chromatography–mass spectrometry for the quantitation of dethiobiotin and biotin in biological samples. *Journal of Chromatography A,* **303**, 272–276.

30 Junko, I. and Takeshi, M. (1990) Application of thermospray liquid chromatography/mass spectrometry to the analysis of water-soluble vitamins. *Analytical Sciences,* **6**, 273–276.

31 Careri, M., Cilloni, R., Lugari, M.T., and Manini, P. (1996) Analysis of water-soluble vitamins by high-performance liquid chromatography–particle beam-mass spectrometry. *Analytical Communications*, **33**, 159–162.

32 De Hoffmann, E. and Stroobant, V. (2003) *Mass Spectrometry: Principles and Applications*, 3rd edn., John Wiley & Sons, Ltd., Chichester

33 Ardrey, R.E. (2003) *Liquid Chromatography–Mass Spectrometry: an Introduction*, Analytical Techniques in the Sciences (AnTS), (ed. D.J. Ando), John Wiley & Sons, Ltd., Chichester.

34 Cole, R.B. (ed.) (1997) *Electrospray Ionization Mass Spectrometry – Fundamentals, Instrumentation and Applications*, John Wiley & Sons, Inc., New York.

35 Niessen, W.M.A. (1998) Advances in instrumentation in liquid chromatography–mass spectrometry and related liquid-introduction techniques. *Journal of Chromatography A*, **794**, 407–435.

36 Niessen, W.M.A. (2003) Progress in liquid chromatography–mass spectrometry instrumentation and its impact on high-throughput screening. *Journal of Chromatography A*, **1000**, 413–436.

37 Robb, D.B., Covey, T.R., and Bruins, A.P. (2000) Atmospheric pressure photoionization: an ionization method for liquid chromatography–mass spectrometry. *Analytical Chemistry*, **72**, 3653–3659.

38 Gentili, A., Caretti, F., D'Ascenzo, G., Marchese, S., Perret, D., Di Corcia, D., and Maniero Rocca, L. (2008) Simultaneous determination of water-soluble vitamins in selected food matrices by liquid chromatography/ electrospray ionization tandem mass spectrometry. *Rapid Communications in Mass Spectrometry*, **22**, 2029–2043.

39 Leporati, A., Catellani, D., Suman, M., Andreoli, R., Manini, P., and Niessen, W.M.A. (2005) Application of a liquid chromatography tandem mass spectrometry method to the analysis of water-soluble vitamins in Italian pasta. *Analytica Chimica Acta*, **531**, 87–95.

40 Hager, J.W. (2004) Recent trends in mass spectrometer development. *Analytical and Bioanalytical Chemistry*, **378**, 845–850.

41 European Commission (2002) Commission Decision 2002/657/CE Implementing Council Directive 96/23/ EC Concerning the Performance of Analytical Methods and the Interpretation of Results. *Official Journal of the European Communities*, **L221**, 8–36.

42 Rychlik, M. (2003) Pantothenic acid quantification by stable isotope dilution assays based on liquid chromatography– tandem mass spectrometry. *Analyst*, **128**, 832–837.

43 Pawlosky, J. and Flanagan, V.P. (2001) A quantitative stable-isotope LC–MS method for the determination of folic acid in fortified foods. *Journal of Agricultural and Food Chemistry*, **49**, 1282–1286.

44 Rychlik, M. (2003) Simultaneous analysis of folic acid and panthothenic acid in foods enriched with vitamins by stable isotope dilution assays. *Analytica Chimica Acta*, **495**, 133–141.

45 Rychlik, M., Englert, K., Kapfer, S., and Kirchhoff, E. (2007) Folate contents of legumes determined by optimized enzyme treatment and stable isotope dilution assays. *Journal of Food Composition and Analysis*, **20**, 411–419.

46 Luo, X., Chen, B., Ding, L., Tang, F., and Yao, S. (2006) HPLC–ESI-MS analysis of vitamin B_{12} in food products and multivitamin–multimineral tablets. *Analytica Chimica Acta*, **562**, 185–189.

47 Garrido French, A., Hernández Torres, M.E., Belmonte Vega, A., Martínez Vidal, J.L., and Plaza Bolaños, P. (2005) Determination of ascorbic acid and carotenoids in food commodities by liquid chromatography with mass spectrometry detection. *Journal of Agricultural and Food Chemistry*, **53**, 7371–7376.

48 Chen, P. and Wolf, W.R. (2007) LC/UV/ MS-MRM for the simultaneous determination of water-soluble vitamins in multi-vitamin dietary supplements. *Analytical and Bioanalytical Chemistry*, **387**, 2441–2448.

49 Barker, S.A. (2000) Matrix solid-phase dispersion. *Journal of Chromatography A*, **885**, 115–127.

50 Barker, S.A. (2000) Matrix solid-phase dispersion. *Journal of Chromatography A*, **880**, 63–68.

51 Ferrer, I., Thurman, E.M., and Zweigenbaum, J.A. (2007) Screening and confirmation of 100 pesticides in food samples by liquid chromatography/tandem mass spectrometry. *Rapid Communications in Mass Spectrometry*, **21**, 3869–3882.

52 Ndaw, S., Bergaentzle, M., Aoudé-Werner, D., and Hasselmann, C. (2002) Enzymatic extraction procedure for the liquid chromatographic determination of niacin in foodstuffs. *Food Chemistry*, **78**, 129–134.

53 Swartz, M.E. (2005) Ultra performance liquid chromatography (UPLC): an introduction. Separation Science Redefined, pp. 8–14. http://chromatographyonline. findanalytichem.com/lcgc/data/ articlestandard//lcgc/242005/164646/ article.pdf (accessed 23 December 2010).

54 Barceló-Barrchina, E., Moyano, E., Galceran, M.T., Lliberia, J.L., Bagó, B., and Cortes, M.A. (2006) Ultra-performance liquid chromatography–tandem mass spectrometry for the analysis of heterocyclic amines in food. *Journal of Chromatography A*, **1125**, 195–203.

Part III
Analysis of Fat-Soluble Vitamins

15
Analysis of Carotenoids

Volker Böhm

15.1
Introduction

Carotenoids are one of the most important classes of plant pigments. Today, about 700 carotenoids are known, which are divided into carotenes (e.g., α-carotene, β-carotene, lycopene) and the oxygenated xanthophylls (e.g., lutein, zeaxanthin, β-cryptoxanthin). Among them, α-carotene and β-carotene, and also β-cryptoxanthin, are the major provitamin A active carotenoids. In plants, carotenoids play a crucial role in light harvesting for photosynthesis and in the protection of chlorophyll against oxidative damage. The latter functions are also considered to be the reason for their health-related effects in humans. Epidemiological studies have shown associations between intake of fruits and vegetables rich in carotenoids and reduced risks of different types of cancer, cardiovascular diseases, and age-related macular degeneration (AMD) [1]. For these reasons, many dietary supplements containing carotenoids are now commercially available. Supplements, for example, with β-carotene or lycopene or lutein in various doses, are produced by several suppliers. In contrast to essential nutrients (vitamins, minerals), recommendations for the dietary intake of carotenoids do not exist. In 1995, a consensus meeting agreed on the advice of an intake of 2–4 mg of β-carotene per day to achieve preventive contents of this carotenoid in plasma [2]. Two large intervention trials with smokers and asbestos workers, investigating the effect of 20 [3] and 30 mg per day [4], respectively, of β-carotene on the incidence of lung cancer, surprisingly showed a higher risk of lung cancer for those volunteers ingesting the β-carotene supplement. Consequently, international and national organizations advised smokers not to ingest high doses of β-carotene. However, many preparations containing more than 10 mg of β-carotene per capsule are commercially available without any warning notice. Following these developments, the analysis of carotenoids is becoming more and more important, not only to determine provitamin A carotenoids. The first crucial step is the extraction of the lipophilic compounds from food matrices, followed by the choice of the appropriate analytical method.

15.2
Materials and Methods

15.2.1
Chemicals

All chemicals for extraction were of analytical-reagent grade and solvents for the chromatography were of high-performance liquid chromatography (HPLC) quality. The carotenoid standards [except (15Z)-β-carotene] were purchased from Carote-Nature (Lupsingen, Switzerland). (15Z)-β-Carotene was isolated from a mixture of β-carotene isomers prepared by photoisomerization [5]. All carotenoids were dissolved in cyclohexane–toluene (4:1, v/v) and stored in the dark at −30 °C. Concentrations of stock standard solutions were calculated periodically using their absorption maxima and appropriate absorptivity values (Table 15.1). For preparing working standard solutions, the stock standard solutions were diluted daily 1:50–1:100 with methanol–methyl tert-butyl ether (1:1, v/v).

15.2.2
Instrumentation

The spectrophotometer, centrifuge, rotary evaporator, Ultra-Turrax, pipettes, etc., used are not explained in detail, as various instruments normally used in a laboratory can be applied.

Table 15.1 Wavelength maxima and absorptivity values of carotenoid standards used for photometric determination of concentrations of stock standard solutions [6–10].

Carotenoid[a]	Solvent	Wavelength (nm)	Absorptivity ($E_{1\%, 1cm}$)
β-apo-8'-Carotenal (IS)	Ethanol	457	2640
(all-E)-Echinenone (IS)	Ethanol	458	2158
(all-E)-Violaxanthin	Ethanol	440	2550
(all-E)-Lutein	Ethanol	445	2550
(all-E)-Zeaxanthin	Ethanol	450	2540
(all-E)-Canthaxanthin	Ethanol	466	2220
(all-E)-β-Cryptoxanthin	Ethanol	452	2386
(all-E)-α-Carotene	n-Hexane	444	2800
(all-E)-Rubixanthin	Petroleum ether	460	2750
(all-E)-β-Carotene	n-Hexane	453	2592
(9Z)-β-Carotene	n-Hexane	445	2550
(13Z)-β-Carotene	n-Hexane	443	2090
(15Z)-β-Carotene	n-Hexane	447	1820
(all-E)-Lycopene	n-Hexane	472	3450

a)   IS, internal standard.

15.2.3
HPLC Equipment

The HPLC instruments used are mentioned in detail, as some differences in the chromatograms might be dependent on the instruments used.

15.2.3.1 Separation of Carotenoids
The equipment consisted of a Model L-7100 HPLC gradient pump (VWR, Darmstadt, Germany), Model L-7200 autosampler (VWR), degasser, Jetstream column oven (Jasco, Gross-Umstadt, Germany), and Model L-7450 diode-array detector (VWR).

15.2.3.2 Separation of Xanthophylls
The equipment consisted of a Model PU-1580 HPLC pump with Model LG-1580-02 ternary gradient unit (Jasco), Model AS-1555-10 autosampler (Jasco), Model DG-1580-53 degasser (Jasco), Jetstream column oven (Jasco), and Model MD-1510 diode-array detector (Jasco).

15.3
Results and Discussion

15.3.1
Extraction

For fruits and vegetables, an extraction procedure using methanol and tetrahydrofuran [8] was successfully tested within an interlaboratory study [11]. This method was used several times in our laboratory, for example, for the extraction of tomato products [12]. Briefly, 200 mg of MgO, 500 μl of apo-8′-carotenal or echinenone (internal standard solution), and 35 ml of methanol–tetrahydrofuran (1:1, v/v) containing 0.1% butylated hydroxytoluene (BHT) (to prevent isomerization and oxidation) were added to 2 g of sample. The mixture was homogenized on ice for 5 min using an Ultra-Turrax. The resulting solution was filtered through 390 paper in a Büchner funnel. The extraction was repeated twice or more often until the extract was colorless. The combined extracts were dried under vacuum at 30 °C in a rotary evaporator. The residue was dissolved in methanol–tetrahydrofuran (1:1, v/v), containing 0.1% BHT, until the solution reached the defined volume of 10 ml. The solution was centrifuged (14 000 rpm, 5 min) and then used for analysis. All extraction steps were carried out under subdued light.

However, this extraction procedure did not result in the quantitative extraction of carotenoids from durum wheat. Finally, it was shown that soaking of samples with water for 5 min prior to extraction with organic solvents had the strongest impact on the extraction yield and led to the most rapid and mild method. The contents of carotenoids in the extracts of several durum wheat and corn samples were doubled by soaking with water before extracting with methanol–tetrahydrofuran (1:1, v/v) [13]. Ongoing investigations should clarify the reasons for the limited extractability of carotenoids from durum wheat.

15.3.2
Saponification

The presence of xanthophyll esters in several samples requires saponification. Often, the sample is mixed with methanolic KOH solution prior to solvent extraction, using different concentrations mainly at room temperature for short or longer periods of time [14–18]. However, in addition to the release of xanthophylls from their esters, degradation of carotenoids was observed during saponification [15, 19]. Therefore, different concentrations (5, 10, 20, and 40 g per 100 ml) of a methanolic KOH solution were tested in order to achieve the best cleavage of esters and the lowest degradation of carotenoids. Saponification was performed with food extracts instead of saponifying the food samples directly to obtain a much more convenient procedure.

Optimized conditions, which were later only used for foods containing xanthophyll esters, were achieved by using KOH solution (10 g per 100 ml). A 1 ml volume of that solution was mixed with 2 ml of the food extract for 15–60 min (depending on the food extracted) at room temperature under subdued light. Subsequently, 2 ml of petroleum ether and 0.5 ml of water were added followed by vigorous mixing for 1 min and centrifugation (5000 rpm, 5 min). The upper layer, containing the carotenoids, was transferred in another test-tube. The lower hydrophilic layer was extracted several times with 2 ml of petroleum ether each time until the organic layer was colorless, followed by washing the combined organic phases several times with water to remove the KOH. The petroleum ether solution was dried under vacuum at 30 °C in a rotary evaporator. The residue was dissolved in methanol–tetrahydrofuran (1:1, v/v), containing 0.1% BHT, until the solution reached a defined volume. Prior to analysis, the solution was centrifuged (14 000 rpm, 5 min).

The increasing demand for natural food ingredients as additives and also their use as supplements have inspired discussions on possible solvents for the extraction of carotenoids. Application of extraction techniques with supercritical carbon dioxide has been widely studied in recent years owing to the various advantages of carbon dioxide as a solvent, such as low toxicity, low cost, and easy separation of the extracted product. Optimizations of the extraction of carotenoids from carrots and also from a marine cyanobacterium by using supercritical carbon dioxide extraction have been reported recently [20, 21]. However, this extraction technique is more suited to obtain carotenoids for further use in other foods rather than for analytical purposes.

15.3.3
Analysis

HPLC procedures are often used to analyze carotenoids in food samples. In addition, some screening methods using spectrophotometric procedures [22] or working with color measurement [23] have been published.

15.3.4
HPLC

Reversed-phase materials and lime are the stationary phases mostly investigated for the separation of carotenoids. However, some interesting separations have also been achieved by using normal-phase columns. Carotenoid geometric isomers were mainly investigated with β-carotene. About 40 years ago, Sweeney and Marsh [24] analyzed isomers of β-carotene to evaluate their provitamin A activity, using a column with calcium hydroxide. They separated three β-carotene isomers, without an exact structural knowledge on the two Z-isomers. More recently, calcium hydroxide was often used to separate geometric isomers of carotenes [25–28]. However, the majority of the carotenoid separations reported in the literature used reversed-phase materials, either C_{18} or C_{30}. For C_{18} columns, polymeric materials (e.g., VYDAC TP columns) were often used to separate carotenoids [29–32]. C_{30} columns were found to give a better resolution, particularly for the separation of geometric isomers [33–35]. Hence C_{30} columns are nowadays state of the art for the HPLC analysis of carotenoids, although in some cases a C_{18} column could be a good choice, depending on the separation needed. Elution was often carried out isocratically with solvent mixtures, avoiding time-consuming re-equilibration of the system. Gradient elution gives a better resolution of different geometric isomers.

A chromatographic separation of 14 carotenoids (11 all-E)-isomers and three (Z)-isomers) by using a C_{30} column eluted with a gradient consisting of methanol and methyl *tert*-butyl ether, using a column temperature of 17°C, is shown in Figure 15.1. To achieve a better separation of the xanthophylls, isocratic elution with methanol–water (97 : 3, v/v) at 21°C was used (Figure 15.2). As internal standard, β-apo-8′-carotenal or echinenone is a good choice. Both are well separated from the most relevant carotenoids present in food samples. The right choice and also control of the column temperature are essential to achieve a good separation of carotenoids. Figure 15.3 shows the separation of six carotenoids in a dietary supplement (capsules) containing an algae extract from *Dunaliella salina*. For HPLC, the same conditions as for Figure 15.1 were used without adding an internal standard.

Laboratories planning to determine carotenoids in food samples by using HPLC with a C_{30} column should be aware that the resolution often changes from column to column even if they are from the same supplier. Poorer separation of several carotenoids or co-elution of some compounds can result when using a different batch of stationary phase. In these cases, changing the column temperature is one possibility to improve the separation [36], followed by modification of the mobile phase gradient if needed.

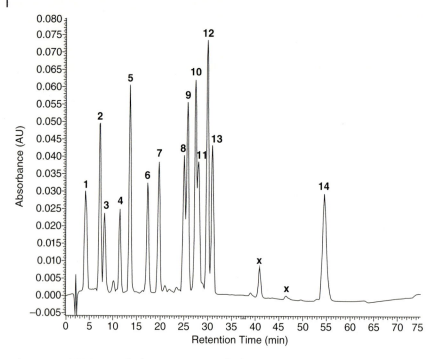

Figure 15.1 HPLC trace (diode-array detector, 450 nm) of 14 carotenoids. C$_{30}$ column, 250 × 4.6 mm i.d., 5 μm (Trentec, Rutesheim, Germany); column temperature, 17 °C; mobile phase, methyl *tert*-butyl ether (solvent A) and methanol (solvent B) using a gradient procedure as follows: (1) initial conditions 10% solvent A and 90% solvent B, (2) a 45 min linear gradient to 55% solvent A, (3) 55% solvent A and 45% solvent B for 15 min, and (4) a 10 min linear gradient to 10% solvent A; flow-rate, 1.3 ml min^{-1}. For further HPLC conditions, see text. Peaks: 1 = (all-*E*)-violaxanthin, 2 = (all-*E*)-lutein, 3 = (all-*E*)-zeaxanthin, 4 = (all-*E*)-canthaxanthin, 5 = (all-*E*)-β-apo-8′-carotenal (internal standard), 6 = (all-*E*)-β-cryptoxanthin, 7 = (all-*E*)-echinenone (internal standard), 8 = (15*Z*)-β-carotene, 9 = (13*Z*)-β-carotene, 10 = (all-*E*)-α-carotene, 11 = (all-*E*)-rubixanthin, 12 = (all-*E*)-β-carotene, 13 = (9*Z*)-β-carotene, x = lycopene *Z*-isomers, 14 = (all-*E*)-lycopene.

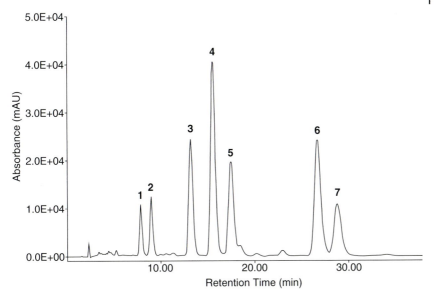

Figure 15.2 HPLC trace (diode-array detector, 450 nm) of xanthophylls. C_{30} column, 250 × 4.6 mm i.d., 5 μm (Trentec); column temperature, 21 °C; mobile phase, methanol–water (97 : 3, v/v); flow-rate, 1.3 ml min^{-1}. For further HPLC conditions, see text. Peaks: 1 = (all-*E*)-violaxanthin, 2 = (all-*E*)-luteoxanthin, 3 = (all-*E*)-capsanthin, 4 = (all-*E*)-lutein, 5 = (all-*E*)-zeaxanthin, 6 = (all-*E*)-β-apo-8′-carotenal (internal standard), 7 = (all-*E*)-canthaxanthin.

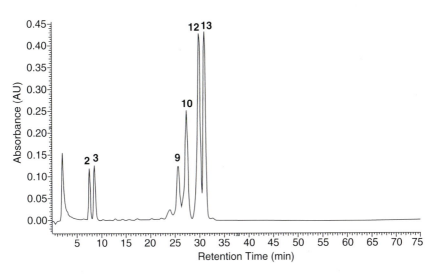

Figure 15.3 HPLC trace (diode-array detector, 450 nm) of carotenoids in a dietary supplement (capsules). C_{30} column, 250 × 4.6 mm i.d., 5 μm (Trentec); column temperature, 17 °C; mobile phase, methyl *tert*-butyl ether (solvent A) and methanol (solvent B) using a gradient procedure as described in Figure 15.1. For further HPLC, conditions see text. For peak identification, see Figure 15.1.

References

1 Cooper, D.A., Elridge, A.L., and Peters, J.C. (1999) Dietary carotenoids and certain cancers, heart disease, and age-related macular degeneration: a review of recent research. *Nutrition Reviews*, **57**, 201–214.

2 Biesalski, H.K., Böhles, H., Esterbauer, H., Fürst, P., Gey, F., Hunsdörfer, K.H., Kasper, H., and Sies, H. (1995) Antioxidative Vitamine in der Prävention. *Deutsches Ärzteblatt*, **18**, 1316–1321.

3 Heinonen, O.P. and Albanes, D. (1994) The effect of vitamin E and beta carotene on the incidence of lung cancer and other cancers in male smokers. *New England Journal of Medicine*, **330**, 1029–1035.

4 Omenn, G.S., Goodman, G.E., Thornquist, M.D., Balmes, J., Cullen, M.R., Glass, A., Keogh, J.P., Meyskens, F.L., Valanis, B., Williams, J.H., Barnhart, S., and Hammar, S. (1996) Effects of a combination of beta carotene and vitamin A on lung cancer and cardiovascular disease. *New England Journal of Medicine*, **334**, 1150–1155.

5 Böhm, V., Puspitasari-Nienaber, N.L., Ferruzzi, M.G., and Schwartz, S.J. (2002) Trolox equivalent antioxidant capacity of different geometrical isomers of α-carotene, β-carotene, lycopene and zeaxanthin. *Journal of Agricultural and Food Chemistry*, **50**, 221–226.

6 Craft, N.E., Brown, E.D., and Smith, J.C. (1988) Effects of storage and handling conditions on concentrations of individual carotenoids, retinol, and tocopherol in plasma. *Clinical Chemistry*, **34**, 44–48.

7 Olmedilla, B., Granado, F., Rojas-Hidalgo, E., and Blanco, I. (1990) A rapid separation of ten carotenoids, three retinoids, alpha-tocopherol and D-alpha-tocopherol acetate by high performance liquid chromatography and its application to serum and vegetable samples. *Journal of Liquid Chromatography and Related Technologies*, **13**, 1455–1483.

8 Hart, D.J. and Scott, K.J. (1995) Development and evaluation of an HPLC method for the analysis of carotenoids in foods, and the measurement of the carotenoid content of vegetables and fruits commonly consumed in the UK. *Food Chemistry*, **54**, 101–111.

9 Schierle, J., Härdi, W., Faccin, N., Bühler, I., and Schüep, W. (1995) Example 8: geometrical isomers of β,β-carotene, in *Carotenoids, Vol. 1 A: Isolation and Analysis* (eds. G. Britton, S. Liaaen-Jensen, and H. Pfander), Birkhäuser, Basel, pp. 265–272.

10 Britton, G., Liaaen-Jensen, S., and Pfander, H. (2004) *Carotenoids Handbook*, Birkhäuser, Basel.

11 Scott, K.J., Finglas, P.M., Seale, R., Hart, D.J., and de Froidment-Görtz, I. (1996) Interlaboratory studies of HPLC procedures for the analysis of carotenoids in foods. *Food Chemistry*, **57**, 85–90.

12 Seybold, C., Fröhlich, K., Bitsch, R., Otto, K., and Böhm, V. (2004) Changes in contents of carotenoids and vitamin E during tomato processing. *Journal of Agricultural and Food Chemistry*, **52**, 7005–7010.

13 Burkhardt, S. and Böhm, V. (2007) Development of a new method for the complete extraction of carotenoids from cereals with special reference to durum wheat (*Triticum durum* Desf.). *Journal of Agricultural and Food Chemistry*, **55**, 8295–8301.

14 Müller, H. (1997) Determination of the carotenoid content in selected vegetables and fruit by HPLC and photodiode array detection. *Zeitschrift für Lebensmitteluntersuchung und -Forschung A*, **204**, 88–94.

15 Riso, P. and Porrini, M. (1997) Determination of carotenoids in vegetable foods and plasma. *International Journal for Vitamin and Nutrition Research*, **67**, 47–54.

16 Breithaupt, D.E. and Bamedi, A. (2002) Carotenoids and carotenoid esters in potatoes (*Solanum tuberosum* L.): new insights into an ancient vegetable.

Journal of Agricultural and Food Chemistry, **50**, 7175–7181.

17 Updike, A.A. and Schwartz, S.J. (2003) Thermal processing of vegetables increases cis isomers of lutein and zeaxanthin. *Journal of Agricultural and Food Chemistry*, **51**, 6184–6190.

18 Cortés, C., Esteve, M.J., Frígola, A., and Torregrosa, F. (2004) Identification and quantification of carotenoids including geometrical isomers in fruit and vegetable juices by liquid chromatography with ultraviolet-diode array detection. *Journal of Agricultural and Food Chemistry*, **52**, 2203–2212.

19 Kimura, M., Rodriguez-Amaya, D.B., and Godoy, H.T. (1990) Assessment of the saponification step in the quantitative determination of carotenoids and provitamins A. *Food Chemistry*, **35**, 187–195.

20 Sun, M. and Temelli, F. (2006) Supercritical carbon dioxide extraction of carotenoids from carrot using canola oil as a continuous co-solvent. *Journal of Supercritical Fluids*, **37**, 397–408.

21 Macías-Sánchez, M.D., Mantell, C., Rodríguez, M., Martínez de la Ossa, E., Lubían, L.M., and Montero, O. (2007) Supercritical fluid extraction of carotenoids and chlorophyll a from *Synechoccus* sp. *Journal of Supercritical Fluids*, **39**, 323–329.

22 Hulshoff, P.J.M., Kosmeijer-Schuil, T., West, C.E., and Hollmann, P.C.H. (2007) Quick screening of maize kernels for provitamin A content. *Journal of Food Composition and Analysis*, **20**, 655–661.

23 Meléndez-Martínez, A.J., Vicario, I.M., and Heredia, F.J. (2007) Rapid assessment of vitamin A activity through objective color measurement for the quality control of orange juices with diverse carotenoids profiles. *Journal of Agricultural and Food Chemistry*, **55**, 2808–2815.

24 Sweeney, J.P. and Marsh, A.C. (1970) Separation of carotene stereoisomers in vegetables. *Journal of AOAC International*, **53**, 937–940.

25 Pettersson, A. and Jonsson, L. (1990) Separation of *cis–trans* isomers of alpha- and beta-carotene by adsorption HPLC and identification with diode array

detection. *Journal of Micronutrient Analysis*, **8**, 23–41.

26 Rodriguez-Amaya, D.B. and Tavares, C.A. (1992) Importance of *cis*-isomer separation in determining provitamin A in tomato and tomato products. *Food Chemistry*, **45**, 297–302.

27 Schmitz, H.H., Schwartz, S.J., and Catignani, G.L. (1994) Resolution and quantitation of the predominant geometric β-carotene isomers present in human serum using normal-phase HPLC. *Journal of Agricultural and Food Chemistry*, **42**, 2746–2750.

28 Schmitz, H.H., Emenhiser, C., and Schwartz, S.J. (1995) HPLC separation of geometric carotene isomers using a calcium hydroxide stationary phase. *Food Chemistry*, **43**, 1212–1218.

29 Schierle, J., Bretzel, W., Bühler, I., Faccin, N., Hess, D., Steiner, K., and Schüep, W. (1997) Content and isomeric ratio of lycopene in food and human blood plasma. *Food Chemistry*, **59**, 459–465.

30 Nyambaka, H. and Ryley, J. (1996) An isocratic reversed-phase HPLC separation of the stereoisomers of the provitamin A carotenoids (α- and β-carotene) in dark green vegetables. *Food Chemistry*, **55**, 63–72.

31 Saleh, M.H. and Tan, B. (1991) Separation and identification of *cis/trans* carotenoid isomers. *Journal of Agricultural and Food Chemistry*, **39**, 1438–1443.

32 Quackenbush, F.W. (1987) Reverse phase HPLC separation of *cis*- and *trans*-carotenoids and its application to β-carotenes in food materials. *Journal of Liquid Chromatography*, **10**, 643–653.

33 Sharpless, K.E., Brown Thomas, J., Sander, L.C., and Wise, S.A. (1996) Liquid chromatographic determination of carotenoids in human serum using an engineered C_{30} and a C_{18} stationary phase. *Journal of Chromatography B*, **678**, 187–195.

34 Emenhiser, C., Simunovic, N., Sander, L.C., and Schwartz, S.J. (1996) Separation of geometrical carotenoid isomers in biological extracts using a polymeric C_{30} column in reversed-phase liquid chromatography. *Journal of*

Agricultural and Food Chemistry, **44**, 3887–3893.

35 Lessin, W.J., Catigani, G.L., and Schwartz, S.J. (1997) Quantification of *cis–trans* isomers of provitamin A carotenoids in fresh and processed fruits and vegetables. *Journal of Agricultural and Food Chemistry*, **45**, 3728–3732.

36 Böhm, V. (2001) Use of column temperature to optimize carotenoid isomer separation on C_{30} high performance liquid chromatography. *Journal of Separation Science*, **24**, 955–959.

16
HPLC Determination of Vitamin E in Fortified Foods

Afaf Kamal-Eldin and Jelena Jastrebova

16.1
Introduction

Vitamin E is a generic name for tocol derivatives having the biological function of α-tocopherol. Four tocopherols and four tocotrienols (Figure 16.1) are generally reported, although other tocol derivatives with limited distribution have also been found in plant sources [1].

The most common natural E vitamers in the diet are α- and γ-tocopherols, which are available from green leaves and vegetables (α-tocopherol) and from oilseeds, vegetable oils, and nuts (both α- and γ-tocopherols). Tocotrienols (α-, β- and γ-) are found in cereal grains and palm oil. The main recognized biochemical effects of the tocopherols relate to their ability to act as antioxidants and to protect the polyunsaturated fatty acids in foods and in cell membranes and other compartments from oxidative damage caused by over-production of free radicals [2, 3]. Other physiological effects of vitamin E such as inhibition of protein kinase-C or smooth muscle cell proliferation [4] might or might not relate to its antioxidant function [3]. Food fortification with vitamin E is performed for two main reasons:

- as a vitamin; to meet the vitamin E requirements of the population
- as an antioxidant; to protect unstable polyunsaturated fatty acids in certain foods against oxidation by free radicals.

The purpose of this chapter is to describe the foods used for vitamin E fortification, the chemical nature and levels of fortificants, and the methods used for their analysis.

16.2
Fortification of Foods with Vitamin E

Dietary reference intakes (DRIs) are used to specify reference values used for planning and assessing nutrient intake for healthy people based on α-tocopherol. Table 16.1 presents the DRIs for vitamin E developed by the US Institute of

Fortification of Foods with Vitamins, First Edition. Edited by Michael Rychlik.
© 2011 Wiley-VCH Verlag GmbH & Co. KGaA. Published 2011 by
Wiley-VCH Verlag GmbH & Co. KGaA.

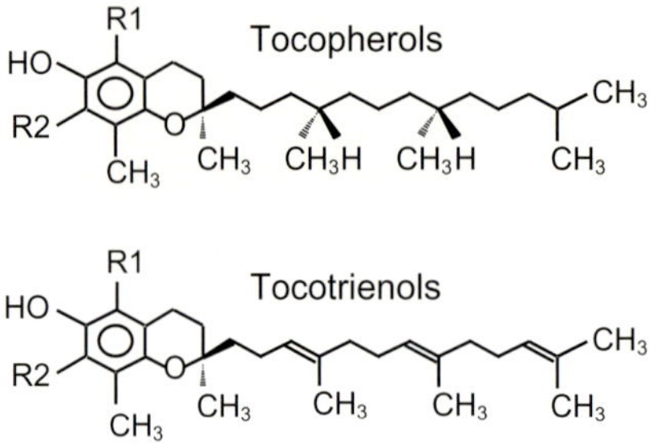

Figure 16.1 Structures of tocopherols and tocotrienols. The different vitamers include α- (R1 = R2 = methyl), β- (R1 = methyl, R2 = hydrogen), γ- (R1 = hydrogen, R2 = methyl), and δ- (R1 = R2 = hydrogen). The systematic names of tocopherols and tocotrienols were devised by the IUPAC–IUB Joint Commission on Biochemical Nomenclature (1982).

Table 16.1 Recommended dietary allowances (RDAs) and tolerable upper intake levels (ULs) for vitamin E for infants, children and adults in mg per day (IU per ay)[a].

Age group	Age	RDA for R,R,R-α-tocopherol (d-α-tocopherol)		UL for d-α-tocopherol
		Males	Females	All
Infants	0–6 months	4 (6)	4 (6)	Not possible to establish
	7–12 months	5 (7.5)	5 (7.5)	
Children	1–3 years	6 (9)	6 (9)	200 (300)
	4–8 years	7 (10.5)	7 (10.5)	300 (450)
	9–13 years	11 (16.5)	11 (16.5)	600 (900)
Adolescents	14–18 years	15 (22.5)	15 (22.5)	800 (1200)
Adults	>19 years	15 (22.5)	15 (22.5)	1000 (1500)
Pregnant women	All ages	–	15 (22.5)	
Breastfeeding women	All ages	–	19 (28.5)	

a) 1 mg α-tocopherol = 1.49 IU.

Medicine [5], including (i) the recommended dietary allowances (RDA), setting the recommended average daily dietary intake level sufficient to meet the nutrient requirements of healthy individuals in each age and gender group, and (ii) the tolerable upper intake levels (ULs), setting the maximum daily intake unlikely to result in adverse health effects.

The average daily intake of α-tocopherol from foods in the United States is estimated to be about 8 mg for men and 6 mg for women [6], which is much lower than the RDA of 15 mg per day of (R,R,R)-α-tocopherol. It is in fact very difficult to meet the 15 mg per day recommendation of α-tocopherol with unsupplemented foods [7]. The United States Department of Agriculture (USDA) list of vitamin E contents in foods is available at http://www.nal.usda.gov/fnic/foodcomp/Data/ SR20/nutrlist/sr20w323.pdf. The RDA of vitamin E can be met only by consumption of supplements or fortified foods. Foods generally fortified with vitamin E include breakfast cereals, milk, some juices, and canned foods. Because of the high relative biopotency of α-tocopherol compared with other E vitamers (Table 16.2), only α-tocopherol is considered when the vitamin E content of foods is calculated.

The tocopherol molecule has three chiral centers in its phytyl tail, making a total of eight (2^3) stereoisomeric forms possible. All naturally occurring tocopherols (α-, β-, γ-, and δ-) have the same molecular configurations (R,R,R) in their phytyl

Table 16.2 Relative vitamin E biopotency of different forms of tocopherols and tocotrienols [8].

Form	Vitamin E activity based on the rat fetal resorption test	
	IU[a]/mg	Relative biopotency (%)
d-α-Tocopherol ($2R,4'R,8'R$)	1.49	100
l-α-Tocopherol ($2S,4'R,8'R$)	0.46	31
α-Tocopherol ($2R,4'R,8'S$)	1.34	90
α-Tocopherol ($2R,4'S,8'S$)	1.09	73
α-Tocopherol ($2R,4'S,8'R$)	0.85	57
α-Tocopherol ($2S,4'S,8'S$)	1.10	60
α-Tocopherol ($2S,4'R,8'S$)	0.55	37
α-Tocopherol ($2S,4'S,8'R$)	0.31	21
all-rac-α-Tocopherol, or dl-α-tocopherol (equi-mixture of R,R,R, S,R,R, R,R,S, R,S,S, R,S,R, S,S,S, S,R,S, S,S,R)	0.74	50
d-β-Tocopherol ($2R,4'R,8'R$)	0.75	50
d-γ-Tocopherol ($2R,4'R,8'R$)	0.15	10
d-δ-Tocopherol ($2R,4'R,8'R$)	0.05	3
d-α-Tocotrienol	0.75	50
d-β-Tocotrienol	0.08	5
d-γ-Tocotrienol	Unknown	Unknown
d-δ-Tocotrienol	Unknown	Unknown

a) The IU for vitamin E is defined as 1 mg of dl-α-tocopheryl acetate.

groups. Synthetic α-tocopherol (all-*rac*-α-tocopherol) is a mixture of approximately equal amounts of the eight stereoisomers [2D,4′D,8′D (*R,R,R*), 2L,4′D,8′D (*S,R,R*), 2D,4′D,8′L (*R,R,S*), 2L,4′D,8′L (*S,R,S*), 2D,4′L,8′D (*R,S,R*), 2L,4′L,8′D (*S,S,R*), 2D,4′L,8′L (*R,S,S*), and 2L,4′L,8′L (*S,S,S*)] [9]. Since the natural tocopherol forms present in foods are the *R,R,R*-isomers, fortificants obtained from natural sources contain only *R,R,R*-forms (also named *d*-tocopherols). These, usually obtained from the deodorizer distillates of vegetable oil refining, are generally mixtures of α- and γ-tocopherols and in the case of soybean oil also δ-tocopherol. Synthetic α-tocopherol, containing the eight isomers (*S,S,R, S,S,S, S,R,S, S,R,R, R,S,S, R,R,S, R,R,R, R,S,R*) labeled all-*rac*-α-tocopherol or *dl*-α-tocopherol, is often found in fortified foods and nutritional supplements as the free phenolic compound or as ester forms including acetate, succinate, nicotinate, and, in limited cases, phosphates. The all-*rac*-α-tocopherol isomeric mixture has lower vitamin E potency (1 IU = 1.32 mg) than the natural and most potent (*R,R,R*)-α-tocopherol (1 IU = 0.67 mg). When tocopherol fortificants are added to food to supply the RDA of vitamin E, they are added in ester forms because these are stable and resistant to oxidation.

Tocopherols are sometimes added to foods to provide antioxidant protection to polyunsaturated fatty acids in foods, for example, functional foods enriched with *n*-3 fatty acids. In this case, the free forms of tocopherols are needed since the ester forms cannot function as antioxidants. Supplementation of foods with *n*-3 fatty acids, for example, fish oils, increase requirements for vitamin E, which can be met by the added tocopherols. Most of these foods contain added tocopherols obtained from deodorizer distillates of vegetable oil, but synthetic tocopherols are also used [10].

Vitamin E supplements containing higher than the recommended intake levels of tocopherols and tocotrienols are also available for individuals wishing to take supplementary levels for possible protection against degenerative diseases including cardiovascular diseases, neurodegenerative diseases, and cancer. In addition, vitamin E supplements may be medically prescribed for premature infants, and for people having anemia, fat absorption disorders including intestinal disorders such as Crohn's disease, liver disorders including biliary cirrhosis and abetalipoproteinemia, cystic fibrosis, and genetic ataxia with vitamin E deficiency (AVED). In these cases, vitamin E supplements containing 100–1000 IU of α-tocopherol are generally consumed per day.

16.3
Experimental Procedures Used to Determine Vitamin E in Fortified Foods by HPLC

Tocopherols in foods can be analyzed by gas chromatography, high-performance liquid chromatography (HPLC), and micellar electrokinetic chromatography (MEKC). Of these methods, HPLC is the mostly used for the analysis of E vitamins as it provides a versatile and reliable technique with high sensitivity, selectivity, reproducibility, and accuracy. The versatility of HPLC is provided by the possible

Table 16.3 Molar extinction coefficients for UV absorption by tocopherols and tocotrienols [11].

Compound	CAS Registry No.	Molecular mass	λ_{max} (nm)	ε_{max} (l mol^{-1} cm^{-1})
α-Tocopherol	59-02-9	430.71	292	3265
β-Tocopherol	148-03-8	416.69	296	3725
γ-Tocopherol	7616-22-0	416.69	298	3809
δ-Tocopherol	119-13-1	402.66	298	3515
α-Tocotrienol	2265-13-4	424.67	292	3652
β-Tocotrienol	49-23-3	410.64	296	3540
γ-Tocotrienol	14101-61-2	410.64	297	3737
δ-Tocotrienol	2561-59-3	396.61	297	3403

use of both normal- and reversed-phase columns together with ultraviolet, fluorescence, electrochemical, and mass spectrometric detectors.

16.3.1
Chemicals

The tocopherol and tocotrienol analytes can be purchased in an isomer kit from Merck (Darmstadt, Germany) and their concentrations can be determined from their molar extinction coefficients (Table 16.3). Chemicals suitable for use as internal standards include tocol [2-methyl-2-(4,8,12-trimethyltridecyl)chroman-6-ol], 5,7-dimethyl tocol, 2,2,5,7,8-pentamethyl-6-hydroxychroman (PMHC) (CAS Registry No. 950-99-2), β-tocopherol, δ-tocopherol, α-tocopherol acetate, and α-tocopherol succinate. The free phenols can be used as standards in different chromatographic systems whereas the ester forms of α-tocopherol are only suitable for the analysis of unfortified foods when saponification will not be performed. Deuterium isotope-labeled forms of α-and γ-tocopherols can also be used as internal standards in stable isotope dilution assays (SIDAs) requiring liquid chromatography–mass spectrometry (LC–MS). Chemicals suitable for use as antioxidants during saponification include ascorbic acid (AA), butylated hydroxytoluene (BHT), and pyrogallol (PG), which can be used separately or combined.

16.3.2
Extraction Procedures

Sample preparation is an important step, which is often time consuming and most critical in HPLC analysis. The extraction procedure for the analysis of E vitamers depends on the nature of sample and the HPLC setup, most specifically the chromatographic mode. When normal-phase HPLC is used, the E vitamers are extracted as part of a lipid extract, which can be dissolved in hexane or the mobile phase and injected directly after suitable dilution. However, when samples are analyzed by reversed-phase HPLC, the whole sample or the lipid extracts obtained therefrom need to be saponified to remove the bulk of lipids, the presence of which is not compatible with the column material.

Depending on the sample type, lipids can be extracted with *n*-hexane by Soxhlet extraction, or with a mixture of hexane or heptane and an alcohol, generally 2-propanol, according to the method of Hara and Radin employing *n*-hexane–2-propanol (3:2, v/v) [12]. For more difficult samples, extraction of lipids is achieved by the Folch method, which employs chloroform–methanol (2:1, v/v) [13] or by the Bligh and Dyer method employing chloroform–methanol–water [14].

The E vitamers can also be extracted from complex samples (e.g., cereals and cereal products, milk, and animal and plant tissue samples) with *n*-hexane or diethyl ether after direct alkaline hydrolysis (saponification) with alcoholic sodium hydroxide or potassium hydroxide (~60%). Saponification is not recommended for samples fortified with tocopherol esters (e.g., acetates or succinates) when the aim is to determine the level of fortification. The E vitamers need to be protected against oxidation by the addition of ample amounts of antioxidants such as ascorbic acid, BHT, pyrogallol, and so on. Despite protection with antioxidants, tocopherol losses are frequently encountered when saponification is used [15]. Saponification is essential, however, for the efficient extraction of tocopherols and tocotrienols from cereal samples [16]. When tocols are easily extracted without saponification, but elimination of bulk lipids is needed before reversed-phase chromatography, solid-phase extraction can be used [17]. In summary, the extraction and sample preparation methods depend on the complexity of the sample matrix [18].

16.3.3
Instrumentation

16.3.3.1 **Columns**
HPLC analysis of tocol derivatives can be performed on normal-phase or reversed-phase columns depending on the samples to be analyzed and the detector used. Normal-phase columns are compatible with ultraviolet (UV)/diode-array detectors (DADs) and fluorescence detectors and are more efficient in the separation of the isomeric pairs β- and γ-tocopherols and β- and γ-tocotrienols, which are not separable by C_{18} reversed-phase columns. Normal-phase columns require longer equilibration times and give less reproducible results than reversed-phase columns. On the other hand, reversed-phase columns are much easier to use and are compatible with the above two detectors and also with electrochemical and electrospray mass spectrometric detectors, and are useful for the separation of samples of animal origin mainly containing α-tocopherol. The disadvantage of using reversed-phase columns is that samples need to be saponified for the elimination of the triacylglycerols and phospholipids. Although saponification is necessary for complex samples, it is not required for the analysis of tocopherols and tocotrienols in lipid extracts, vegetable oils, and oilseeds, for which normal-phase HPLC would be preferable.

Normal-phase HPLC columns used for tocopherol analysis include silica, amino, diol, and cyano columns. The mobile phases are generally based on hexane, but

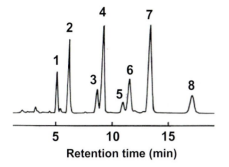

Figure 16.2 Normal-phase HPLC trace of tocopherols and tocotrienols on a silica column with 4% 1,4-dioxane in hexane as mobile phase (2 ml min^{-1}) and with fluorescence detection ($\lambda_{exc.}$ 295 nm, $\lambda_{em.}$ 326 nm). Peaks: (1) α-tocopherol, (2) α-tocotrienol, (3) β- tocopherol, (4) γ-tocopherol, (5) β-tocotrienol, (6) γ-tocotrienol, (7) δ-tocopherol, and (8) δ-tocotrienol.

heptane and isooctane are sometimes used. The elution power of the mobile phase is modulated with the addition of 1–4% of polar solvents, for example, diethyl ether, diisopropyl ether, *tert*-butyl methyl ether, methanol, 2-propanol, and 1,4-dioxane. Separations are performed under isocratic conditions and the columns, especially silica columns, have a very long lifetime. Tocopherols and tocotrienols are separated by adsorption in normal-phase HPLC columns and they elute in order of increasing polarity, mainly depending on the number of methyl substituents in the chromanol head and to a minor extent on the position of the methyl substituents, which induce different steric effects on the phenolic group, influencing its interaction with the polar groups of the stationary phase. In addition, the unsaturation in the side chain also influences the molecular polarity, making the tocotrienols slightly more polar than their corresponding tocopherols. Figure 16.2 shows that the retention of these tocol derivatives on silica columns follows the order α-tocopherol < α-tocotrienol < β-tocopherol < γ-tocopherol < β-tocotrienol < γ-tocotrienol < δ-tocopherol < δ-tocotrienol [19]. The α-, β-, γ-, and δ-tocopherols and 5,7-dimethyltocol were separated on β- and γ-cyclodextrin-bonded silica [20]. Separation of α-tocopherol stereoisomers as methyl ethers can be achieved by chiral HPLC, which separates the eight isomers into five peaks: (i) the four 2S-isomers (S,S,S, S,S,R, S,R,R, S,R,S) together, (ii) R,S,S, (iii) R,R,S, (iv) R,R,R, and (v) R,S,R [21].

Reversed-phase HPLC columns used for tocopherol analysis include C$_8$, C$_1$8, C$_{30}$, and pentafluorophenylsilica (PFP) columns. They are commonly run with aqueous methanol or aqueous acetonitrile and cannot tolerate a heavy lipid load, as discussed above. The more polar tocotrienols elute before less polar tocopherols, each series in the order δ-, followed by β-/γ-, followed by α-. Reversed-phase HPLC columns can also be run under non-aqueous conditions, for example, with acetonitrile–methanol (3:1, v/v). C$_8$ and C$_{18}$ reversed-phase columns are not able to separate the β- and γ-tocopherol and tocotrienol pairs, but separation of all

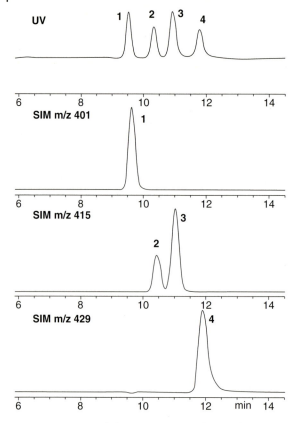

Figure 16.3 Reversed-phase HPLC traces of tocopherols on a pentafluorophenylsilica column with 5% water in methanol as mobile phase (0.5 ml min⁻¹). Detectors: UV (292 nm) and APCI-MS (negative ion mode) – selected ion monitoring (SIM) at m/z 401, 415, and 429. Peaks: (1) δ-tocopherol, (2) β- tocopherol, (3) γ- tocopherol, and (4) α-tocopherol.

tocopherols and tocotrienols and α-tocopherol acetate can be achieved using a C_{30} stationary phase and methanol as mobile phase [22], a PFP column and methanol and water as mobile phase [23] (Figure 16.3), or an octadecylpoly(vinyl alcohol) (ODVA) column [20]. C_{18} reversed-phase HPLC can also be performed on nano-columns of 3 or 5 μm diameter [24].

16.3.3.2 Detectors

Tocol derivatives eluting from HPLC column can be detected using a wide range of detectors, including evaporative light scattering, UV/DAD, fluorescence, electrochemical (EC), NMR, and mass spectrometric detectors [25]. The evaporative light scattering detector is non-selective and has poor sensitivity. Tocols absorb UV light at about 290 nm (Table 16.3) and fluoresce with an excitation wavelength of ~296 nm and an emission wavelength of ~320 nm. The optimal redox potential for the detection of α-tocopherol by electrochemical detectors is around 550 mV. The presence of water in the mobile phase is essential in the case of electrochemical and mass spectrometric detection.

HPLC with UV/DADs is the most commonly used method in the analysis of tocopherols and its sensitivity is good for most samples. The fluorescence detector is about 100 times more sensitive than the UV detector and is widely used with normal-phase columns. Amperometric and coulometric electrochemical detectors, approximately 10–20 times more sensitive than fluorescence detectors, are also the most selective, but their use is limited to reversed-phase HPLC separations because of the need for an aqueous mobile phase to provide sufficient conductivity. Enhancement of electrochemical detector sensitivity can be achieved by adding certain perchlorate or acetate salts.

LC–MS methods reported for tocopherol analysis of foods include the use of normal- and reversed-phase HPLC columns with atmospheric pressure chemical ionization (APCI) (LC–APCI-MS) [26–30] or the use of reversed-phase columns with electrospray ionization (ESI) (LC–ESI-MS) [28, 31]. APCI and ESI can be used in both the positive and negative ion modes using single-quadrupole [26–28], triple-quadrupole [29, 30], or ion-trap [31] mass separation. Ionization in ESI or APCI positive ion modes provide the protonated molecular ions $[M + H]^+$ of tocopherols and also abundant characteristic fragment ions $[m/z$ 165.3 (α-tocopherol), 151.3 (β- and γ-tocopherols), and 137.3 (δ-tocopherol)] [28] due to the elimination of a methyl radical followed by elimination of a hydrocarbon moiety including the phytyl chain by retro-Diels–Alder processes [29]. Minor formation of Na adduct ions occurs in the positive ion mode of ESI but not APCI [28]. In addition to the main $[M + H]^+$ ions and the fragment ions, positive ESI or APCI of tocopherols yields molecular cations $[M]^+$ and $[M - H]^+$ fragments. The latter ion was observed only for α- and β-tocopherols and is probably a result of dehydrogenation of the $[M + H]^+$ ions. Ionization of tocols with ESI can be improved several-fold by adding silver ions, which form adducts with the tocols with minimal fragmentation [32]. On the other hand, negative ion APCI and ESI produce only target deprotonated molecular ions and solvent adduct ions, which results in higher sensitivity compared with positive ionization [28]. APCI analysis of tocopherols provides a wider linearity range and lower detection limits and is less sensitive to differences in the chemical structure of the analytes and solvents [28]. NMR detectors are much less sensitive than UV/DAD and the other more sensitive detectors, but have the advantage of providing structural confirmation [31].

16.4
Conclusions

Several HPLC methods can be used for tocopherol analysis, including normal- and reversed-phase separation followed by UV, fluorescence, electrochemical, and mass spectrometric detection. The method of choice for the analysis of tocol derivatives in fortified foods depends on the type of fortificant (free α-tocopherol, mixtures of α-, γ-, and/or δ-tocopherols extracted from vegetable oils, or synthetic tocopherol esters) and the sample matrix. If the nature of the fortificant is not known, HPLC–NMR can be used to identify the fortificant(s) and decide on a reasonable analytical method. Oils and lipid extracts can be analyzed directly by normal-phase HPLC after dilution in hexane or the mobile phase. After adding an internal standard and a protective antioxidant (BHT, AA, and/or PG), samples fortified with α-tocopherol acetate or other esters can be extracted with methanol without saponification and analyzed by normal-phase HPLC. This method might lead to lower recovery of the tocols. Samples can also be saponified, which provides a higher recovery for total (endogenous + fortified) α-tocopherol.

References

1 IUPAC–IUB Joint Commission on Biochemical Nomenclature (JCBN) (1982) Nomenclature of tocopherols and related compounds: recommendations 1981. *Pure and Applied Chemistry*, **54**, 1507–1510.

2 Kamal-Eldin, A. and Appelqvist, L.Å. (1996) The chemistry and antioxidant properties of the tocopherols and tocotrienols. *Lipids*, **31**, 671–701.

3 Traber, M.G. and Atkinsson, J. (2008) Vitamin E, antioxidant and nothing more. *Free Radical Biology and Medicine*, **43**, 610–618.

4 Zingg, J.-M. and Azzi, A. (2004) Non-antioxidant activities of vitamin E. *Current Medicinal Chemistry*, **11**, 1113–1133.

5 Food and Nutrition Board, Institute of Medicine. Vitamin E (2000) *Dietary Reference Intakes for Vitamin C, Vitamin E, Selenium, and Carotenoids*, National Academies Press, Washington, DC, pp. 186–283.

6 Ahuja, J.K., Goldman, J.D., and Moshfegh, A.J. (2004) Current status of vitamin E nutriture. *Annals of the New York Academy of Sciences*, **1031**, 387–390.

7 Maras, J.E., Bermudez, O.I., Qiao, N., Bakun, P.J., Boody-Alter, E.L., and Tucker, K.L. (2004) Intake of alpha-tocopherol is limited among US adults. *Journal of the American Dietetic Association*, **104**, 567–575.

8 VERIS, the Vitamin E Research and Information Service (1993) *Vitamin Research Summary*, VERIS, LaGrange, IL.

9 Burton, G.W. and Traber, M.G. (1990) Vitamin E: antioxidant activity, biokinetics, and bioavailability. *Annual Review of Nutrition*, **10**, 357–382.

10 Sally, S.F., Schakel, F., and Pettit, J. (2004) Expansion of a nutrient database with the "new" vitamin E. *Journal of Food Composition and Analysis*, **17**, 1371–1378.

11 Eitenmiller, R.E. and Landen, W.O., Jr. (1998) Vitamin E, in *Vitamin Analysis for the Health and Food Sciences*, CRC Press, Boca Raton, FL, pp. 109–148.

12 Hara, A. and Radin, N.S. (1978) Lipid extraction of tissues with a low-toxicity solvent. *Analytical Biochemistry*, **90**, 420–426.

13 Folch, J., Lees, M., and Stanley, G.H.S. (1957) A simple method for the isolation and purification of total lipids from animal tissues. *Journal of Biological Chemistry*, **226**, 497–509.

14 Bligh, E.G. and Dyer, W.J. (1959) A rapid method of total lipid extraction and purification. *Canadian Journal of Biochemistry and Physiology*, **37**, 911–917.

15 Czauderna, M. and Kowalczyk, J. (2007) Alkaline saponification results in decomposition of tocopherols in milk and ovine blood plasma. *Journal of Chromatography B*, **858**, 8–12.

16 Ryynänen, M., Lampi, A.-M., Salo-Väänänen, P., Ollilainen, V., and Piironen, P. (2004) A small-scale sample preparation method with HPLC analysis for determination of tocopherols and tocotrienols in cereals. *Journal of Food Composition and Analysis*, **17**, 749–765.

17 Lechner, M., Reiter, B., and Lorbeer, E. (1999) Determination of tocopherols and sterols in vegetable oils by solid-phase extraction and subsequent capillary gas chromatographic analysis. *Journal of Chromatography A*, **857**, 231–238.

18 Rupérez, F.J., Martín, D., Herrera, E., and Barbas, C. (2001) Chromatographic analysis of α-tocopherol and related compounds in various matrices. *Journal of Chromatography A*, **935**, 45–69.

19 Kamal-Eldin, A., Görgen, S., Pettersson, J., and Lampi, A.-M. (2000) Normal-phase high-performance liquid chromatography of tocopherols and tocotrienols. Comparison of different chromatographic columns. *Journal of Chromatography A*, **881**, 217–227.

20 Abidi, S.L. and Mounts, T.L. (1994) Separations of tocopherols and methylated tocols on cyclodextrin-bonded silica. *Journal of Chromatography A*, **670**, 67–75.

21 Jensen, S.K. and Lauridsen, C. (2007) α-Tocopherol stereoisomers. *Vitamins and Hormones*, **76**, 281–308.

22 Strohschein, S., Pursch, M., Lubda, D., and Albert, D. (1998) Shape selectivity of C_{30} phases for RP-HPLC separation of tocopherol isomers and correlation with MAS NMR data from suspended stationary phase. *Analytical Chemistry*, **70**, 13–18.

23 Richheimer, S.L., Kent, M.C., and Bernart, M.W. (1994) Reversed-phase high-performance liquid chromatographic method using a pentafluorophenyl bonded phase for analysis of tocopherol. *Journal of Chromatography A*, **677**, 75–80.

24 Fanali, S., Camera, E., Chankvetadze, B., D'Orazio, G., and Quaglia, M.G. (2004) Separation of tocopherols by nano-liquid chromatography. *Journal of Pharmaceutical and Biomedical Analysis*, **35**, 331–337.

25 Abidi, S.L. (2000) Chromatographic analysis of tocol-derived lipid antioxidants. *Journal of Chromatography A*, **881**, 197–216.

26 Heudi, O., Trisconi, M.J., and Blake, C.J. (2004) Simultaneous quantification of vitamins A, D-3 and E in fortified infant formulae by liquid chromatography–mass spectrometry. *Journal of Chromatography A*, **1022**, 115–123.

27 Kalman, A., Mujahid, C., Mottier, P., and Heudi, O. (2003) Determination of alpha-tocopherol in infant foods by liquid chromatography combined with atmospheric pressure chemical ionisation mass spectrometry. *Rapid Communications in Mass Spectrometry*, **17**, 723–727.

28 Lanina, S., Toledo, P., Samples, S., Kamal-Eldin, A., and Jastrebova, J. (2007) Comparison of reversed-phase liquid chromatography–mass spectrometry with electrospray and atmospheric pressure chemical ionization techniques for determination of tocopherols in foods. *Journal of Chromatography A*, **1157**, 159–170.

29 Perri, E., Mazzotti, F., Raffaelli, A., and Sindona, G. (2000) High-throughput screening of tocopherols in natural extracts. *Journal of Mass Spectrometry*, **35**, 1360–1361.

30 Stöggl, W.M., Huck, C.W., Scherz, H., Popp, M., and Bonn, G.K. (2001) Analysis of vitamin E in food and phytopharmaceutical preparations by HPLC and HPLC–APCI-MS–MS. *Chromatographia*, **54**, 179–185.

31 Krucker, M., Lienau, A., Putzbach, K., Grynbaum, M.D., Schuler, P., and Albert, K. (2004) Hyphenation of capillary HPLC to microcoil ^{1}H NMR spectroscopy for the determination of tocopherol homologues. *Analytical Chemistry*, **76**, 2623–2628.

32 Rentel, C., Strohschein, S., Albert, K., and Bayer, E. (1998) Silver-plated vitamins: a method of detecting tocopherols and carotenoids in LC/ESI-MS coupling. *Analytical Chemistry*, **70**, 4394–4400.

17
Determination of Vitamin D by LC–MS/MS

Dorit Kern

17.1
Introduction

As early as the sixteenth century, rickets had been diagnosed among the impoverished rural population in the course of industrialization. Today, it is known under the name "rachitis" as a deficiency symptom of an inadequate vitamin D intake. Typical disease patterns (e.g., distorted bones, weakened musculature, poor dental forming, reduced immune defense, and stronger preparedness for spasms of the musculature) were the results of the air pollution in the growing cities, the frequent work in mines – including by children – and the resultant deficiency of sunlight, and ultraviolet (UV) radiation needed to activate vitamin D precursors [1]. Today, it is impossible to imagine primary health care without rachitic prophylaxes with vitamin D supplements for infants. Because a lack of vitamin D in adulthood can lead to osteomalacia [2, 3], elderly people have an increased demand.

The term "vitamin D" describes a number of steroid-like compounds. Of these, only vitamin D_2 and D_3, in the form of bioactive calciferol, are important for the human organism and occur in both vegetable (D_2) and animal foods (D_3). Strictly, the D vitamins do not match the definition of vitamins, because they are formed in the human metabolism, especially in the skin, if the solar radiation is sufficient. Furthermore, calcitriol (1,25-dihydroxycholecalciferol) is the active metabolite, which is considered to be a hormone [4].

High concentrations of vitamin D_3 in food are found only in animal foods such as herring (27 µg per 100 g), salmon (16 µg per 100 g), sardines (11 µg per 100 g), egg (2.9 µg per 100 g), and butter (1.2 µg per 100 g). In contrast, vitamin D_2 is found in vegetable foods, such as mushrooms (1.9 µg per 100 g) [5].

Vitamin D_4 is the saturated form of vitamin D_2 (22,23-dihydroergochalciferol), and is used in pharmaceutical products in exchange for vitamin D_2 or D_3 to treat medically hyperparathyroidism (a dysfunction of the parathyroids) [6], whereas the other forms given in Table 17.1 play a minor role.

In Germany, the fortification of food to fulfill the daily demand for vitamin D is regulated by its equality under the law to the additives and a special admission procedure. Only ergocalciferol (vitamin D_2) and cholecalciferol (vitamin D_3) are

Fortification of Foods with Vitamins, First Edition. Edited by Michael Rychlik.
© 2011 Wiley-VCH Verlag GmbH & Co. KGaA. Published 2011 by
Wiley-VCH Verlag GmbH & Co. KGaA.

Table 17.1   Overview of the classes of vitamin D molecules.

Vitamin D_1	A mixture of ergocalciferol and lumisterol (1:1)
Vitamin D_2	Ergocalciferol (made from ergosterol, or pro-vitamin D_2)
Vitamin D_3	Cholecalciferol (made from 7-dehydrocholesterol, or pro-vitamin D_3
Vitamin D_4	Dihydrotachysterol (vitamin D_2 without the 22,23-double bond)
Vitamin D_5	Sitocalciferol (made from 7-dehydrositosterol)

Source: modified from [7].

allowed to be added to food as vitamin D compounds. For margarine, the maximum concentration of vitamin D is stated as 2.5 µg per 100 g and in infant formulas 3 µg per 100 kcal per ready-to-eat product, corresponding to 100 and 120 international units (IU), respectively. Regulation 1925/2006 of the European Union allows fortification with vitamins in all foods in the future, and maximum levels are currently under discussion in European boards.

The concentration of vitamin D in enriched food follows the demand of the various age groups. In Germany, it has been statistically proven that the vitamin D level in the blood serum of some children is too low [8]. Therefore, the consumption of vitamin D-rich food is necessary and, for this reason, has a preventive character.

For the determination of vitamin D in food, analysis is focused only on vitamin D_2 and D_3, as a result of the above-mentioned legal situation. Therefore, the purpose of this study was to develop a method for simultaneous determination of both vitamin D_2 and D_3 by liquid chromatography–tandem mass spectrometry (LC–MS/MS) with previous saponification and enrichment on a solid-phase extraction (SPE) unit.

17.2
Materials and Methods

17.2.1
Chemicals

Vitamin D_2 (ergocalciferol), vitamin D_3 (cholecalciferol), vitamin D_4 (dihydrotachysterol), and sodium ascorbate were obtained from Fluka (Buchs, Switzerland). Benzoyl and acetyl esters of vitamin D_3 were synthesized by Dr. A. Barthel at the Chair of Organic and Bioorganic Chemistry (Professor R. Csuk) of the Martin-Luther-Universität Halle-Wittenberg, Germany. Methanol, chloroform, formic acid, sodium heptanesulfonate, sodium sulfide, and potassium hydroxide were obtained from Merck (Darmstadt, Germany), and citric acid from Riedel-de Haën (Seelze, Germany). Deionized water was purified using a Millipore system.

17.2.2
Preparation of Standards

Amounts of 10 mg of vitamins D_2, D_3, and D_4 were dissolved in methanol and diluted to 100 ml to give stock standard solutions. Because the vitamins are light sensitive, amber-colored glassware was used.

In case the standard substances were not dissolved completely [9], the concentrations of the stock standard solutions were determined by UV spectrometry using the molar extinction coefficients $\varepsilon = 475 \, l \, mol^{-1} \, cm^{-1}$ for vitamin D_2 and $\varepsilon = 480 \, l \, mol^{-1} \, cm^{-1}$ for vitamin D_3.

The operating range was selected as 5–250 $\mu g \, l^{-1}$ to measure both the low concentrations in margarine and the relatively high concentrations in food supplements or fish oil in one measurement. The concentration of the internal standard vitamin D_4 was 1000 $\mu g \, l^{-1}$. The high concentration of the internal standard was chosen because its signal is less intense than those of the analytes.

17.2.3
Samples and Reference Materials

Dietetic infant formulas, dietetic food for reduction in weight, margarine, fish oil, cereals, dietary supplements, and animal food were analyzed. Some products were also analyzed by the Gesellschaft für Bioanalytik Hamburg mbH using the §64 LFGB reference method [semipreparative high-performance liquid chromatography (HPLC), analytical HPLC with diode-array detection [9]]. Because of the instability of the vitamins, no certified reference material exists in Europe that contains a known vitamin D concentration.

17.2.4
Sample Preparation

17.2.4.1 Saponification
After homogenization of food samples in a blender, 1–10 g of the homogenized samples were weighed in a 500 ml amber-colored two-necked Erlenmeyer flask and dissolved in 30 ml of water. Then, the internal standard vitamin D_4, 1 g of sodium ascorbate, 2 ml of aqueous sodium sulfide (4%), 25 ml of methanolic potassium hydroxide solution [5 mol l^{-1} potassium hydroxide in methanol–water (9 : 1)] and 100 ml of methanol were added successively.

Saponification was performed under reflux at 85 °C for 30 min with nitrogen as inert gas. If the fat content of the sample was higher than 25%, the saponification time was increased to 60 min. Then, 20 ml of methanol were rinsed through the cooler into the flask and the latter was cooled to room temperature. By rinsing with portions of methanol, the sample was quantitatively transferred into a beaker, and the pH was adjusted to 9 using citric acid (1 mol l^{-1}). Subsequently, 5 ml of a methanolic sodium heptanesulfonate solution (0.3 mol l^{-1}) were added. Then, the sample solution was transferred into a 250 ml volumetric flask and diluted to volume with methanol. An aliquot of the prepared sample solution was transferred into a centrifuge tube and centrifuged for 15 min with 4800 rpm. The supernatant liquid was used for the following SPE.

Losses during SPE were minimized by working at an alkaline pH and by the use of an ion-pair reagent [10].

17.2.4.2 Solid-Phase Extraction (SPE)

Liquid–liquid extraction with nonpolar solvents (e.g., hexane or diethyl ether) is used in official methods [9, 11] for the separation of vitamin D from the saponifiable part. These methods are both solvent and time consuming. Therefore, they have been replaced by SPE in many areas in recent years. Another advantage is that SPE can often be automated.

In this work, a Strata-X SPE cartridge based on a polymer from Phenomenex (Aschaffenburg, Germany) was employed. After conditioning with 6 ml of methanol and 6 ml of water adjusted to pH 9 with aqueous sodium hydroxide, the cartridge was loaded with 10–50 ml of the above-prepared solution. The elution rate should not exceed 2 ml min^{-1}.

The analytes were then eluted four times with 3 ml of chloroform and evaporated to dryness under a stream of nitrogen at 40 °C using a Turbovap® (Zymark, Rüsselsheim, Germany). The residue was dissolved in 0.5–1 ml of methanol and filtered through a membrane filter. A washing step was not carried out because the SPE cartridge became clogged on washing with methanolic solutions [10].

17.2.4.3 Equipment

17.2.4.3.1 Liquid Chromatography

An LC 10 series HPLC system (Shimadzu Deutschland, Duisburg, Germany) was used. A 20 µl volume of the sample were injected followed by an isocratic separation on a reversed-phase column (Reprosil-Pur 120 C_{18} AQ, 250 × 4.0 mm i.d., 3 µm, from Trentec Analysentechnik, Gerlingen, Germany) at a flow-rate of 0.6 ml min^{-1} with 0.1% formic acid in methanol as eluent. The temperature of the column oven was 40 °C.

17.2.4.3.2 Mass Spectrometry

LC–MS/MS analysis was carried out using an API 3200 QTrap mass spectrometer (Applied Biosystems, Applera Deutschland, Darmstadt, Germany). For ionization, an atmospheric pressure chemical ionization (APCI) and an electrospray ionization (ESI) source were used alternately. The APCI source was operated in the positive mode using gas pressures of 69, 448, 448, and 34 kPa for curtain gas, gas 1, gas 2, and collision gas, respectively, at a temperature of 250 °C and a nebulizer current of 5 µA. The voltages of the mass spectrometer were as follows: declustering potential, 30 V; entrance potential, 4 V; collision exit potential for quantifier and qualifier, 4 V; collision exit potential for quantifier/qualifier, of vitamin D_2 21.26/21.27 V, of vitamin D_3 22.00/20.93 V, and of vitamin D_4 21.33 V; and collision energy for quantifier/qualifier, of vitamin D_2 37/17 V, of vitamin D_3 19/33 V, and of vitamin D_4 22 V.

17.3
Results and Discussion

17.3.1
Analytical Methods

In recent years, some methods for the determination of vitamin D and its metabolized products were developed for a couple of matrices. Saponification followed by the liquid–liquid extraction with nonpolar solvents is the most commonly used sample preparation method for matrices with a high fat content [9, 11–15]. Because of the low fat content in pharmaceuticals, animal foods, and human serum, saponification is not necessary here. Instead, extraction with supercritical carbon dioxide, extraction with hexane, or simply acidification of the sample is applied [16–20]. The determination of vitamin D is often performed by UV detection at 265 ± 1 nm after a semipreparative step of normal-phase chromatography, in order to remove compounds with similar analytical properties such as sterols or vitamin A and E which are excessively available in food [9, 11, 21]. Further, a cleanup with kieselguhr cartridges or reversed-phase cartridges can additionally be applied [22, 23]. Furthermore, the purification of the sample solutions and the enrichment of vitamin D are necessary because UV detection is not sufficiently specific or sensitive. However, electrochemical detection has been reported to be successfully used for its determination in fish [23]. The use of mass spectrometry to identify unequivocally fat-soluble vitamins in foods, pharmaceuticals, human serum, and cosmetics is continuously increasing, but became important for vitamin D quantitation only in the last few years [24–28]. However, in almost all of the methods reported, vitamin D_2 is used as an internal standard for the determination of vitamin D_3 and provitamin D_2 is used for provitamin D_3, and so on. For the production of foods or supplements, more and more plant extracts, phytosterols, and other novel food ingredients are added. Hence foods or supplements can contain both vitamin D_2 from vegetable ingredients and also vitamin D_3 from animal ingredients. Therefore, another internal standard must be used when vitamin D_2 and D_3 have to be determined simultaneously. Isotope-labeled vitamin standards are not available so far, and custom syntheses for routine analysis are too expensive.

The aim of this study was the development of an LC–MS/MS method with internal standard for the simultaneous determination of vitamin D_2 and D_3 in different foods with various fat contents and different matrices.

17.3.2
Development of the Mass Spectrometry Method

Vitamin D is neither clearly polar nor non-polar, but something in between. Therefore, both ESI and APCI were applied. Before the quantification could be carried out in the multiple reaction monitoring (MRM) mode, definite fragment ions had to be identified and focusing and accelerating voltages within the ion path of the mass spectrometer had to be optimized. Acceptable results were obtained only by positive ionization because of the structure of the analytes. In addition, the signal intensity was four times higher in APCI than in ESI [10]. Fragment ions of the analytes are dependent on the type of mass spectrometer. The most intense fragments for vitamin D_2 were m/z 69.1 and 125.4 when using an API 2000 instrument (Applied Biosystems) [10], whereas the most intense fragments with an API 3200 QTrap were m/z 69.1 and 379.4. In order to increase the reliability of the qualitative results, at least one additional fragment (qualifier) should be determined along with that used for quantification (quantifier). The peak area ratio of these fragments is a criterion for correct quantitation. Substances such as vitamin D show only few fragment ions because the energy of the collision cell is not sufficient to cleave the molecule into a number of fragments. According to this, the molecular peak itself (m/z 397.4) and the signal after elimination of water (m/z 379.4) are the only specific fragments for vitamin D_2 (Figures 17.1 and 17.2).

In the lower mass-to-charge region (below m/z 100), fragment ions are often nonspecific and susceptible to interferences. Only for vitamin D_2 it was decided to use m/z 69.1 for quantification because of its notably higher intensity than m/z 379.4, which was used as a qualifier (Table 17.2).

17.3.3
Development of the Liquid Chromatographic Method

Usually, fat-soluble vitamins are analyzed by liquid chromatography using normal-phase columns and nonpolar solvents, such as hexane, chloroform, or diethyl ether. In this study, we aimed to use a reversed-phase column because of more convenient in-house handling. Several columns with different diameters and different stationary phases were tested along with various polar solvents. The best results for the signal intensity were achieved with a Luna C_{18} column (150 × 2.0 mm i.d., 5 μm, Phenomenex) and a Synergy fusion RP-80 column (50 × 2.0 mm i.d., 4 μm, Phenomenex) with methanol as the mobile phase.

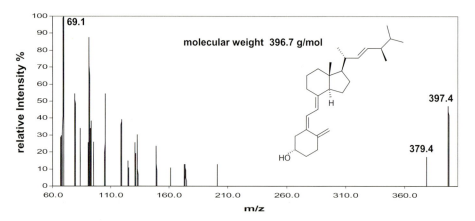

Figure 17.1 Mass spectrum of the protonated molecule of vitamin D_2 in the positive mode.

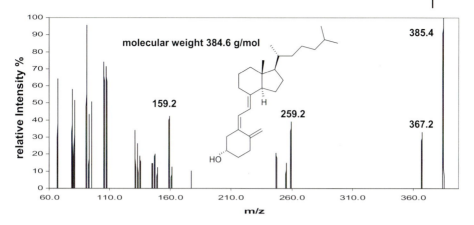

Figure 17.2   Mass spectrum of the protonated molecule of vitamin D₃ in the positive mode.

Table 17.2   Fragments in positive ionization mode from API 3200 QTrap.

Vitamin	Product ion (m/z)	Quantifier (m/z)	Qualifier (m/z)	Dwell time (m)
Vitamin D_2	397.4	69.1	379.4	150
Vitamin D_3	385.4	259.2	159.2	150
Vitamin D_4 (internal standard)	399.4	381.4		150

As the generation of ions in the ion source of the mass spectrometer is highly dependent on the composition of the solvent, various additives (modifiers) were tested. The best results were obtained with formic acid. For the initial analysis of food samples, a mobile phase of methanol, water and formic acid (99% methanol and 1% of a solution with the composition 99.9% water + 0.1% formic acid) at a flow-rate of 0.4 ml min⁻¹ was used [10].

However, the separation of the vitamins was poor for samples with complex matrices, such as cereals and margarine. Despite the high specificity of tandem mass spectrometric detection, there were interferences in the MRM traces. Therefore, a new HPLC method had to be developed with special attention to the separation of analytes and matrices. This could be achieved using a Reprosil Pur 120 C₁₈ AQ column (250 mm × 4 mm i.d., Trentec Analysentechnik) at a flow–rate of 0.6 ml min⁻¹. However, the better separation resulted in a decrease in signal intensity (by a factor of 3) and peak broadening.

To compensate for this decrease in intensity, the mobile phase was modified [methanol–formic acid (99.9 : 0.1, v/v)]. Because of the higher flow-rate, the gas pressure in the ion source was increased from 276 to 448 kPa in order to obtain stable signals. Interestingly, reducing the inner diameter of the Reprosil Pur 120

C_{18} AQ column to 2 mm did not improve the results. The presented method is clearly a compromise in favor of a better separation of the analytes to exclude false-positive results (Figure 17.3).

17.3.4
Choice of Internal Standard

Internal standards are used to compensate for losses of analytes during sample preparation and by matrix effects. Isotopically labeled standards are preferred when working with mass spectrometry as they show nearly the same behavior as the analyte, regarding chromatography and ionization. Currently, there are no commercially available isotopically labeled vitamin standards, hence compounds with a similar structure to vitamin D had to be used.

First, compounds related to sterol were tested, such as cholesterol, cholestane, and β-estradiol. However, none of these substances showed sufficient sensitivity. Therefore, vitamin D acetate ($[M + H]^+$ m/z 425.5 → 259.3 and 367.3) and vitamin D benzoate ($[M + H]^+$ m/z 489.4 → 259.2 and 367.4) were synthesized by Dr. A. Barthel, as mentioned earlier. The advantages of the synthesized chemicals include their similar behavior during chromatography and ionization. However, a major disadvantage is their instability during alkaline saponification. For this reason, vitamin D esters were added after saponification [10].

To monitor the entire sample preparation process, vitamin D_4 was tested. Since it is not allowed to use vitamin D_4 as a food additive, vitamin D_4 is a suitable internal standard for the determination of vitamin D_2 and vitamin D_3.

The mass spectrum of vitamin D_4 showed a protonated molecule peak at m/z 399.4 and intense fragments at m/z 381.4, 273.4, and 95.4. Therefore, vitamin D_4 was used as an internal standard for all measurements in this study (Figure 17.4).

17.3.5
Calibration and Validation

The method for the determination of vitamin D_2 along with vitamin D_3 by LC–MS/MS was linear in the range 5–250 µg l^{-1} (Mandel test of linearity).

The limit of determination (LOD) and limit of quantification (LOQ) were determined by external calibration using an internal standard [29]. For vitamin D_2, the LOD and LOQ were 2.1 and 7.6 µg l^{-1}, respectively, whereas for vitamin D_3 1.9 and 6.9 µg l^{-1}, respectively, were obtained.

To determine the recoveries, samples were spiked with vitamin D. The spiking range was 10–100 µg l^{-1}, depending on the total amount of vitamin D in the sample. The recovery for vitamin D_3 was 71–114% and that of vitamin D_2 was 70–113%.

17.3.6
Analysis of Real Samples and Conclusion

Twenty-nine different foods and one animal feed with indications of vitamin D were examined. The foods chosen represent the most important matrices of today

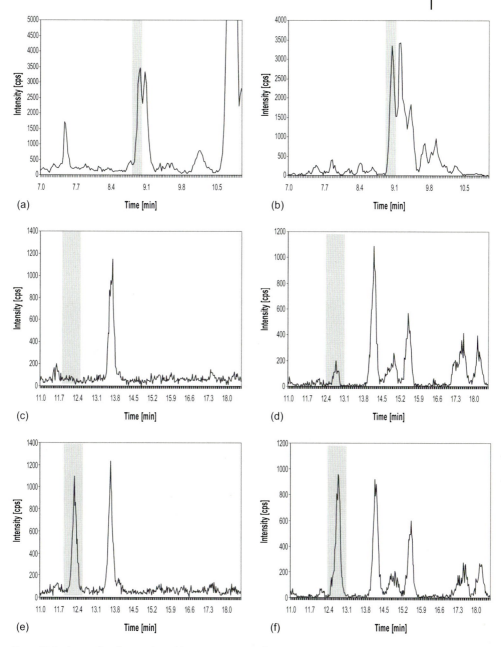

Figure 17.3 A sample of margarine with butterfat, analyzed using LUNA C_{18} (a, b) and Reprosil-Pur 120 C_{18} AQ (c–f) columns. (a) Sample spiked with $50 \mu g \, l^{-1}$ of vitamin D_2 m/z 397.4 → 69.1. (b) Sample spiked with $50 \mu g \, l^{-1}$ of vitamin D_3 m/z 385.4 → 259.2.

(c) Sample Vitamin D_2 m/z 397.4 → 69.1.
(d) Sample vitamin D_3 m/z 385.4 → 259.2.
(e) Sample spiked with $50 \mu g \, l^{-1}$ of vitamin D_2 m/z 397.4 → 69.1. (f) Sample spiked with $50 \mu g \, l^{-1}$ vitamin D_3 m/z 385.4 → 259.2.

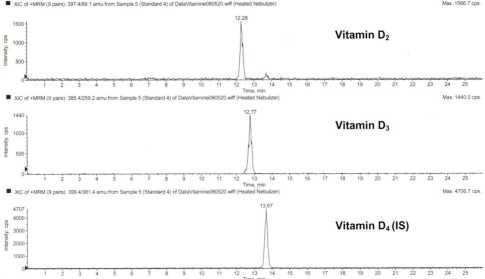

Figure 17.4 MRM of vitamin D_2 and D_3 ($50\,\mu g\,l^{-1}$) and vitamin D_4 as an internal standard ($1000\,\mu g\,l^{-1}$).

and also various fat contents and a wide range of vitamin D additions. The results are presented in Table 17.3.

Vitamin D was always determined as vitamin D, except for one sample of soy margarine, in which vitamin D_2 was determined because it was explicitly highlighted on the label. So far, the labeling "vitamin D" corresponds to vitamin D_3. In the near future, it cannot be ruled out that vitamin D_2 will be found originating from plant extracts or novel food ingredients in supplements or foods enriched with phytosterins, for example.

Comparing the declared and determined values of vitamin D, a deviation of 51–173% was observed. However, this deviation is acceptable when considering the following criteria. The vitamin D content decreases during storage, hence the declared vitamin D value represents the average content during the shelf-life. This value is set up by parameters such as formulation, shelf-life trials, seasonal influences, or analysis. It is the food producer's responsibility to guarantee the average content of vitamin D in the product until the end of the shelf-life.

The coefficient of variation (CV) as an indicator of the analytical deviation of the method is 13% on average, with one CV being 30%. It is generally known that small values represent a large analytical error. Therefore, this error must be taken into consideration during the evaluation of the results.

The German working group "Fragen der Ernährung" ("Questions on Nutrition") suggests a fluctuation range of ±30% for declared vitamin D contents in order to

Table 17.3 Results obtained for vitamin D in foods by LC–MS/MS – comparison in part with the results of the reference HPLC–UV method [9][a].

Sample	Declared (µg per 100 g)	n	HPLC–UV (µg per 100 g)	LC–MS/MS (µg per 100 g)	CV (%)
Dietary supplement 1	71.4 D$_3$	3	71.9	101.4	19
Dietary supplement 2	375.1 D$_3$	2	301	365.0	3
Milk pudding powder 1	6.5 D	2		9.4	16
Milk pudding powder 2	5 D$_3$	2		6.5	23
Milk pudding powder 3	6.6 D	2		7.9	2
Cereal 1	5 D	2	3.9	3.3	2
Cereal 2	5 D	2	2.6	2.6	19
Cereal 3	5 D	2	3.5	5.1	17
Margarine 1	2.5 D	2		2.1	24
Margarine 2	2.5 D	2		2.7	21
Margarine 3	–	1		–	–
Margarine 4	2.5 D	2		1.8	1
Margarine 5	2.3 D	2		2.3	6
Margarine with sterine	7.5 D	2		8.5	9
Soy margarine 1	2.5 D	2		2.3	30
Soy margarine 2	6.7 D$_2$	4		9.5 D$_2$	9
Margarine with butterfat	1.25 D	10	0.6	0.8	11
Fishoil 1	69 D$_3$	2		58.9	3
Fishoil 2	–	2	158	122.7	1
Fishoil 3	–	2	1.4	1.0	22
Animal feed	3.75 D$_3$	2		6.5	15
Slim drink powder 1	4.7 D	2	5.5	5.4	11
Slim drink powder 2	2,4 D	2		2.3	9
Food for special medical purposes 1	8.1 D	4		7.9	18
Food for special medical purposes 2	6.6 D	2		5.1	6
Infant formula 1	7.6 D	4		10.1	22
Infant formula 2	7.6 D	6	7.9	6.7	13
Infant formula 3	9.7 D	2		9.9	2
Infant formula 4	9.3 D	4	7.0	6.8	11
Infant formula 5	9.3 D	4	11.3	6.8	11

a) Declared, declared content of vitamin D; *n*, number of independent repetitions; vitamin D, determined mean content of vitamin D$_3$; CV, coefficient of variation.

take into account natural and technological fluctuations [30]. Regarding the two aspects above, the variation for vitamin D should be at least ±30% for all foods.

Because of changing consumer habits and adaptation of European legislation, the enrichment of foods with vitamin D will become even more important in the future. With the presented LC–MS/MS method, vitamins D$_2$ and D$_3$ can be determined routinely in all important matrices. Therefore, the labeling of vitamin D can be controlled.

Acknowledgments

Special thanks are due to Udo Hartmann for the fundamental method development and to Gabriela Jacob for technical support with sample preparation. Dr. A. Barthel at the Chair of Organic and Bioorganic Chemistry (Prof. R. Csuk) of the Martin-Luther-Universität Halle-Wittenberg, the Gesellschaft für Bioanalytik Hamburg mbH, Phenomenex and Trentec Analysentechnik are also acknowledged for their fruitful cooperation. Dr. Petra Uchida and Kerstin Mägel are thanked for linguistic corrections.

References

1 Holick, M.F. (2006) Resurrection of vitamin D deficiency and rickets. *Journal of Clinical Investigation*, **116** (8), 2062–2072.

2 Kraenzlin, M. (2003) Osteomalazie. *Swiss Medical Forum*, **32/33**, 754–763.

3 Domke, A., Grossklaus, R., Niemann, B., Przyrembel, H., Richter, K., Schmidt, E., Weissenborn, A., Wörner, B., and Ziegenhagen, R. (2004) Verwendung von Vitaminen in Lebensmitteln. Available at http://www.bfr.bund.de/cm/238/verwendung_von_vitaminen_in_lebensmitteln.pdf (accessed 31 May 2008).

4 Biesalski, H.K., Köhrle, J., and Schümann, K. (2002) *Vitamine, Spurenelemente und Mineralstoffe*, Georg Thieme, Stuttgart.

5 Souci, S.W., Fachmann, H., and Kraut, H. (2002) *Food Composition and Nutrition Tables*, 6th edn., Medpharm Scientific, Stuttgart.

6 European Directorate for the Quality of Medicines (2006) Pharmacopoea Europaea Ph. Eur. 5.3, *Dihydrotachysterol*, 2014, European Directorate for the Quality of Medicines (EDQM), Strasbourg.

7 Byrdwell, W.C. (2006) LC-MS of Vitamin D. Available at http://www.LCMSofVitaminD.com (accessed 25 May 2008).

8 Hintzpeter, B., Scheidt-Nave, C., and Mensink, G.B.M. (2007) Vitamin D Status von Kindern und Jugendlichen in Deutschland. Poster presented at the *44th Wissenschaftlichen DGE-Tagung*, Halle/Saale, Germany.

9 Bundesamt für Verbraucherschutz und Lebensmittelsicherheit (2001) Bestimmung von Vitamin D in Lebensmitteln mittels HPLC, §64 LFGB-Methode, Bundesamt für Verbraucherschutz und Lebensmittelsicherheit, Braunschweig.

10 Hartmann, U. (2007) Erstellung einer Methode zur Bestimmung von Vitamin D in Lebensmitteln mittels LC–MS/MS, Diploma thesis, Martin-Luther Universität Halle-Wittenberg.

11 Comité Européen de Normalisation (2000) BS EN 12821. Foodstuffs. Determination of Vitamin D by High Performance Liquid Chromatography. Measurement of Cholecalciferol (D3) and Ergocalciferol (D2), Comité Européen de Normalisation (CEN), Brussels.

12 Thompson, J.N. and Plouffe, L. (1993) Determination of cholecalciferol in meat and fat from livestock fed normal and excessive quantities of vitamin D. *Food Chemistry*, **46** (3), 313–318.

13 Mattila, P.H., Piironen, V.I., Uusi Rauva, E.J., and Koivistoinen, P.E. (1995) Contents of cholecalciferol, ergocalciferol, and their 25-hydroxylated metabolites in milk products and raw meat and liver as determined by HPLC. *Journal of Agricultural and Food Chemistry*, **43** (9), 2394–2399.

14 Faulkner, H., Hussein, A., Foran, M., and Szijarto, L. (2000) A survey of vitamin A and D contents of fortified fluid milk in Ontario. *Journal of Dairy Science*, **83** (6), 1210–1216.

15 Staffas, A. and Nyman, A. (2003) Determination of cholecalciferol (vitamin D_3) in selected foods by liquid chromatography: NMKL collaborative study. *Journal of AOAC International*, **86** (2), 400–406.

16 Qian, H. and Sheng, M. (1998) Simultaneous determination of fat-soluble vitamins A, D and E and pro-vitamin D_2 in animal feeds by one-step extraction and high-performance liquid chromatography analysis. *Journal of Chromatography A*, **825** (2), 127–133.

17 Gámiz-Gracia, L., Jiménez-Carmona, M.M., and Luque de Castro, M.D. (2000) Determination of vitamins D_2 and D_3 in pharmaceuticals by supercritical-fluid extraction and HPLC separation with UV detection. *Chromatographia*, **51** (7–8), 428–432.

18 Iwase, H. (2000) Determination of vitamin D_2 in emulsified nutritional supplements by solid-phase extraction and column-switching high-performance liquid chromatography with UV detection. *Journal of Chromatography A*, **881** (1–2), 189–196.

19 Luque-Garcia, J.L. and Luque de Castro, M.D. (2001) Extraction of fat-soluble vitamins. *Journal of Chromatography A*, **935** (1–2), 3–11.

20 Bustamante-Rangel, M., Delgado-Zamarreno, M.M., Sanchez-Perez, A., and Carabias-Martinez, R. (2006) Microemulsion electrokinetic chromatography for the separation of retinol, cholecalciferol, delta-tocopherol and alpha-tocopherol. *Journal of Chromatography A*, **1125** (2), 270–273.

21 Gomis, D.B., Fernandez, M.P., and Gutierrez Alvarez, M.D. (2000) Simultaneous determination of fat-soluble vitamins and provitamins in milk by microcolumn liquid chromatography. *Journal of Chromatography A*, **891** (1), 109–114.

22 Sliva, M.G. and Sanders, J.K. (1996) Vitamin D in infant formula and enteral products by liquid chromatography: collaborative study. *Journal of AOAC International*, **79** (1), 73–80.

23 Ostermeyer, U. and Schmidt, T. (2005) Vitamin D and provitamin D in fish. *European Food Research and Technology*, **222** (3–4), 403–413.

24 Careri, M., Lugari, M.T., Mangia, A., Manini, P., and Spagnoli, S. (1995) Identification of vitamins A, D and E by particle beam liquid chromatography–mass spectrometry. *Fresenius' Journal of Analytical Chemistry*, **351** (8), 768–776.

25 Josephs, J.L. (1999) Analysis of Vitamin D_3 in Feed Extract by MS^3. Thermo Finnigan Application Report, pp. 1–8.

26 Lopez, L.L. and Goodley, P.C. (2003) Rapid Determination of Vitamin D_3 in Poultry Feed Supplements. Application Note, Food and Flavours, Agilent Technologies, pp. 1–4.

27 Heudi, O., Trisconi, M.-J., and Blake, C.-J. (2004) Simultaneous quantification of vitamins A, D_3 and E in fortified infant formulae by liquid chromatography–mass spectrometry. *Journal of Chromatography A*, **1022** (1–2), 115–123.

28 Murao, N., Ohishi, N., Nabuchi, Y., Ishigai, M., Kawanishi, T., and Aso, Y. (2005) The determination of 2beta-(3-hydroxypropoxy)-1alpha,25-dihydroxy vitamin D_3 (ED-71) in human serum by high-performance liquid chromatography–electrospray tandem mass spectrometry. *Journal of Chromatography B*, **823** (2), 61–68.

29 DIN (1994) 32645. Chemical Analysis, Decision Limit, Detection Limit and Determination Limit, Estimation in Case of Repeatability, Terms, Methods, Evaluation. Beuth Verlag, Berlin.

30 Arbeitsgruppe "Fragen der Ernährung" der Lebensmittelchemischen Gesellschaft (1998) Empfehlungen zu Toleranzen für Nährstoffschwankungen bei der Nährwertkennzeichnung. *Lebensmittelchemie*, **52** (25), 132–133.

18
Quantitation of Vitamin K in Foods

Sameh Ahmed, Naoya Kishikawa, Kaname Ohyama, and Naotaka Kuroda

18.1
Introduction

Vitamin K is the family name for a series of fat-soluble compounds that have a common 2-methyl-1,4-naphthoquinone nucleus but differ in the structure of the side chain at the 3-position. There are two major forms of vitamin K in Nature (Figure 18.1): vitamin K_1 (phylloquinone, PK) that have a phytyl side chain and the vitamin K_2 series (menaquinones, MKs) that have a side chain with repeated isoprenoid units. The length of the isoprenoid side chain in MKs is defined by its carbon number or the number of isoprenoid units. The major dietary form of vitamin K has been considered to be PK, which occurs in green and leafy vegetables. In contrast, MKs are found in fermented foods such as natto (fermented soyabeans) and in the colon, where they are synthesized by the intestinal microflora [1]. PK exists naturally only in the *trans* form, and the all-*trans* configuration is also the most common one for MKs. *Cis–trans* isomers, which are formed by exposure to ultraviolet (UV) light or in the synthetic production of vitamin K, are considered to have low bioactivity [2]. Menadione, formerly known as vitamin K_3, is a synthetic form and may be regarded as a provitamin because vertebrates can convert it to MK-4 by adding a 4-prenyl side chain at the 3-position. Further, hydrogenated PK-rich vegetable oils are used widely in the food industry because of their physical characteristics and oxidative stability. During the commercial hydrogenation of PK-rich oils, there is some conversion of PK to 2′,3′-dihydrophylloquinone (dihydro-PK) [3].

18.2
Biological Role of Vitamin K

The biological role of vitamin K is to act as a cofactor for an enzyme that converts specific glutamyl residues in several proteins such as plasma clotting factors II (prothrombin), VII, IX, and X, protein C, and protein S and also non-coagulation proteins such as osteocalcin to γ-carboxyglutamyl residues. These

Fortification of Foods with Vitamins, First Edition. Edited by Michael Rychlik.
© 2011 Wiley-VCH Verlag GmbH & Co. KGaA. Published 2011 by
Wiley-VCH Verlag GmbH & Co. KGaA.

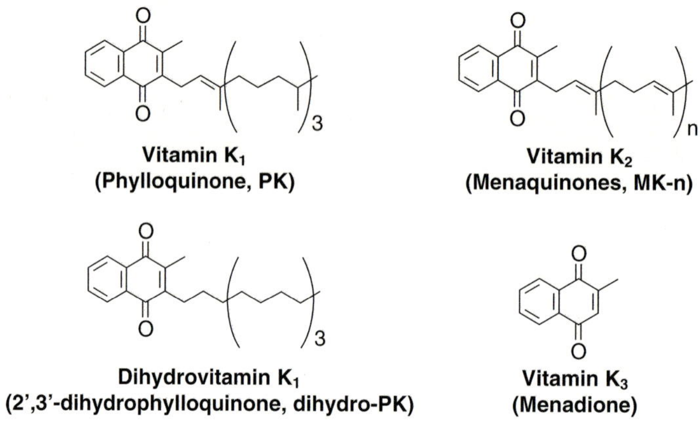

Vitamin K₁
(Phylloquinone, PK)

Vitamin K₂
(Menaquinones, MK-n)

Dihydrovitamin K₁
(2',3'-dihydrophylloquinone, dihydro-PK)

Vitamin K₃
(Menadione)

Figure 18.1 Structures of different forms of vitamin K.

vitamin K-dependent proteins play crucial roles in homeostasis and calcification [1, 4]. There is evidence for the role of PK as a protective dietary factor against hip fracture in the elderly. It is assumed that the beneficial effects of PK on bone are mediated through the carboxylation of γ-carboxyglutamic acid residues of vitamin K-dependent proteins in bone, including osteocalcin, matrix γ-carboxyglutamic acid protein, and protein S [5]. In addition, bone resorption is inhibited by MK-4 through targeting osteoclasts to undergo programmed cell death [6]. A number of studies have claimed beneficial results using synthetic MK-4 in the treatment of osteoporosis [7, 8].

In a metabolic study with human participants, dihydro-PK was less effective in carboxylating glutamic acid residues of the hepatic vitamin K-dependent protein prothrombin. Dihydro-PK also had no measurable effect on biochemical markers of bone turnover [9].

18.3
Dietary Vitamin K Sources

Most of the published studies of vitamin K have concentrated on plant foods, which naturally contain only phylloquinone. On the other hand, during food processing, other vitamin K compounds can be formed, including dihydro-PK in hydrogenated oils and MK-4 in fermented soybean products. A recommended dietary allowance (RDA) for vitamin K ranging from 10 µg/day for infants aged 6 months to 80 µg/day for adults was reported [10]. Booth *et al.* [11] tabulated the PK contents of 261 foods using high-performance liquid chromatography (HPLC) as the method of analysis [12]. Representative data from this survey are presented in Table 18.1. The foods selected for analysis were from theTotal Diet Study (TDS) of the US Food and Drug Administration (FDA), which lists the main foods in the American food supply. Based on their PK content, it was concluded that green

Table 18.1 Phylloquinone content of some foods from the US FDA TDS.

Food	Phylloquinone content (mean ± SD) (µg per 100g)	Food	Phylloquinone content (mean ± SD) (µg per 100g)
Vegetables			
Spinach, fresh frozen, boiled	360 ± 70	Cabbage, fresh, boiled	98 ± 10
Collards, fresh frozen, boiled	440 ± 85	Asparagus, fresh frozen, boiled	80 ± 2.9
Broccoli, fresh frozen, boiled	113 ± 2.5	Cauliflower, fresh frozen, boiled	20 ± 4.6
Tomato, raw	3 ± 0.3	Corn, fresh frozen, boiled	0.3 ± 0.2
Tomato, juice, bottled	2.3 ± 0.1	Squash, fresh frozen, boiled	4.4 ± 0.6
Green beans, fresh frozen, boiled	16 ± 6.6	Iceberg lettuce, raw	31 ± 8.6
Brussels sprouts, fresh frozen, boiled	289 ± 55	Green pepper, raw	2.5 ± 0.2
Fruits			
Orange, raw	<0.01	Grapefruit, raw	<0.01
Apple, raw	1.8 ± 0.09	Pear, raw	4.9 ± 0.5
Apricot, raw	3.3 ± 0.4	Strawberry, raw	1.5 ± 0.3
Avocado, raw	14 ± 0.7	Orange juice, from frozen concentrate	<0.01
Banana, raw	0.2 ± 0.02	Grape juice, from frozen concentrate	0.4 ± 0.04
Milk and cheese			
Whole milk, fluid	0.3 ± 0.02	Cheddar cheese	2.1 ± 0.2
Plain yogurt, low fat	0.1 ± 0.01	Swiss cheese	2.8 ± 0.6

(Continued)

Table 18.1 (*Continued*)

Food	Phylloquinone content (mean ± SD) (µg per 100 g)	Food	Phylloquinone content (mean ± SD) (µg per 100 g)
Meat and fish			
Beef steak, pan cooked	1.8 ± 0.2	Fried chicken	24.5 ± 0.3
Pork bacon, pan cooked	0.1 ± 0.05	Fish sticks, frozen, heated	6.8 ± 0.2
Pork roast, baked	<0.01	Tuna, canned in oil, drained	24 ± 1.2
Beverages			
Tea, from tea bag	0.08 ± 0.02	Beer	<0.01
Cola, carbonated beverage	0.02 ± 0.01	Whiskey	<0.01
Others			
White sauce, home-made	6.9 ± 1.1	Milk-baked infant formula	13 ± 0.6
Mayonnaise, regular, bottled	41 ± 1.2	Rice cereal, strained/junior	0.3 ± 0.02
Potato chips	15 ± 3.8	White bread	1.9 ± 0.1
Tomato catsup (ketchup)	3.6 ± 0.4	Olive/safflower oil	28 ± 1.0

Reproduced from reference [11] with permission.

leafy vegetables were the primary dietary source of vitamin K in the form of PK. This has been confirmed in dietary surveys and observational studies [13, 14], as the highest concentrations of PK are found in cabbage, broccoli, Brussels sprouts, spinach, and collards, with values ranging between 98 and 440 µg of PK per 100 g. PK contents of five different vegetables grown at two different locations were compared and significant differences were observed. It was suggested that possible reasons for variations are differences in climate, soil, and growing conditions [15]. In addition, the PK content of green leafy vegetables increases during plant maturation.

Animal products (meat, fish, milk, and eggs) contain low concentrations of PK, whereas appreciable amounts of MKs are present in liver [16].

Some vegetable oils, including canola, soybean, and olive oils, are rich sources of PK, whereas peanut oil and maize oil contain low concentrations of PK. Soybean oil is the most common vegetable oil in the American diet [17]. The addition of vegetable oils rich in PK during food cooking modifies poor sources of vitamin K to important dietary sources. Also, margarines, mayonnaises, sauces, and regular-calorie salad dressings that are derived from vegetable oils come second to green leafy vegetables in their PK content. The addition of these fats and oils to mixed dishes and desserts has an important impact on the amount of vitamin K in the American diet. Infant formulas are supplemented with PK and, therefore, contain significantly more vitamin K than human milk [10].

MKs seem to be less distributed than PK in the diet. In Western diets, significant amounts of MKs were found in animal livers and fermented foods. MK-rich foods are those with a bacterial fermentation stage. The Japanese food natto (fermented soyabeans) has an MK content even higher than that of PK in green leafy vegetables [18].

During hydrogenation of vegetable oils, a percentage of PK is converted to dihydro-PK, the amount of which increases with higher levels of hydrogenation, whereas the amount of PK decreases [3]. Dihydro-PK was quantified in 261 prepared foods with a high fat content among a total of 261 foods from the FDA TDS. Of these foods, only 36 contained dihydro-PK [19].

18.4
Stability of Vitamin K

Vitamin K_1 is fairly stable to most food processing and food preparation procedures. It is unstable to light and strongly alkaline conditions. The instability of vitamin K to alkalinity prevents the use of saponification for sample extraction [2]. Ferland and Sadowski studied the effect of heating and light exposure on the PK content of vegetable oils [17]. This study demonstrated that PK was slightly unstable during heating and the content decreased by 11% after heating at 185–190 °C for 40 min. PK is also sensitive to both fluorescent light and sunlight. After 2 days of exposure to fluorescent light and sunlight, the PK content of rapeseed and safflower oils decreased by 46–59 and 87–94%, respectively. Therefore, it is necessary to work in subdued light when foods are being analyzed. Amber-colored glass bottles were effective in protecting PK in oils from the destructive effects of light [17].

Studies concerned with the effects of freezing, heat processing, and γ-irradiation indicated that there is no decrease in PK content during deep-freezing, cooking, or γ-irradiation [20]. The PK content of commercially available vegetable products in cans or glass containers, and dried or deep-frozen products, did not differ significantly from the content of fresh vegetables [10].

Hydrogenation is a common process used in the food industry to increase the oxidative stability of polyunsaturated oils and to convert liquid oils into semi-solid fats. Dihydro-PK is usually formed during the hydrogenation of vegetable oils [3]. The biological activity of vitamin K_1 was significantly decreased by hydrogenation [9].

18.5
Bioavailability of Vitamin K from Foods

The bioavailability of vitamin K from foods varies widely, depending on the source of the vitamin and the amount of fat in the meal. It has been estimated that the efficiency of absorption of PK from boiled spinach is not greater than 10%, compared with 80% when PK is given in its free form [21]. The absorption of PK was six times higher after the ingestion of a PK tablet compared with after the ingestion of raw spinach containing the same amount of PK. The poor bioavailability of vitamin K in green leafy vegetables was explained by its location in chloroplasts and the tight binding of PK to the thylakoid membranes of chloroplasts [22]. The free PK in vegetable oils, margarines, and dairy products is well absorbed, owing to the transporting effect of fat. The PK from a PK-fortified oil was better absorbed compared with that from an equivalent amount from cooked broccoli, regardless of the triglyceride concentrations [23]. In comparison, the bioavailability of PK from spinach enriched with butter was more than double that of PK from spinach [24]. The bioavailability of dihydro-PK that is produced by the commercial hydrogenation of vegetable oils is lower than that of PK [9, 19].

18.6
Dietary Vitamin K Deficiency

Vitamin K deficiency is uncommon in healthy humans because of the wide distribution of PK in foods and the synthesis of MKs in the gut by the intestinal microflora. Vitamin K deficiency was reported only in neonates and in patients with low dietary intakes of vitamin K who are also receiving antibiotics [4]. The risk of serious vitamin K deficiency is greatest among newborn infants, because their vitamin K stores are low, their gut does not produce MKs, and the vitamin K content in human milk is low. Neonatal and infantile vitamin K deficiency causes melena neonatorum and intracranial hemorrhagic disorders. This deficiency is prevented in many countries by a prophylactic dose of vitamin K at birth [25].

The prothrombin time, the classical measure of vitamin K deficiency, offers a relatively insensitive insight into vitamin K nutritional status and the detection of subclinical vitamin K-deficient states. A more sensitive measure of vitamin K sufficiency can be obtained from tests that detect undercarboxylated species of vitamin K-dependent proteins such as under-γ-carboxylated prothrombin (PIVK-II) and under-γ-carboxylated osteocalcin (% ucOC). In the case of vitamin K deficiency, PIVK-II is released from the liver into the blood and its level increases with the degree of severity of vitamin K deficiency. The measurement of PIVKA-II is the most useful and sensitive homeostatic marker of subclinical vitamin K deficiency [26]. In the same way, vitamin K deficiency in bone causes the osteoblasts to secrete ucOC into the bloodstream. The concentration of circulating ucOC reflects vitamin K deficiency in bone tissue [27]. Most assays for ucOC are indirect and depend on the difference in the affinity of carboxylated and undercarboxylated forms to hydroxyapatite or barium sulfate, so it is difficult to interpret [28]. Other markers for vitamin K deficiency are plasma measurements of vitamin K homologs [29] and the measurement of urinary γ-carboxyglutamate excretion [27].

18.7
Analytical Methods for the Determination of Vitamin K in Foods

The determination of vitamin K in foods presents severe challenges to most of conventional methods due to the complexity of the matrices and the low vitamin K contents in foods. Therefore, the methods used for their determination in food must be of high sensitivity and selectivity. The methods used for the determination of vitamin K in foods can be classified into biological, spectroscopic, and chromatographic methods. However, only HPLC methods are nowadays regarded as reliable techniques for the determination of vitamin K in foods [30].

18.7.1
Official and Regulatory Methods

The *Official Methods of Analysis* of AOAC International [31] provide a gas chromatography (GC) procedure with flame ionization detection for analyzing menadione sodium bisulfate in food premixes. For the analysis of PK in milk and infant formulas, AOAC International suggests an HPLC method with fluorescence (FL) detection. The method includes digestion with lipase from *Candida cylindracea*, extraction of the digest with hexane, reversed-phase chromatography, and fluorescent detection of the analytes after postcolumn reduction with zinc. The procedure shows high reproducibility with a relative standard deviation (RSD) of 6.53%.

The European Committee for Standardization [32] presented the procedure reported by Woollard *et al.* [33] as the official method for the determination of vitamin K in food. The method involved the resolution of vitamin K on a C_{30}

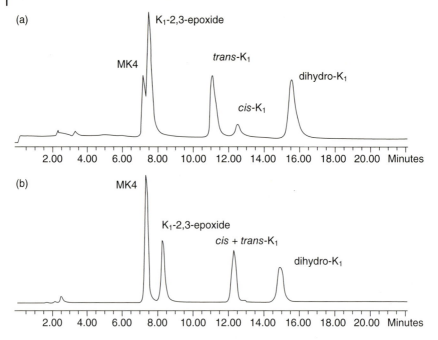

Figure 18.2 Liquid chromatograms of vitamin K standards on (a) a C_{30} 3 μm YMC column and (b) a C_{18} column. Reproduced with permission from reference [33].

column followed by postcolumn reduction and FL detection. The resolution of *cis* and *trans* isomers of PK in foods was not possible on a C_{18} column. Recently, a C_{30} column provided high efficiency in the separation of vitamin K isomers. Figure 18.2 shows the comparative resolution of vitamin K standards on C_{30} and C_{18} columns. The ability to resolve the *cis* and *trans* isomers is necessary to assess accurately the true levels of the biologically active *trans* isomer of PK. In addition, quantification of MKs makes the method valuable for general food analysis.

18.7.2
Recent Methods for the Determination of Vitamin K in Foods

18.7.2.1 Biological Methods
Until the beginning of the 1980s, bioassays with chicks were the only practical methods to determine the vitamin K content of foods [2]. In this test, chicks first receive a vitamin K-free diet to deplete their body stores. The chicks are then divided into two groups and fed with diets supplemented with a graded series of known amounts of vitamin K to establish a response curve or to determine the unknown vitamin K content of a test sample. After 18–24 h or 2 week test periods, the clotting times or prothrombin times of the animals are compared. Many variables affect the results in the chick bioassay, such as test animal, differences in food intakes between test groups, the length of the test period and the standard

used for comparison. In addition, the low sensitivity of bioassay methods limits their use.

18.7.2.2 Spectroscopic Methods

Excellent spectroscopic methods exist for the determination of vitamin K in foods using flow injection analysis (FIA) with spectrofluorimetry [34, 35]. FIA methods are sensitive, reproducible, usually simple, and provide high sample throughput.

Perez-Ruiz *et al.* reported an FIA method for the determination of PK in foods after extraction with a mixture of dichloromethane and isooctane [34]. The method was based on the photosensitization reaction of PK with glucose to generate hydrogen peroxide. Hydrogen peroxide reacted with Fe(II) to generate hydroxyl radicals and these radicals were further scavenged with benzoic acid to generate fluorescent hydroxybenzoic acid. The generated hydroxybenzoic acid was measures using FL detection. Another method based on FIA with FL detection for the analysis of PK in fruit and vegetables [35] involved an on-line photochemical reaction to reduce non-fluorescent PK to a fluorescent hydroquinone form. Vegetables and fruits were extracted with hexane and an automated solid-phase extraction (SPE) on silica gel cartridges was used to purify the samples. However, chromatographic methods are still the most convenient approach for the routine analysis of vitamin K in foods.

18.7.2.3 Chromatographic Methods

18.7.2.3.1 GC Methods

Although methods based on GC are available for vitamin K analysis in foods, they are unpopular, because of the long retention times and the possible degradation of vitamin K on the column since its low volatility requires high temperatures (300 °C). Despite these problems, GC was used for analysis of PK in green vegetables after hexane extraction and purification on alumina columns [36]. Recently, a GC method with flame ionization detection was used for the analysis of vitamin K in green tea leaves after a solid-phase micro-extraction of the sample [37]. Moreover, after the introduction of new technologies such as GC couples with mass spectrometry (MS) and special fused-silica columns, the determination of vitamin K with GC has attracted more interest [38, 39].

18.7.2.3.2 HPLC Methods

Nowadays, HPLC is the most popular method for the routine analysis of vitamin K in foods. Its introduction facilitated the analysis of vitamin K in complex matrices by decreasing analytical variations and increasing resolution [40]. The versatility of HPLC analysis originated from the ability to use various stationary phases and detection systems. Other advantages of HPLC methods are the absence of the risk for thermal degradation and protection against light during the chromatographic run. Several HPLC methods for determining the presence of naturally occurring vitamin K, especially for PK, have been developed, more recent examples of which are given in Table 18.2.

Table 18.2 Some recent HPLC methods used for the determination of vitamin K in foods.

Sample matrix Analyte[a)]	Extraction/purification	Column	Mobile phase	Detection	Ref.
Infant formulas and milk PK	Extraction with hexane after enzymatic hydrolysis. Purification with normal-phase semipreparative LC	Resolve C_{18} column (100×8 mm i.d., 5μm)	Methanol–2-propanol–ethyl acetate–water ($450:350:145:135$)	UV detection at 269 and 277 nm	[41]
Olive oil and chard PK, dihydro-PK (IS)	Extraction with hexane and purification with a semipreparative μ-Porasil column	C_{18} μ-Bondapak column (300×3.9 mm i.d., 10μm)	Acetonitrile–dichloromethane–methanol ($60:20:20$)	UV detection at 247 nm	[42]
Infant formula PK	Addition of NH_4OH, methanol, then extraction with dichloromethane–isooctanol ($2:1$) and purification over a silica layer	Apex I, silica column (250×4.6 mm i.d., 5μm)	Dichloromethane–isooctanol ($30:70$) containing 0.02% 2-propanol	UV detection at 254 nm	[43]
Various foods PK, dihydro-PK (IS)	After initial extraction in 2-propanol and hexane, the food extracts were purified by SPE on silica gel.	Hypersil ODS column (150×4.6 mm i.d., 3μm)	Methanol–dichloromethane ($90:10$) containing 10 mM zinc chloride, 5 mM sodium acetate, and glacial acetic acid	FL detection at λ_{ex} 244 nm and λ_{em} 418 nm after postcolumn reduction with zinc	[11]
Vegetables and edible oils PK, dihydro-PK IS	Extraction was performed with 2-propanol–hexane ($3:2$) and the crude lipid extracts were purified by SPE on silica	Hypersil ODS column (250×4.6 mm i.d., 3μm)	Methanol–dichloromethane ($80:20$) containing 10 mM zinc chloride, 5 mM sodium acetate, and glacial acetic acid	FL detection at λ_{ex} 248 nm and λ_{em} 418 nm and reduction with zinc reactor.	[17]

Sample/Analytes	Extraction	Column	Mobile phase	Detection	Reference
Vegetables PK, MK-6 (IS)	Homogenization by means of an Ultra-Turrax mixer then extraction with hexane	Hypersil MOS C_{18} column (100 × 30 mm i.d., 5 μm)	Methanol–water (92.5 : 7.5) containing 0.05 M sodium perchlorate	FL detection at λ_{ex} 320 nm and λ_{em} 430 nm after postcolumn electrochemical reduction	[20]
Various foods PK isomers, MK-4, dihydro-PK	Lipase digestion followed by single extraction into hexane	YMC C_{30} column (250 × 4.6 mm i.d., 3 μm)	Methanol–dichloromethane (920 : 80) containing 0.41 g sodium acetate, 1.37 g zinc chloride, and 0.3 g glacial acetic acid	FL detection at λ_{ex} 243 nm and λ_{em} 430 nm after postcolumn reduction with zinc	[33]
Margarine and margarine-like products PK isomers, dihydro-PK	Enzymatic digestion with lipase, then extraction with hexane, or the sample after homogenization was filtered and extracted directly with hexane	YMC C_{30} column (250 × 4.6 mm i.d., 3 μm)	Methanol–dichloromethane (90 : 10) containing 5 mM sodium acetate, 10 mM zinc chloride, and 5 mM glacial acetic acid	FL detection at λ_{ex} 243 nm and λ_{em} 430 nm after post-column reduction with zinc	[44]
Human milk PK, MK-5 (IS)	Sonication with albumin, sodium taurocholate, CaCl$_2$, and NaCl, then enzymatic hydrolysis followed by extraction with hexane and purification on an adsorption column	Rosil C_{18} HL column (150 × 3.2 mm i.d., 5 μm)	Methanol–ethyl acetate (96 : 4) containing tetramethylammonium octahydridotriborate	FL detection at λ_{ex} 325 nm and λ_{em} 430 nm after online thermally induced postcolumn reduction	[40]

(Continued)

Table 18.2 (*Continued*)

Sample matrix Analyte[a]	Extraction/purification	Column	Mobile phase	Detection	Ref.
Fats and oils PK, dihydro-PK (IS)	Extraction with hexane, then samples were purified by SPE on silica	BDS Hypersil C_{18} column (150 × 3 mm i.d., 5 μm)	Gradient elution with methanol containing 5 mM sodium acetate, 10 mM zinc chloride, and 5 mM glacial acetic acid	FL detection at λ_{ex} 244 nm and λ_{em} 430 nm after post-column reduction	[45]
Emulsified nutritional supplements PK	The nutrient was dissolved in aqueous sodium sulfate containing 1 mM EDTA. The solution was applied to a BondElut C_{18} cartridge followed by elution with ethanol	Intersil ODS-2 column (150 × 4.6 mm i.d., 5 μm)	Methanol–ethanol (50:50)	FL detection at λ_{ex} 254 nm and λ_{em} 400 nm with postcolumn reduction with RC-10 platinum oxide catalyst	[46]
Various foods PK, dihydro-PK (IS)	Extraction with hexane, then liquid–liquid sample cleanup with methanol–water (9:1)	Hypersil ODS column (250 × 4.6 mm i.d., 5 μm)	Methanol–dichloromethane (90:10) containing 10 mM zinc chloride, 5 mM sodium acetate, and 5 mM acetic acid	FL detection at λ_{ex} 243 nm and λ_{em} 430 nm using zinc postcolumn reductor	[47]
Various foods PK, dihydro-PK	Extraction with hexane, then samples were purified by SPE on silica	BDS Hypersil ODS column (150 × 3 mm i.d., 5 μm)	Gradient elution using (A) methanol containing 10 mM zinc chloride, 5 mM sodium acetate, and 5 mM acetic acid–(B) dichloromethane	FL detection at λ_{ex} 244 nm and λ_{em} 430 nm using zinc postcolumn reductor	[48]

Food / Analytes	Sample preparation	Column	Mobile phase	Detection	Reference
Various foods PK, dihydro-PK (IS)	Extraction with hexane, then liquid–liquid sample cleanup	Hypersil ODS column (250 × 4.6 mm i.d., 5 μm)	Methanol–dichloromethane (90:10) containing 10 mM zinc chloride, 5 mM sodium acetate, and 5 mM acetic acid	FL detection at λ_{ex} 243 nm and λ_{em} 430 nm after postcolumn derivatization	[49]
Milk and infant formula PK, MK-4, dihydro-PK	Lipase digestion followed by the extraction with hexane	Resolve C$_{18}$ column (100 × 8 mm i.d., 5 μm)	Methanol–dichloromethane (90:10) containing 5 mM sodium acetate, 10 mM zinc chloride, and 5 mM glacial acetic acid	FL detection at λ_{ex} 243 nm and λ_{em} 430 nm after postcolumn reduction with zinc	[50]
Oil, margarine, butter, animal products PK, MK-4 (IS)	Lipase hydrolysis for fat content, then extraction with hexane–2-propanol (1:1) and purification by semipreparative LC	Vydac 205TP54 column (250 × 4.6 mm i.d., 5 μm)	Methanol–sodium acetate (25 mM, pH 3) (95:5)	EC detection	[51]
Human milk PK	Alkaline hydrolysis for fatty materials, then extraction with hexane	OD-224 C$_{18}$ reversed-phase column (220 × 4.6 mm i.d., 10 μm)	Methanol–water (99:1) containing 2.5 mM sodium acetate–acetic acid	EC detection	[52]
Vegetables PK	Sonication with methanol and centrifugation, then saponification of the extract with sodium carbonate followed by extraction with hexane	LiChrosorb RP-8 column (250 × 4.6 mm i.d., 5 μm)	Methanol (100%)	HPLC with particle beam MS detection with electron ionization and negative chemical ionization modes	[53]

a) IS, internal standard.

Extraction and Purification Procedures for Vitamin K from Foods The selection of convenient extraction and purification procedures is necessary for the accurate quantitation of vitamin K in foods. Although different procedures have been reported, the choice of the procedures depends mainly on the food matrices. Due to instability of vitamin K in alkaline conditions, saponification, which is generally used in the extraction of fat-soluble vitamins, is not suitable for vitamin K analysis. However, mild alkaline hydrolysis has been used successfully [53]. The extraction of vitamin K from foods is usually carried out with common lipid extraction methods with nonpolar solvents such as hexane [47]. Due to the tight bonding of PK to membranes, vortexing with some mechanical apparatus is necessary for efficient extraction [20]. In recent studies, the use of a mixture of polar and non-polar solvents was reported. The most common solvent mixtures were of 2-propanol and hexane [11, 51]. The use of hexane with methanol, ethanol, and acetone has also been reported for extraction from various animal products [54].

Furthermore, removal of lipid components is necessary after the initial extraction of the lipid fraction containing vitamin K components. The crude extracts usually cannot be used directly in HPLC analysis due to the presence of high molecular weight lipids in foods. A purification step was necessary to remove these interferences and also to increase the column lifetime. Several methods have been reported for the removal of these lipid components, including enzymatic hydrolysis, SPE, semipreparative liquid chromatography (LC) and liquid–liquid cleanup. These methods are used either alone or in combination.

Enzymatic hydrolysis by lipase is usually used as the initial step in the extraction of vitamin K from oils, milk, and infant formulas in order to hydrolyze the triglycerides before extraction with a nonpolar solvent. Generally, lipase powder and a buffer such as phosphate buffer are added to the liquid sample, which is then incubated at 37°C for 90–120 min. Subsequently, vitamin K is extracted with hexane [33, 40, 41, 44, 50, 51].

Nowadays, SPE is the most common purification method when analyzing vitamin K in various food matrices. Silica cartridges are highly efficient for the separation of the interfering lipid components in foods. After applying the crude extract to the cartridge, vitamin K is eluted with a polar solvent [11, 17, 45, 48].

Semipreparative LC has been used also in combination with lipase. It is a practical method that removes different classes of lipids from sample extracts [41, 42, 51]. In this method, the fraction containing vitamin K is isolated on the column, detected with a UV detector and collected, then the concentrated fraction is used for analysis. Silica columns have been the most frequently used and many interfering compounds were eliminated.

Liquid–liquid cleanup has also been used for the purification of the crude extract [47, 49]. The hexane extract was purified by using methanol–water (9:1). Although simple, this method still is not sufficient for the removal of all interfering compounds.

Selection of the Internal Standard Since extensive extraction and purification steps are used for vitamin K processing in foods, it is necessary to use an internal standard for its quantitation. An efficient internal standard should have a quinone structure with similar lipophilicity to vitamin K. The most commonly used internal standards in the quantitation of vitamin K in foods are dihydro-PK, MK-5, MK-6, MK-7, and PK-epoxide [20, 40, 47]. Dihydro-PK has been reported to be a product of the hydrogenation of edible oils [19]. Therefore, its use as an internal standard should be limited to foods where dihydro-PK is not present. Also, some MKs are present in food of animal origin, so dihydro-PK is the most preferred in such cases. Recently, vitamin K analogs with a saturated alkyl side chain were synthesized and used as internal standards for the quantitation of vitamin K in human plasma [29, 55]. These internal standards appear to be the most promising compounds for vitamin K analysis in different matrices. In addition, stable isotopically labeled analogs of vitamin K are presented in Chapter 1.

18.7.2.3.3 Detection Methods

Several methods have been used for the detection of PK and MKs, including UV and FL spectrometry, electrochemical (EC) detection, and MS. Among these, FL detection after chemical reduction to the corresponding hydroquinones is the most popular.

UV Detection The UV absorbance characteristics of vitamin K are relatively poor, with four peaks in its UV spectrum. Although the most intense UV absorbance is at 248 nm, this wavelength is not very selective. Therefore, other wavelengths, such as 270 nm, are generally used to obtain better selectivity [41–43]. However, the use of UV absorbance for the determination of vitamin K contents in food samples is hindered by its poor sensitivity and selectivity.

FL Detection FL detection is the most popular method for the detection of vitamin K in various food matrices owing to its high sensitivity and selectivity. Because vitamin K does not exhibit natural fluorescence, several methods have been developed to reduce the non-fluorescent quinone form into the fluorescent hydroquinone form. These reduction methods include chemical [11, 17, 32, 40–46, 50], photochemical [35], and electrochemical reactions [20]. The chemical reduction of vitamin K can be achieved in many ways. The most common method is reduction on a solid-phase reactor in which either zinc or platinum oxide is used as catalyst. When zinc particles are used, zinc chloride, sodium acetate, and acetic acid have to be added to the mobile phase, which is usually 10–20% dichloromethane in methanol. On the other hand, when platinum oxide is used as catalyst in the solid-phase reactor, the reduction system does not need any reagents in the mobile phase [46]. When thermally induced postcolumn reduction was applied [40], tetramethylammonium octahydridotriborate was used as the reductant while the reduction reaction was induced thermally. Electrochemical reduction before FL detection has also been reported [20]. However, the chemical reduction method is preferred over the electrochemical method because of the high reduction efficiency and the simplified instrumental setup.

EC Detection   Although EC detection is regarded as a simple and reproducible detection method for vitamin K analysis, it has not become very popular in food analysis. Vitamin K is usually analyzed by an EC detector in the reductive mode. However, in this mode, dissolved oxygen causes a high background. Therefore, complicated degassing instruments are required. Due to these problems with the reductive mode, a dual-electrode detection system was developed. In this system, vitamin K is first reduced to its hydroquinone form by a negative potential of the first electrode. Subsequently, the hydroquinone form is oxidized back to the quinone form by the second electrode. A further drawback of EC detection is that the sensitivity of detection can easily be reduced to an undetectable level by the absorption of co-eluted compounds on the surfaces of the electrodes. Another disadvantage is that the sample has to be dissolved in a semi-aqueous solvent, in which the required electrolyte is dissolved [51, 52].

MS Detection   There have been few reports on the application of LC–MS methodology to the analysis of vitamin K in food. Careri *et al.* applied MS detection to the identification and quantitation of vitamin K in vegetables [53]. Particle beam mass spectra were obtained in the electron ionization and chemical ionization modes. LC–MS assays were performed with negative-ion chemical ionization by monitoring the molecular ion of PK at m/z 450. In Chapter 19, the MS behavior of vitamin K is described and demonstrates that LC–MS is a promising technique.

References

1 Shearer, M.J. (1995) Vitamin K. *Lancet*, **345**, 229–234.

2 Parrish, D.B. (1980) Determination of vitamin K in foods: a review. *Critical Reviews Food Science and Nutrition*, **13**, 337–352.

3 Davidson, K.W., Booth, S.L., Dolnikowski, G.G., and Sadowski, J.A. (1996) The conversion of phylloquinone to 2′,3′-dihydrophylloquinone during hydrogenation of vegetable oils. *Journal of Agricultural and Food Chemistry*, **44**, 980–983.

4 Suttie, J.W. (1992) Vitamin K and human nutrition. *Journal of the American Dietetic Association*, **92**, 585–590.

5 Feskanich, D., Weber, P., Willett, W.C., Rockett, H., Booth, S.L., and Colditz, G.A. (1999) Vitamin K intake and hip fractures in women: a prospective study. *American Journal of Clinical Nutrition*, **69**, 74–79.

6 Akiyama, Y., Hara, K., Tajima, T., Murota, S., and Morita, I. (1994) Effect of vitamin K_2 (menatetrenone) on osteoclast-like cell formation in mouse bone marrow cultures. *European Journal of Pharmacology*, **263**, 181–185.

7 Shiraki, N., Shiraki, Y., Aoki, C., and Miura, M. (2000) Vitamin K_2 (menatetrenone) effectively prevents fractures and sustains lumbar bone mineral density in osteoporosis. *Journal of Bone and Mineral Research*, **15**, 515–522.

8 Yamaguchi, Y., Taguchi, H., Gao, I.A., and Tsukamoto, Y. (1999) Effect of vitamin K_2 (menaquinone-7) in fermented soyabean (natto) on bone loss in overiectomized rats. *Journal of Bone and Mineral Metabolism*, **17**, 23–29.

9 Booth, S.L., Lichtenstein, A.H., O'Brien-Morse, M., McKeown, N.M., Wood, R.J., and Saltzman, E. (2001) Effects of a hydrogenated form of

vitamin K on bone formation and resorption. *American Journal of Clinical Nutrition*, **74**, 783–790.

10 National Research Council (1989) *Recommended Dietary Allowances*, 10th edn., National Academies Press, Washington, DC.

11 Booth, S.L., Sadowski, J.A., and Pennington, J.A.T. (1995) Phylloquinone (vitamin K_1) content of foods in the U.S. Food and Drug Administration's Total Diet Study. *Journal of Agricultural Food Chemistry*, **43**, 1574–1579.

12 Booth, S.L., Davidson, K.W., and Sadowski, J.A. (1994) Evaluation of an HPLC method for the determination of phylloquinone (vitamin K_1) in various food matrices. *Journal of Agricultural Food Chemistry*, **42**, 295–300.

13 Booth, S.L., Pennington, J.A.T., and Sadowski, J.A. (1996) Food sources and dietary intakes of vitamin K-1 (phylloquinone) in the American diet: data from the FDA Total Diet Study. *Journal of the American Dietetic Association*, **96**, 149–154.

14 Booth, S.L. and Suttie, J.W. (1998) Dietary intake and adequacy of vitamin K. *Journal of Nutrition*, **128**, 785–788.

15 Ferland, G., and Sadowski, J.A. (1992) Vitamin K_1 (phylloquinone) content of green vegetables: effects of plant maturation and geographical growth location. *Journal of Agricultural Food Chemistry*, **40**, 1874–1877.

16 Terhi, J.K., Velimatti, O., and Vieno, I.P. (2000) Determination of phylloquinone and menaquinones in animal products with fluorescence detection after postcolumn reduction with metallic zinc. *Journal of Agricultural Food Chemistry*, **48**, 6325–6331.

17 Ferland, G. and Sadowski, J.A. (1992) Vitamin K_1 (phylloquinone) content of edible oils: effects of heating and light exposure. *Journal of Agricultural Food Chemistry*, **40**, 1869–1873.

18 Ikeda, Y., Iki, M., and Morita, A. (2006) Intake of fermented soybeans, natto, is associated with reduced bone loss in postmenopausal women: Japanese Population-Based Osteoporosis (JPOS) Study. *Journal of Nutrition*, **136**, 1323–1328.

19 Booth, S.L., Davidson, K.W., Lichtenstein, A.H., and Sadowski, J.A. (1996) Plasma concentrations of dihydro-vitamin K_1 following dietary intake of a hydrogenated vitamin K_1-rich vegetable oil. *Lipids*, **31**, 709–712.

20 Langenberg, J.P., Tjaden, U.R., De Vogel, E.M., and Langerak, D.I. (1986) Determination of phylloquinone (vitamin K_1) in raw and processed vegetables using reversed phase HPLC with electrofluorometric detection. *Acta Alimentaria*, **15**, 187–198.

21 Shearer, M.J., McBurney, A., and Barkhan, P. (1974) Studies on the absorption and metabolism of phylloquinone (vitamin K_1) in man. *Vitamins and Hormones*, **32**, 513–542.

22 Garber, A.K., Binkley, N.C., Krueger, D.C., and Suttie, J.W. (1999) Comparison of phylloquinone bioavailability from food sources or a supplement in human subjects. *Journal of Nutrition*, **129**, 1201–1203.

23 Booth, S.L., Lichtenstein, A.H., and Dallal, G.E. (2002) Phylloquinone absorption from phylloquinone-fortified oil is greater than from a vegetable in younger and older men and women. *Journal of Nutrition*, **132**, 2609–2612.

24 Gijsbers, B., Jie, K.-S.G., and Vermeer, C. (1996) Effect of food composition on vitamin K absorption in human volunteers. *British Journal of Nutrition*, **76**, 223–229.

25 Manji, K.P. and Azzopardi, D. (1999) Intracranial haemorrhage due to vitamin K deficiency following gastroenteritis in an infant. *Journal of Tropical Pediatrics*, **45**, 105–106.

26 Widdershoven, J., Munster, P.V., Abreu, R., Bosman, H., Lith, T.M., Meyel, M.P., Motohara, K., and Matsuda, I. (1987) Four methods compared for measuring des-carboxyprothrombin (PIVKA-II). *Clinical Chemistry*, **33**, 2074–2078.

27 Vermeer, C., Jie, K.-S., and Knapen, M.H.J. (1995) Role of vitamin K in bone metabolism. *Annual Review of Nutrition*, **15**, 1–22.

28 Gundberg, C.M. (1998) Vitamin K status and bone health: an analysis of methods for determination of undercarboxylated osteocalcin. *Journal of Clinical*

Endocrinology and Metabolism, **83**, 258–266.

29 Ahmed, S., Kishikawa, N., Nakashima, K., and Kuroda, N. (2007) Determination of vitamin K homologues by high-performance liquid chromatography with on-line photoreactor and peroxyoxalate chemiluminescence detection. *Analytica Chimica Acta*, **591**, 148–154.

30 Kamao, M., Tsugawa, N., Suhara, Y., and Okano, T. (2007) Determination of fat-soluble vitamins in human plasma, breast milk and food samples: application in nutrition survey for establishment of Dietary Reference Intakes for Japanese. *Journal of Health Science*, **53**, 257–262.

31 AOAC International (2005) *Official Methods of Analysis*, 18th edn., AOAC International, Arlington, VA.

32 Comité Européen de Normalisation, Technical Committee (2003) CEN/TC275. Foodstuffs. Determination of Vitamin K by HPLC–EN14148, Comité Européen de Normalisation (CEN), Brussels.

33 Woollard, D.C., Indyk, H.E., Fong, B.Y., and Cook, K.K. (2002) Determination of vitamin K_1 isomers in foods by liquid chromatography with C_{30} bonded-phase column. *Journal of AOAC International*, **85**, 682–687.

34 Perez-Ruiz, T., Martinez-Lozano, C., Tomas, V., and Martin, J. (1999) Flow-injection fluorimetric determination of vitamin K_1 based on a photochemical reaction. *Talanta*, **500**, 49–56.

35 Perez-Ruiz, T., Martinez-Lozano, C., Martin, J., and Garia, M.D. (2006) Automatic determination of phylloquinone in vegetables and fruits using online photochemical reduction and fluorescence detection via solid phase extraction and flow injection. *Analytical and Bioanalalytical Chemistry*, **384**, 280–285.

36 Seifert, R.M. (1979) Analysis of vitamin K_1 in some green leafy vegetables by gas chromatography. *Journal of Agricultural Food Chemistry*, **27**, 1301–1304.

37 Reto, M., Figueira, M.E., Filipe, H.M., and Almeida, M. (2006) Analysis of vitamin K in green tea leaves and infusions by SPME–GC–FID. *Food Chemistry*, **100**, 405–411.

38 Dolnikowski, G., Sun, Z., Grusak, M.A., Peterson, J.W., and Booth, S.L. (2002) HPLC and GC/MS determination of deuterated vitamin K (phylloquinone) in human serum after ingestion of deuterium-labeled broccoli. *Journal of Nutritional Biochemistry*, **13**, 168–174.

39 Osman, A. and Hannestad, U. (2008) A possible ethanol-catalyzed rearrangement of vitamin K(1) detected by gas chromatography/mass spectrometry. *Rapid Communications in Mass Spectrometry*, **22**, 3861–3866.

40 Lambert, W.E., Vanneste, L., and De Leenheer, A.P. (1992) Enzymatic sample hydrolysis and HPLC in a study of phylloquinone concentration in human milk. *Clinical Chemistry*, **38**, 1743–1748.

41 Indyk, H.E., Littlejohn, V.C., Lawrence, R.J., and Woollard, D.C. (1995) Liquid chromatographic determination of vitamin K_1 in infant formulas and milk. *Journal of AOAC International*, **78**, 719–723.

42 Otles, S. and Cagindi, O. (2007) Determination of vitamin K_1 content in olive oil, chard and human plasma by RP-HPLC method with UV–Vis detection. *Food Chemistry*, **100**, 1220–1222.

43 Tanner, J.T., Barnett, S.A., and Mountford, M.K. (1993) Analysis of milk-based infant formula. Phase IV. Iodine, linoleic acid, and vitamins D and K. U.S. Food and Drug Administration–Infant Formula Council: collaborative study. *Journal of AOAC International*, **76**, 1042–1056.

44 Cook, K.K., Mitchell, G.V., Grundel, E., and Rader, J.I. (1999) HPLC analysis for *trans*-vitamin K_1 and dihydro-vitamin K_1 in margarines and margarine-like products using the C30 stationary phase. *Food Chemstry*, **67**, 79–88.

45 Peterson, J.W., Muzzey, K.L., Haytowitz, D., Exler, J., Lemar, L., and Booth, S.L. (2002) Phylloquinone (vitamin K_1) and dihydrophylloquinone content of fats and oils. *JAOCS*, **79**, 641–646.

46 Iwase, H. (2000) Determination of vitamin K_1 in emulsified nutritional supplements by solid-phase extraction

and high performance liquid chromatography with post column reduction on a platinum catalyst and fluorescence detection. *Journal of Chromatography A*, **881**, 261–266.

47 Jakob, E. and Elmadfa, I. (1996) Application of a simplified HPLC assay for the determination of phylloquinone (vitamin K_1) in animal and plant food items. *Food Chemistry*, **56**, 87–91.

48 Jennine, F.D., Peterson, J., Haytowitz, D., and Booth, S.L. (2003) Phylloquinone and dihydrophylloquinone contents of mixed dishes, processed meats, soups and cheeses. *Journal of Food Composition and Analysis*, **16**, 595–603.

49 Jakob, E. and Elmadfa, I. (2002) Rapid and simple HPLC analysis of vitamin K in food, tissues and blood. *Food Chemistry*, **68**, 219–221.

50 Indyk, H.E. and Woollard, D.C. (1997) Vitamin K in milk and infant formulas: determination and distribution of phylloquinone and menaquinone-4. *Analyst*, **122**, 465–469.

51 Piironen, V., Koivu, T., Tammisalo, O., and Mattila, P. (1997) Determination of phylloquinone in oils, margarines and butter by high performance liquid chromatography with electrochemical detection. *Food Chemistry*, **59**, 473–480.

52 Zamarreño, M.M.D., Sanchez Perez, A., Gomez Perez, M.C., Fernandez, M.A., and Mendez, J.H. (1995) Determination of vitamins A, E and K_1 in milk by high-performance liquid chromatography with dual amperometric detection. *Analyst*, **120**, 2489–2492.

53 Careri, M., Mangia, A., Manini, P., and Taboni, N. (1996) Determination of phylloquinone (vitamin K_1) by high performance liquid chromatography with UV detection and with particle beam-mass spectrometry. *Fresenius' Journal of Analytical Chemistry*, **355**, 48–56.

54 Schurgers, L.J., Geleijnse, J.M., Grobbee, D.E., Pols, H.A.P., Hofman, A., Witteman, J.C.M., and Vermeer, C. (1999) Nutritional intake of vitamins K_1 (phylloquinone) and K_2 (menaquinone) in The Netherlands. *Journal of Nutritional and Environmental Medicine*, **9**, 115–122.

55 Kamao, M., Suhara, Y., Tsugawa, N., and Okano, T. (2005) Determination of plasma vitamin K by high-performance liquid chromatography with fluorescence detection using vitamin K analogs as internal standards. *Journal of Chromatography B*, **816**, 41–48.

19

Trace Analysis of Carotenoids and Fat-Soluble Vitamins in Some Food Matrices by LC–APCI-MS/MS

Alessandra Gentili and Fulvia Caretti

19.1
Introduction

Fat-soluble vitamins are classified into four groups of vitamin-active compounds: vitamin A (retinoids and carotenoids as provitamins), vitamin D [ergocalciferol (D_2) and cholecalciferol (D_3)], vitamin E (α-, β-, γ-, and δ-tocopherols and the corresponding tocotrienols), and vitamin K [phylloquinone (K_1) and menaquinones (K_2)]. All these micronutrients are involved in important biological functions, such as vision (vitamin A), calcium adsorption (vitamin D), antioxidative protection of cell membranes (vitamin E and carotenoids), and blood coagulation (vitamin K) [1].

Currently, the conventional methods for the analysis of fat-soluble vitamins in food products require that each vitamin be individually determined by high-performance liquid chromatography (HPLC) with ultraviolet (UV) or fluorescence detection [2–6]. The extraction of fat-soluble vitamins from foods is a very critical step in their analysis [7–9]. It must be quantitative and non-destructive, but the vitamin group heterogeneity and the different stabilities of vitamers are among the main factors making this goal difficult to achieve.

Foods of animal origin contain vitamin A in the form of retinoids [1, 8], whereas in plants it occurs in the form of carotenoids (provitamins A) [10]. These compounds are very sensitive to light, heat, and oxidants. Because of their lability, retinol, retinyl acetate, retinyl palmitate, or β-carotene are frequently added to various processed foods . In order to avoid isomerization and oxidation phenomena, special care has to be taken during sample collection and storage, the extraction procedure, and the final chromatographic separation. The approaches adopted for sample pretreatment consist of (i) saponification followed by solvent extraction, used when the total retinol determination is required, and (ii) direct extraction, applied for the determination of retinyl esters or of specific forms supplemented in foods. Ascorbic acid, hydroquinone, and pyrogallol are some of the antioxidants utilized to prevent analyte degradation.

Only fish liver is an abundant source of vitamin D_3 [1], and most natural foods have only a low content. Vitamin D_2 [1] is mainly found in foods of plant origin.

Fortification of Foods with Vitamins, First Edition. Edited by Michael Rychlik.
© 2011 Wiley-VCH Verlag GmbH & Co. KGaA. Published 2011 by
Wiley-VCH Verlag GmbH & Co. KGaA.

Both forms are employed to fortify milk, yogurt, margarine, cereals, pastries, and bread. Sensitivity to heat, light, and oxygen and the presence of fat and proteins make its analysis complicated [8, 11, 12]. For vitamin A, two procedures can be applied for sample preparation. One includes hot saponification (to release protein-bound forms, to hydrolyze triglycerides, and to dissolve any gelatin) and extraction of the unsaponifiable fraction; the other is based on a direct extraction with organic solvents. Saponification is omitted or carried out at room temperature overnight in order to reduce thermal isomerization, and the addition of an antioxidant is useful to prevent oxidation.

Predominant forms of vitamin E in foods are α- and γ-tocopherols, but for estimation of its total content it is necessary to evaluate all the tocopherols, tocotrienols, and, if supplemented, tocopheryl acetate and tocopheryl acid succinate [8]. During its extraction, vitamin E must be protected from light and oxygen in order to quantify its actual content. It can be analyzed directly in oils after sample dilution with *n*-hexane, but for its determination in other foods it must be concentrated and freed from the matrix. Saponification facilitates its extraction with solvents, hydrolyzing esters linkages of tocopherols and tocotrienols, liberating vitamers E from protein, lipid, and carbohydrate complexes, and destroying the sample matrix. Alkaline conditions must be carefully balanced, as they can cause degradation of vitamin E [13].

Vitamin K is a collective term used when referring to a series of compounds with a common 2-methyl-1,4-naphthoquinone nucleus (menadione moiety) and differing in the isoprenoid side chains at the 3-position. They include natural and synthetic forms [14, 15]. Phylloquinone or vitamin K_1 is the single major form, characterized by the same phytyl side chain as chlorophyll, occurring in green plants, whereas vitamin K_2 is represented by a family of bacterially synthesized menaquinones with multiprenyl side chains. Menadione (vitamin K_3) and menadiol (vitamin K_4) are the synthetic forms. This compound group is relatively stable to heat and oxygen exposure, but is destroyed by light and alkali [8]. Therefore, saponification cannot be used and it has been replaced by enzymatic hydrolysis (lipase) followed by liquid–liquid extraction [16] or, more simply, direct solvent extraction [16, 17].

In this overview, some fundamental difficulties in developing a method to determine simultaneously fat-soluble vitamins in foods are summarized: (i) chemical heterogeneity, which requires specific detection parameters; (ii) occurrence of vitamers and geometric isomers, difficult to separate and identify; (iii) different stabilities, complexity of matrices, and artifact creation, all factors that make extraction a critical step; (iv) high fat content in animal food products, which, if not efficiently removed, could affect the chromatographic efficiency and shorten the column life; and (v) occurrence of bound forms, for example, esters of fatty acids, which are difficult to separate, to identify, and to quantify owing to the unavailability of standards. Saponification is an expedient employed to simplify vitamin analysis since chromatographic separation, identification, and quantification are performed on the free forms. Nevertheless, saponification can cause some problems: first, it is not convenient for K vitamers, which are rapidly decomposed in alkaline media; further, it may cause losses of all those vitamins and carotenoids which, in the free form, are more sensitive to light and oxygen; and finally, it is not always necessary to extract vitamin E.

In the literature, few analytical methods have been proposed for the simultaneous determination of fat-soluble vitamins in foods by chromatographic techniques [18, 19]. So far, only two methods, based on liquid chromatography–tandem mass spectrometry (LC–MS/MS), have been published: one is limited to those analytes that can be extracted by means of the same procedure [20]; the other was addressed to quantify several fat-soluble vitamins in human breast milk [21].

The main purpose of this work was to develop a liquid chromatography–atmospheric pressure chemical ionization tandem mass spectrometry (LC–APCI-MS/MS) method for the simultaneous analysis of 13 compounds, namely carotenoids and fat-soluble vitamins, in some selected foods. Matrix solid-phase dispersion (MSPD) was used as a relatively mild technique for the extraction/cleanup of the target compounds from different foods, with few modifications depending on the matrix. Saponification was omitted in order to preserve retinyl palmitate, vitamin K, vitamin E, and vitamin D and to avoid isomerization of carotenoids. This method was utilized to characterize the vitamin profile of the selected foodstuffs. Moreover, it highlighted some interesting results, such as the occurrence of vitamin K_2 in foods of plant origin and the natural occurrence of geometric isomers of β-carotene in all real samples and of lycopene in tomato pulp.

19.2
Materials and Methods

19.2.1
Chemicals and Materials

The following carotenoids and fat-soluble vitamin standards were purchased from Aldrich-Fluka-Sigma (Milan, Italy): lutein, zeaxanthine, lycopene, β-carotene, retinol, retinyl palmitate, α-tocopherol, δ-tocopherol, γ-tocopherol, ergocalciferol, cholecalciferol, phylloquinone, and menaquinone-4. All chemicals had a purity grade of at least 90% and were used without further purification.

Butylated hydroxytoluene (BHT), provided by Aldrich-Fluka-Sigma, was used as stabilizer both in standard solutions and during the dispersion step of MSPD.

Acetonitrile and methanol were of special grade; 2-propanol, hexane, and chloroform were of RS grade; absolute ethanol was of elevated purity grade or analytical-reagent grade. All these solvents were purchased from Carlo Erba (Milan, Italy).

The following materials were employed for extraction studies: as adsorbents, C_{18}-bonded silica with particles of diameter 35–70 μm and porosity 60 Å, supplied by Alltech Associates (Deerfield, IL, USA), and diatomaceous earth SPE-ED™ MATRIX 38, supplied by Applied Separations (Allentown, PA, USA); glass cartridges (1 cm i.d., capacity 6 ml) with frits made of Teflon, purchased from Supelco

(Bellefonte, PA, USA); and polypropylene tubes (2.5 cm i.d., capacity 20 ml) with polyethylene frits, purchased from Alltech Associates.

19.2.2
Standards

Individual stock standard solutions of each vitamin were prepared monthly at a concentration of 1 μg μl^{-1} using different solvents because of their solubility: chole-calciferol, ergocalciferol, phylloquinone, menaquinones, δ-tocopherol, and γ-tocopherol in acetonitrile containing 0.1% (v/v) BHT; retinol and α-tocopherol in methanol containing 0.1% (v/v) BHT; and retinyl palmitate and β-carotene in chloroform containing 0.1% (v/v) BHT. These solutions were kept in dark bottles at −20 °C to avoid degradation.

Owing to their low stability or solubility, the individual solutions of the other carotenoids were prepared weekly as follows: lycopene at 200 ng μl^{-1} in chloroform containing 0.1% (v/v) BHT with sonication for 6 min.; and lutein and zeaxanthine at 500 ng μl^{-1} in chloroform containing 0.1% (v/v) BHT. Working multistandard solutions were obtained weekly by mixing and diluting individual solutions in methanol containing 0.1% (v/v) BHT at different concentrations, depending on the purpose.

19.2.3
Samples

Tomato pulp and green kiwi were purchased from a supermarket in Rome; maize flour and golden kiwi were obtained from a mill and a producer in central Italy, respectively.

19.2.4
Sample Treatment

The extraction procedures adopted for the different food matrices were rapid, simple, and relatively mild in order to minimize the exposure of vitamins to light, heat, and air and to avoid undesirable losses. All operations were performed in subdued light.

The food matrices chosen for this work were ready to be directly subjected to the simple extraction procedure, with the exception of kiwi, which had to be homogenized. For this purpose, one kiwi fruit was peeled and the edible part was cut into small pieces and homogenized by a rotating blade (~100 rpm for 30 s).

The technique applied to extract analytes from the different foods was MSPD, using a C_{18} sorbent as dispersing tool in the case of maize flour and diatomaceous earth for the other food samples, the water content of which was high.

Before dispersion, the C_{18} sorbent and diatomaceous earth were subjected to a cleaning procedure: 10 g of sorbent were filled into a polypropylene tube (12.5 cm i.d., capacity 20 ml), washed with 30 ml of methanol, and dried with a gentle flow of nitrogen.

19.2.4.1 **Maize Flour**

The sample (2 g) was weighed into a glass mortar, and 15 mg of BHT and 2 g of C_{18} sorbent were added. For recovery studies, the sample was fortified with various volumes of the working standard solution before applying the MSPD technique. At this point, maize flour and the solid support were blended with a glass pestle to obtain complete dispersion, which was verified from the uniform color and consistency of the whole mixture. A 6 ml syringe-like glass tube was filled with the blend, retained by two Teflon frits. The cartridge was connected to a side-arm flask, and the extractant was forced to pass through it by water pump vacuum.

The analytes were extracted from matrix by passing 4 ml of methanol, 4 ml of 2-propanol, and 4 ml of hexane through the cartridge, in sequence. The whole extract was collected into a glass tube with a conical bottom (1.4 cm i.d.), evaporated to 500 μl in a thermostated bath at 30 °C, under a gentle nitrogen flow, and diluted to a final volume of 1 ml with 2-propanol–hexane (50 : 50, v/v) solution. Subsequently, the extract solution was transferred into an Eppendorf tube, centrifuged at 4000 rpm for 5 min, and 5 μl were injected into the LC–MS/MS system.

19.2.4.2 **Kiwi and Tomato Pulp**

The sample (2 g) and 15 mg of BHT, 3 g of diatomaceous earth, and 1 g of Na_2SO_4 were weighed into a glass mortar and dispersed as described for maize flour. The extraction/purification cartridge was prepared by introducing first 0.5 g of the C_{18} sorbent into a 6 ml syringe-like glass tube and then the food sample, treated as described above. Teflon frits were located above and below the sorbent/food matrix bed. The cartridge was connected to a side-arm flask, and the extractant was forced to pass through it by water pump vacuum.

The analytes were extracted from matrix by passing 5 ml of methanol, 10 ml of 2-propanol, and 5 ml of hexane through the cartridge, in sequence. The whole extract was collected into a 50 ml Falcon tube and centrifuged at 6000 rpm for 5 min. The supernatant was transferred into a glass tube with a conical bottom (1.4 cm i.d.), concentrated to 500 μl by evaporation in a thermostated bath at 30 °C, under a mild nitrogen flow, and diluted to a final volume of 1 ml with 2-propanol–hexane (50 : 50, v/v) solution. The extract was then centrifuged at 4000 rpm for 5 min in an Eppendorf tube and 20 μl were injected into the LC–MS/MS system.

19.2.5
Liquid Chromatography and Mass Spectrometry

The analytes were separated on a Prontosil C-30 column (25 cm × 4.6 mm i.d., particle size 3 μm) (Bischoff Chromatography, Leonberg, Germany) equipped with an Alltima guard column (Alltech, Sedriano, Italy), by means of a PE Series 200 micro-HPLC–autosampler–vacuum degasser system (Perkin-Elmer, Norwalk, CT, USA). The mobile phase was also used as washing solution for the autosampler injection system.

The analytes were identified and quantified using a PE Sciex API 3000 triple-quadrupole mass spectrometer (Perkin-Elmer), equipped with an atmospheric pressure chemical ionization (APCI) source operating in the positive ionization mode.

A preliminary mass axis calibration of each quadrupole mass-analyzer, Q_1 and Q_3, was carried out by the infusion of a poly(propylene glycol) solution at $10 \mu l\, min^{-1}$. Unit mass resolution was established and kept in each mass-resolving quadrupole by maintaining a full width at half-maximum (FWHM) of approximately 0.7 ± 0.1 Da.

For separating the analytes, phase A was methanol and phase B was 2-propanol–hexane (50 : 50, v/v) solution. Elution was performed for the first minute in the isocratic mode (100% A) followed by a linear increase of phase B from 0 to 75% in 15 min and then from 75 to 99.5% in 0.1 min; finally, phase B was kept at 99.5% for 15 min (the 0.5% percentage of methanol was maintained in this final part of the chromatographic run to support APCI of analytes). The flow-rate of the mobile phase was $1\, ml\, min^{-1}$ and was introduced entirely into the APCI source, operating with a needle current (NC) of $3 \mu A$ and a probe temperature of $450\,°C$. High-purity nitrogen was used as a curtain gas ($2\, l\, min^{-1}$) and collision gas ($4\, mTorr$), and air was used as nebulizer gas ($1.5\, l\, min^{-1}$) and make-up gas ($2\, l\, min^{-1}$).

For each analyte, two multi-reaction monitoring (MRM) transitions were selected for quantitative analysis, after having observed the collision-induced dissociation (CID) spectra obtained by full-scan product ion experiments; the instrumental parameters of the APCI source and the triple quadrupole were optimized for each vitamin by flow-injection analysis (10 ng injected; flow-rate $1\, ml\, min^{-1}$).

The most important liquid chromatography–electrospray ionization mass spectrometry (LC–ESI-MS) parameters for the MRM acquisition of 13 fat-soluble vitamins in the chosen food matrices are summarized in Table 19.1. The parameters used for confirming each analyte in the real samples were the retention time, the two MRM transitions, and their relative abundance.

19.2.6
Quantitative Analysis

Quantitative analysis was carried out on the samples of maize flour, kiwi, and tomato pulp. Since blank samples were not available, quantitative analysis of fat-soluble vitamins in the different matrices was performed by means of the standard additions method. Further method validation parameters were determined after having quantified the natural levels of each vitamin in the different food matrices.

Regarding the two selected MRM transitions, that most intense one (quantifier transition) was used for quantitative analysis, and the least intense one (qualifier transition) was employed for identification purposes and to define the method limits.

19.2.7
Recovery and Precision

After a preliminary determination of the natural vitamin content in a specific food, analyte recovery was assessed by dividing 4 g of each food sample into two aliquots.

Table 19.1 LC–MS/MS parameters for the fat-soluble vitamins selected in this study.

Analyte	Retention time (min)	Qualifier and quantifier MRM transitions[a]	Declustering potential (V)	Collision potential (V)	Relative abundance (mean ± SD)[b]
Retinol	4.48	269/93	25	33	0.54 ± 0.02
		269/95		15	
δ-Tocopherol	7.52	402/177	40	34	0.89 ± 0.02
		402/137		47	
γ-Tocopherol	8.09	416/151	57	40	0.66 ± 0.01
		417/151	35	35	
Ergocalciferol	8.16	397/159	36	30	0.85 ± 0.05
		397/125		22	
Cholecalciferol	8.37	385/367	35	16	0.68 ± 0.03
		385/259		22	
α-Tocopherol	8.75	431/111	20	20	0.03 ± 0.01
		431/165		25	
Lutein	8.83	551/175	45	25	0.39 ± 0.02
		551/459		17	
Menaquinone	8.95	445/149	45	30	0.45 ± 0.02
		445/187		35	
Zeaxanthine	9.23	569/551	40	20	0.46 ± 0.05
		569/135		30	
Phylloquinone	10.80	451/197	45	35	0.18 ± 0.04
		451/187		38	
Retinyl palmitate	13.81	269/213	25	15	0.51 ± 0.04
		269/95		15	
β-Carotene	15.80	537/137	47	25	0.64 ± 0.02
		537/177		28	
		537/537[c]		9	
Lycopene	23.00	537/137	47	25	0.83 ± 0.03
		537/177		28	
		537/537[c]		9	

a) The first line reports the least intense MRM transition (qualifier) and the second line the most intense transition (quantifier).
b) The relative abundance is calculated as ratio of qualifier intensity to quantifier intensity; the results are reported as arithmetic average of six replicates plus the corresponding standard deviation (SD).
c) This pseudo-/pseudomolecular ion transition was monitored due to its intensity, especially helpful for the detection of lycopene.

One was spiked with known amounts of the analytes and subjected to the extraction process (spiked sample). The second aliquot was extracted directly and fortification was applied on the final extract with the same nominal amounts of analytes (control sample). It should be pointed out that the spiking level was chosen so as to increase the original contents of the vitamins in the specific food sample by a factor of 2–3. The recovery of each analyte was calculated as follows:

$$R\,(\%) = \frac{\text{analyte area}_{\text{spiked sample}}}{\text{analyte area}_{\text{sontrol sample}}} \times 100$$

$$= \frac{\text{total amount of analyte recovered from sample spiked}}{\text{analyte amount extracted from the sample control } + \text{ analyte amount added to the final extract of the sample control}} \times 100$$

The mean recovery was calculated for each vitamin by averaging six replicates at the applied spike level, while the corresponding relative standard deviation was representative for intra-day precision.

The recoveries and precision achieved with the described procedures are reported in Table 19.2.

The inter-day precision was estimated as the relative standard deviation [coefficient of variation (CV)] of six replicates performed within 2 weeks and the respective CV was less than 12% for all analytes.

19.2.8
Memory Effect and Standard Additions Method

A possible memory effect was checked by recording calibration curves in the solvent for each analyte. The first series of points were obtained by injecting

Table 19.2 Recoveries (R) (with relative standard deviations, RSDs) of fat-soluble vitamins from selected food matrices.

Analyte	Maize flour		Green kiwi		Golden kiwi		Tomato pulp	
	Spiked level ($\mu g\,g^{-1}$)	R (%) (RSD)	Spiked level ($\mu g\,g^{-1}$)	R (%) (RSD)	Spiked level ($\mu g\,g^{-1}$)	R (%) (RSD)	Spiked level ($\mu g\,g^{-1}$)	R (%) (RSD)
Retinol	1.0[a]	97 (5)	1.0	65 (7)	1.5	63 (5)	1.0	80 (6)
δ-Tocopherol	0.5	100 (4)	1.0	65 (6)	1.0	65 (5)	1.0	100 (3)
γ-Tocopherol	0.3	100 (5)	0.5	63 (6)	0.3	63 (6)	1.0	94 (5)
Ergocalciferol	0.02	100 (3)	0.2	64 (5)	1.0	65 (4)	0.1	98 (4)
Cholecalciferol	0.06	100 (4)	0.15	64 (5)	0.15	64 (5)	0.1	99 (3)
α-Tocopherol	1.0	100 (3)	30	85 (4)	100	82 (7)	50	100 (3)
Lutein	10	78 (7)	10	73 (6)	5	70 (5)	5	93 (5)
Menaquinone	0.03	100 (5)	0.3	75 (4)	0.3	75 (5)	0.3	99 (3)
Zeaxanthine	10	84 (6)	0.5	65 (5)	1.0	63 (6)	0.3	99 (4)
Phylloquinone	0.01	100 (5)	2.5	62 (7)	1.5	60 (4)	0.3	100 (4)
Retinyl palmitate	0.15	95 (4)	0.3	57 (5)	0.15	57 (6)	0.3	98 (5)
β-Carotene	0.2	82 (6)	1.5	60 (5)	1.5	60 (5)	10	97 (4)
Lycopene	0.5	79 (7)	15	61 (5)	15	58 (7)	100	96 (5)

a) An arbitrary fortification level at 2–3 times the LOQ was chosen for those analytes which were not detected by a preliminary estimation in the analyzed food samples (see Table 19.5).

increasing concentrations, and the second series by injecting decreasing concentrations, for a total of four replicates for each concentration value; by comparing the areas of the four replicates, the occurrence of a memory effect could be excluded.

The linear dynamic range, sensitivity, and goodness of fit of the linear model were evaluated by applying the standard additions method. For this purpose, the extraction/purification cartridge was prepared by using 5 g of food sample and maintaining the same ratios between g sample:g sorbent and ml extractant:g solid blend as described in Section 19.2.4. The eluate from the cartridge was subdivided into five aliquots and four of these were spiked with appropriate volumes of working standard solution; each one was evaporated to 250 μl, the final volume was adjusted to 1 ml with 2-propanol–hexane (50:50, v/v) solution, and 20 μl were injected into the LC–MS/MS system. The linear dynamic range was investigated up to 50 ng injected for all analytes, except for: (i) α-tocopherol in kiwi and tomato pulp (up to 1000 ng injected); (ii) lycopene in tomato pulp (up to 1000 ng injected); (iii) β-carotene in tomato pulp (up to 200 ng injected); and (iv) lutein in kiwi and maize flour and zeaxanthine in maize flour (up to 200 ng injected).

All these regression parameters are reported in detail in Table 19.3. Good linearity was verified over two or three orders of magnitude, with all correlation coefficients exceeding 0.9934. Angular coefficients or slopes, which are representative of the method sensitivity, were very similar in the different matrices; this is an advantage of the standard additions method when used in LC–MS, since it permits the avoidance of different calibrations due to matrix effects.

Table 19.3 Angular coefficients (m) and correlation coefficients (R^2) of fat-soluble vitamins in analyzed food matrices using an APCI source.

Analyte	Quantifier transition	Maize flour		Green kiwi		Golden kiwi		Tomato pulp	
		$m \times 10^3$	R^2	$m \times 10^3$	R^2	$m \times 10^3$	R^2	$m \times 10^3$	R^2
Retinol	269/95	9.50	0.9992	9.50	0.9994	8.90	0.9980	9.60	0.9996
δ-Tocopherol	402/137	1.10	0.9971	1.40	0.9996	1.70	0.9995	1.50	0.9967
γ-Tocopherol	417/151	16.5	1.0000	14.6	0.9991	16.5	0.9997	17.3	0.9996
Ergocalciferol	397/125	23.5	0.9996	21.5	0.9998	21.7	0.9997	20.5	0.9995
Cholecalciferol	385/259	60.1	1.0000	61.1	0.9994	65.7	0.9994	62.1	0.9994
α-Tocopherol	431/165	15.6	1.0000	16.0	0.9995	15.3	0.9964	17.0	0.9963
Lutein	551/459	0.20	0.9983	0.20	0.9978	0.20	0.9993	0.20	0.9979
Menaquinone	445/187	38.6	0.9998	41.0	0.9995	46.3	0.9983	42.4	0.9988
Zeaxanthine	569/135	9.60	0.9996	9.10	0.9996	9.70	0.9988	9.80	0.9999
Phylloquinone	451/187	37.6	0.9995	36.7	0.9992	44.1	0.9971	37.2	0.9966
Retinyl palmitate	269/95	65.6	0.9998	51.0	0.9997	62.1	0.9998	66.2	0.9999
β-Carotene	537/177	10.7	0.9960	11.6	0.9994	13.8	0.9992	13.7	0.9941
Lycopene	537/177	57.2	0.9934	69.0	0.9995	66.0	0.9993	61.4	0.9952

19.2.9
Limits of Detection and Quantitation

The limit of detection (LOD) was calculated as the quantity of analyte able to produce a chromatographic peak three times higher than the noise of the baseline in a chromatogram [signal-to-noise ratio (S/N) = 3] of a non-fortified sample, after having estimated naturally occurring quantities. The limit of quantitation (LOQ) was set at three times the LOD. The noise level depended on the matrix and, therefore, the same analyte was characterized by different LODs in the several foods. LODs and LOQs are reported in Table 19.4 as means of six replicates.

19.3
Results and Discussion

19.3.1
Fragmentation Study and Optimization of MS/MS Conditions

Structural information to confirm the identity of analytes were obtained by means of a preliminary fragmentation study, which was intended to optimize the sensitivity by tuning the voltages of several instrumental components during the tandem mass spectrometry (MS/MS) experiments.

Q_1 scan spectra of almost all of the fat-soluble vitamins and carotenoids showed the protonated molecule $[M + H]^+$ as the base peak, but some exceptions were

Table 19.4 Limits of detection (LOD) and limits of quantitation (LOQ) of fat-soluble vitamins in the analyzed food matrices.

Analyte	Quantifier transition	Maize flour		Green kiwi		Golden kiwi		Tomato pulp	
		LOD (ng g^{-1})	LOQ (ng g^{-1})	LOD (ng g^{-1})	LOQ (ng g^{-1})	LOD (ng g^{-1})	LOQ (ng g^{-1})	LOD (ng g^{-1})	LOQ (ng g^{-1})
Retinol	269/93	121	363	151	453	230	690	105	315
δ-Tocopherol	402/177	58.6	176	151	453	108	324	132	396
γ-Tocopherol	416/151	18.0	54.0	20.1	60.3	20.5	61.5	12.7	38.1
Ergocalciferol	397/159	2.35	7.05	31.8	95.4	163	489	15.6	46.8
Cholecalciferol	385/367	9.40	28.2	17.8	53.4	20.8	62.4	14.1	42.3
α-Tocopherol	431/111	25.5	76.5	27.2	81.6	24.6	73.8	20.1	60.3
Lutein	551/175	75.4	226.2	221	663	162	486	138	414
Menaquinone	445/149	4.70	14.1	1.57	4.71	38	114	5.25	15.7
Zeaxanthine	569/551	49.0	147	83.0	249	76.0	228	15.3	45.9
Phylloquinone	451/197	0.39	1.17	18.0	54.0	13.1	39.3	10.0	30.0
Retinyl palmitate	269/213	18.6	55.8	34.0	102	24.8	74.4	29.0	87.0
β-Carotene	537/137	0.96	2.88	30.1	90.3	22.9	68.7	13.2	39.6
Lycopene	537/137	85.0	255	2507	7521	1988	5964	1132	3396

observed. The base peak of retinol was detected at m/z 269, as a consequence of water loss from the less intense $[M + H]^+$ ion at m/z 287. Also for retinyl palmitate the most abundant ion was observed at m/z 269 due to elimination of palmitic acid from $[M + H]^+$ at m/z 525.5 in the APCI source; furthermore, the Q_1 full-scan spectrum of the ester showed a group of peaks at m/z 523.6, 524.5, and 525.6, attributable to $[M - H]^+$, $[M]^{+\bullet}$, and $[MH]^+$, respectively, since they exhibited analogous fragment ions (for example. at m/z 267, 268, and 269, respectively) when subjected to product ion scan experiments (Figure 19.1b). Vitamins D_2 and D_3 also generated an $[MH - H_2O]^+$ ion, but it was not more intense than $[M + H]^+$. In their Q_1 full-scan spectra, each of the three tocopherols showed three ion species: $[M + H]^+$ (base peak for α- and γ-tocopherol), $[M]^{+\bullet}$ (base peak for δ-tocopherol) and $[M - H]^+$. Differently from what van Breemen *et al.* reported [22], we did not observe the radical ion $[M]^{+\bullet}$ for β-carotene and lycopene, but only $[MH]^+$ at m/z 537 as the base peak and $[M - H]^+$ at m/z 535; the latter very low-intensity ion species was also confirmed by chromatographic analysis. The Q_1 full-scan spectrum of lutein presented two very abundant ions, $[M + H]^+$ at m/z 569 and $[MH - H_2O]^+$ at m/z 551 (base peak), and three other less intense ions, $[M]^{+\bullet}$ at m/z 568, $[MH - 2H_2O]^+$ at m/z 533 and $[MH - H_2O - 92]^+$ at m/z 459 for toluene loss from m/z 551. Zeaxanthine had $[M + H]^+$ at m/z 569 as the base peak and the least intense $[MH - H_2O]^+$ ion at m/z 551; no other ion species were detected as for lutein.

The following MS/MS experiments were performed by selecting as precursor ions the base peak and other especially intense ions that were observed in the Q_1 full-scan spectrum of each vitamin, in order to study the fragmentation of each analyte and to choose its most intense MRM transitions.

Product ion scan mass spectra and the hypothesized fragmentation schemes of the studied vitamins are reported in Figures 19.1–19.3.

Upon CID, retinol and retinyl palmitate showed the same fragment ions in the product ion scan mode when dehydrated retinol (m/z 269) was selected as the precursor ion. Its fragmentation led to numerous product ions; the most intense ions were detected at m/z 213, presumably for butene loss (−56 Da) from $[MH - H_2O]^+$, and at m/z 95 and 93 (Figure 19.1a).

The water loss from $[M + H]^+$ of vitamin D_2 and D_3 gave the fragment ions at m/z 379.6 and 367.0, respectively, and the fragment ions at m/z 271 and 259 were due to the cleavage of their side chain (−126 Da). Two other ions were detected at m/z 159 for D_3 and at m/z 125 for D_2 (Figure 19.1c and d).

Fragmentation of tocopherols caused the cleavage of the chromanol ring, generating abundant product ions at m/z 165 (from $[M + H]^+$, $[M]^{+\bullet}$ and $[M - H]^+$ of α-tocopherol), at m/z 151 and at m/z 137 (from $[M + H]^+$, $[M]^{+\bullet}$ and $[M - H]^+$ of γ- and δ-tocopherol, respectively) (Figure 19.2a–c). The isoprenoid chain rupture generated an intense ion at m/z 177 for δ-tocopherol only.

Protonated molecular ions of both K vitamers lost their phytyl chain, giving an intense product ion at m/z 187 (Figure 19.2d and e). Fragmentation of vitamin K_2 was more pronounced and resulted in fragments of appreciable intensity at m/z 427 ($[MH - H_2O]^+$), 363, 341, and 149.

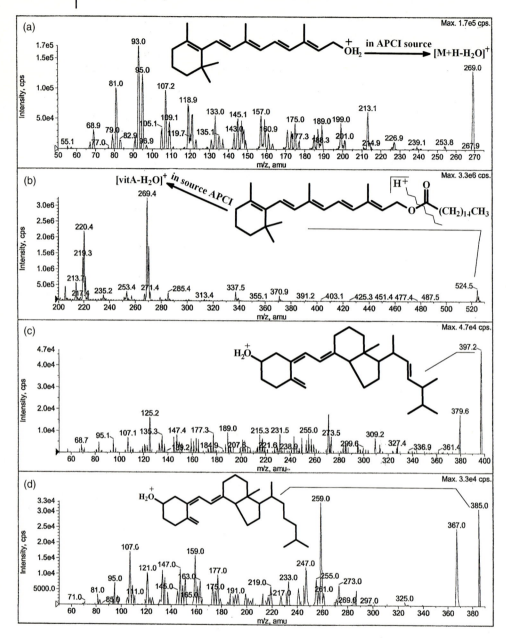

Figure 19.1 Product ion scan mass spectra and corresponding fragmentation schemes for retinol (a), ergocalciferol (c), cholecalciferol (d), and ergocalciferol (e). In (b) is depicted the Q_1 full-scan spectrum of retinyl palmitate and its scission in dehydrated retinol.

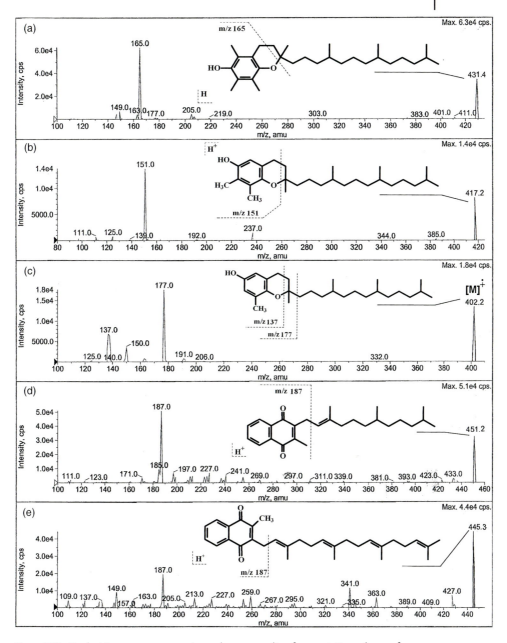

Figure 19.2 Product ion scan mass spectra and corresponding fragmentation schemes for α-tocopherol (a), γ-tocopherol (b), δ-tocopherol (c), phylloquinone (d), and menaquinone (e).

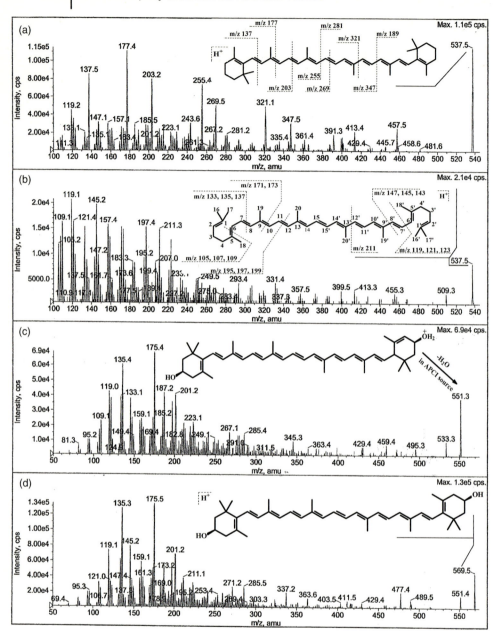

Figure 19.3 Product ion scan mass spectra and corresponding fragmentation schemes for β-carotene (a), lycopene (b), lutein (c), and zeaxanthine (d).

The product ion scan spectrum of β-carotene highlighted two series of fragment ions (Figure 19.3a). The first consisted of those at m/z 137, 177, 203, 243, and 269, which probably originated from cleavage of the C7–C8, C9–C10, C11–C12, C13–C14, and C15–C15′ double bonds, respectively; the most intense ion at m/z 177 might be stabilized by the formation of a ring structure. The second series was due to the cleavage of C6–C7 (fragments at m/z 123 or 413), C10–C11 (m/z 189 or 347), C12–C13 (m/z 321), and C14–C15 (m/z 255 or 281) single bonds.

Lycopene fragmentation generated groups of product ions (related to sequential losses of H_2) less intense than those observed for β-carotene (Figure 19.3b).

At low collision energy, $[MH - H_2O]^+$ of lutein lost a second water molecule (fragment at m/z 533), toluene (fragment at m/z 459), and the terminal ring (fragment at m/z 429). At higher collision energies, a pattern of fragmentation analogous to that described for β-carotene and lycopene was observed; abundant ions were detected at m/z 109, 119, 135, 145, 175, 187, and 201 (Figure 19.3c).

Product ions of zeaxanthine ($[M + H]^+$ at m/z 569 as precursor ion) at m/z 551, 477, and 429 resulted from loss of water, of toluene, and of water plus the terminal ring, respectively. In contrast to from lutein, the second molecule of water produced a peak of poor intensity. As for the other carotenoids, chain fragmentation generated intense product ions at m/z 119, 135, 145, 175, and 201 (Figure 19.3d).

On the basis of this fragmentation study, the two most intense MRM transitions of each analyte (listed in Table 19.1) were taken into consideration for performing quantitative and confirmative analysis on the selected food matrices.

In summary, the precursor ions were selected as follows:

- $[M + H]^+$ for α-tocopherol, γ-tocopherol (for the most intense MRM transition), vitamin D_2, vitamin D_3, vitamin K_1, vitamin K_2, β-carotene, lycopene, and zeaxanthine

- $[MH - H_2O]^+$ for retinol, and lutein

- $[MH - palmitic acid - H_2O]^+$ for retinyl palmitate

- $[M]^{+\bullet}$ for δ-tocopherol and γ-tocopherol (for the less intense MRM transition).

19.3.2
Optimization of the LC–MS Conditions

Chromatographic conditions used for separating nine fat-soluble vitamins and four carotenoids were defined after a series of preliminary trials addressed at the choice of the chromatographic column, mobile phase, and gradient elution.

Several normal-bore reversed-phase columns were tested: Alltima C$_{18}$ (250 × 4.6 mm i.d.; 5 μm) (Alltech, Sedriano, Italy); TSK Gel Super-ODS (100 × 4.6 mm i.d.; 2 μm) (Tosoh Bioscience, Stuttgart, Germany); Ultracarb ODS(30) (150 × 4.6 mm i.d.; 5 μm) (Phenomenex, Torrence, CA, USA); and Prontosil 120-3-C$_{30}$ (250 × 4.6 mm i.d.; 3 μm) (Bischoff Chromatography, Leonberg, Germany).

Owing to the hydrophobicity of the analytes, methanol was chosen as the polar component of the mobile phase (phase A), and 2-propanol was selected as modifier (phase B). Although different gradients were tested, we were not able to avoid two coelutions of isomeric compounds (zeaxanthine plus lutein and β-carotene plus lycopene) on the conventional Alltima and Tosoh C_{18} columns. Owing to its high carbon load (31%), Phenomenex Ultracarb ODS(30) succeeded in separating lyco- pene and β-carotene, but their peaks were broad and showed long retention times; moreover, zeaxanthine and lutein were not separated. Replacing 2-propanol with 2-propanol–hexane (50:50, v/v) solution, the eluotropic power of phase B was increased; furthermore, on setting up a combined elution mode (phase A was kept at 100% for the first 4 min, then phase B was increased linearly from 0 to 100% in 16 min), the retention times of lycopene (t_r = 14.4 min) and β-carotene (t_r = 14.81 min) were shorter as anticipated, but lutein (t_r = 5.6 min) and zeaxan- thine (t_r = 5.8 min) were only partially resolved.

In order to separate the two xanthophylls completely, we tested the Prontosil C_{30} column, which is characterized by superior apolarity and had already been suc- cessfully applied by other workers to the analysis of carotenoids [23] and vitamin K_1 isomers [16, 24]. After having optimized the elution conditions (see Section 19.2.5), all analytes were separated, with the exception of vitamin K_2 and α- tocopherol, and the elution order of lycopene and β-carotene were inverted (see Figure 19.4). At the end of the chromatographic run, phase B was maintained at 99.5% and not at 100%, since we verified that 0.5% of methanol was able to support APCI of analytes that elute at the end, thus increasing their S/N.

19.3.3
Recovery Studies

MSPD is a relatively new extraction technique [25, 26], mainly applied for extract- ing target analytes from solid or semisolid samples. It involves the use of an abrasive dispersing agent, blended with the sample by mortar and pestle. By adopt- ing an adequate dispersing agent, this technique was also applied to liquid food- stuffs such as milk and fruit juice. The shearing forces developed during the dispersion process disrupt the sample matrix and disperse its components on the solid support, thus generating finely divided material suitable for a chromatographic- like extraction. In fact, the solid blend achieved at the end of the MSPD process is transferred into a column and retained between two frits; thereafter, a suitable solvent has to be chosen to extract the analytes from the dispersed sample. Due to its mildness and rapidity, MSPD was used for extracting carotenoid stereoi- somers from spinach samples without inducing analyte isomerization and/or oxidation [27].

For maize flour, a C_{18} sorbent proved to be the best dispersing agent. After having packed the cartridge as described in the Section 19.2, different eluents were tested. Using methanol (12 ml) as extractant, recoveries ranged between 70 and 90% for all analytes, with the exception of lycopene (65%). A 12 ml volume of methanol–2-propanol (50:50, v/v) solution allowed recoveries exceeding 74% to be obtained, with a slight improvement for lycopene (68%). The best results were achieved by eluting sequentially with methanol (4 ml), 2-propanol (4 ml) and hexane (4 ml). In this way, the lycopene recovery was close to 80%.

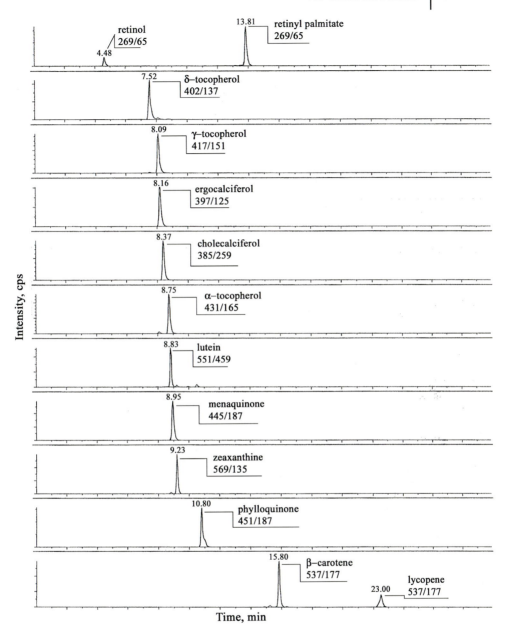

Figure 19.4 LC–MRM chromatogram, acquired by APCI in the positive ionization mode, of the 14 fat-soluble vitamins (100 ng injected for all analytes; 250 ng were injected for lycopene only); for each analyte is shown the most intense MRM transition.

A C_{18} sorbent could not be utilized for the dispersion of kiwi and tomato pulp owing to their high content of water and to the formation of breaks and preferential flow paths of the solid blend during the elution step with the same solvents as employed for maize flour. For these reasons, we replaced C_{18} by diatomaceous earth, also assisted by a drying agent such as anhydrous Na_2SO_4, in order to obtain a fairly dry solid blend suitable for the subsequent extraction. This solution led to recoveries of more than 80% from tomato pulp and 60–85% from the two kinds of kiwis.

19.3.4
Application of the Method to the Analysis of Food Products

The main goal of this work was to demonstrate the practicality of the developed analytical method and its use for the direct determination of the fat-soluble vitamin profile of selected foods. Its application to real samples allowed the detection of some vitamins whose occurrence in foods, on which this study was focused, was not expected. The results of this small monitoring exercise are summarized in Table 19.5.

The analysis of foodstuffs of plant origin such as maize flour, tomato pulp, and kiwi confirmed β-carotene (with the highest concentrations in tomato pulp and golden kiwi) as a provitamin A source and the absence of retinol and/or retinyl palmitate. In these foods, we always found the two xanthophylls zeaxanthine and lutein, of which maize was especially rich, whereas lycopene was determined in tomato pulp at a level of ~43 µg g^{-1}.

Table 19.5 Estimated quantities of fat-soluble vitamins in the different foods analyzed.

Analyte	Maize flour (µg g^{-1})	Green kiwi (µg g^{-1})	Golden kiwi (µg g^{-1})	Tomato pulp (µg g^{-1})
Retinol	n.d.[a]	n.d.	n.d.	n.d.
Retinyl palmitate	n.d.	n.d.	n.d.	n.d.
β-Carotene	0.076	0.51	0.75	4.54
Lycopene	n.d.	n.d.	n.d.	42.6
Lutein	3.16	3.40	2.11	1.55
Zeaxanthine	2.88	<LOQ	0.56	0.10
α-Tocopherol	0.32	15.9	48.4	24.5
γ-Tocopherol	0.12	0.23	0.12	0.31
δ-Tocopherol	n.d.	n.d.	n.d.	<LOQ
Ergocalciferol	<LOQ	n.d.	n.d.	n.d.
Cholecalciferol	n.d	n.d.	n.d.	n.d.
Phylloquinone	0.005	0.86	0.53	0.14
Menaquinone	<LOQ	0.10	0.16	0.097

a) Not detected.

D vitamers were not found in any sample, except ergocalciferol, the presence of which was confirmed at a trace level in maize but not in tomato pulp, for which only the most intense MRM transition of vitamin D_2 was observed in the MRM chromatogram. Both extracted ion currents of ergocalciferol showed a series of interfering peaks at characteristic retention times (7.5, 11.7, 13.2, and 17.7 min) in all matrices, but they had a different relative abundance to that of vitamin D_2.

Additional peaks were also observed in the MRM traces for other vitamins such as K_1 (8.2 and 9.2 min) and tocopherols (e.g., at 8.8 min for δ-tocopherol and 8.7 min for γ-tocopherol, except for maize flour). In these cases, since the relative abundance of the two ion currents coincided with that of the free vitamin, we supposed that these peaks could relate to bound forms broken in the APCI source giving the same MRM transitions as the free form, such as retinyl palmitate and retinol, and/or isomeric forms. When tomato pulp was analyzed, the MRM transitions of β-carotene and lycopene showed several additional peaks ascribable to their geometric isomers, the identification of which could not be achieved by the use of only LC–MS (Figure 19.5).

The major homologs of vitamin E in Nature are α- and γ-tocopherol, and in fact they were found in all of the matrices analyzed. The α form was the dominant species and was determined in golden kiwi at a level three times higher than that in green kiwi.

K vitamers occurred in all samples. According to the literature, vitamin K_2 is produced by bacteria and it has been found in certain animal food products such as eggs and meat. The high sensitivity of the developed method allowed the detection and confirmation of menaquinone-4 in the analyzed samples of maize flour, kiwi, and tomato pulp (Figure 19.6). Its occurrence in foods of plant origin could be explained by the microbial origin of vitamin K_2, that is, as a consequence of food contamination with bacteria.

19.4
Conclusions

An MSPD–LC–APCI-MS/MS method for the simultaneous determination of 13 fat-soluble vitamins and carotenoids was developed and applied to the analysis of some foods. Reversed-phase chromatography performed on a C_{18} column with a high carbon load (31%) was able to separate lycopene and β-carotene but failed in the case of the xanthophylls lutein and zeaxanthine. A C_{30} column proved to be indispensable for achieving this goal, but its apolar nature required the use of a mobile phase with an high eluotropic power, which is unusual when working in the reversed-phase mode. However, this was indispensable for eluting β-carotene and lycopene and for restricting the broadening of their peaks. A small percentage of methanol was also maintained at the end of the chromatographic run to support especially the APCI of β-carotene and lycopene.

The sensitivity and selectivity of the proposed method allow a simple and mild extraction procedure to be applied, thus reducing the time necessary for sample

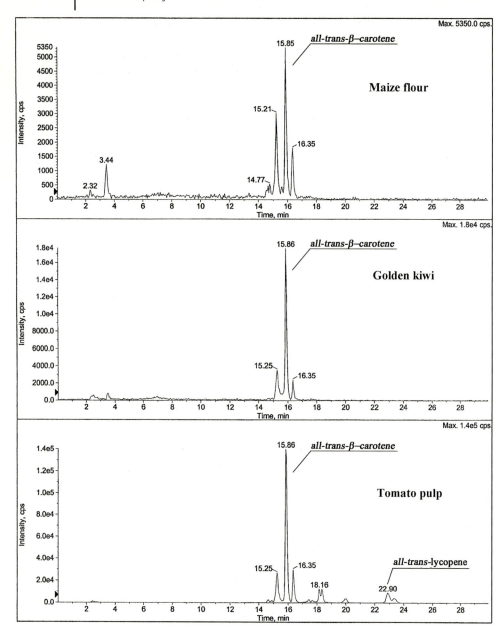

Figure 19.5 Extracted ion chromatograms for the most intense transition of β-carotene and lycopene. In addition to all-*trans* isomers, a series of peaks attributable to other geometric isomers are visible.

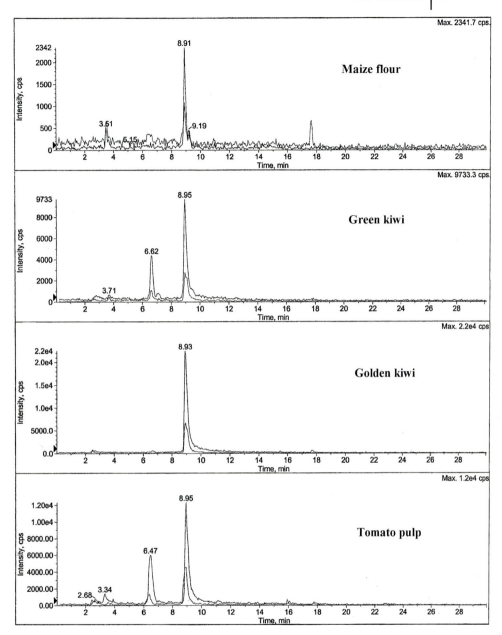

Figure 19.6 Extracted ion chromatograms of menaquinone-4 in the selected food matrices (unspiked extract). It was identified by its characteristic retention time, the two MRM transitions, and their relative abundance.

preparation, preserving all vitamers, and precluding the creation of artifacts. The direct analysis of all forms of a specific vitamin, naturally occurring in a food, can also be performed thanks to the ability of LC–MS to carry out multi-analyte determinations.

This method could be improved by increasing the number of vitamers to be analyzed (e.g., other vitamin esters) in order to justify the omission of the saponification step. After suitable modification, it could also be applied to the direct determination of those vitamin forms utilized for food fortification.

Acknowledgment

This work was supported by the Ministero Italiano delle Politiche Agricole e Forestali (MIPAF), "Qualità Alimentare" project.

References

1 Belitz, H.-D., Grosch, W., and Schieberle, P. (2004) Vitamins, in *Food Chemistry*, 3rd edn. Springer, Berlin, pp. 409–426.

2 Comité Européen de Normalisation (2000) BS EN 12823-1. *Foodstuffs. Determination of Vitamin A by High Performance Liquid Chromatography. Part 1: Measurements of All-trans-Retinol and 13-cis-Retinol*. Comité Européen de Normalisation (CEN), Brussels.

3 Comité Européen de Normalisation (2000) BS EN 12823-2. *Foodstuffs. Determination of Vitamin A by High Performance Liquid Chromatography. Part 2: Measurements of Beta-Carotene*. Comité Européen de Normalisation (CEN), Brussels.

4 Comité Européen de Normalisation (2000) BS EN 12821. *Foodstuffs. Determination of Vitamin D by High Performance Liquid Chromatography. Measurement of Cholecalciferol (D3) and Ergocalciferol (D2)*. Comité Européen de Normalisation (CEN), Brussels.

5 Comité Européen de Normalisation (2000) BS EN 12822. *Foodstuffs. Determination of Vitamin E by High Performance Liquid Chromatography. Measurement of Alpha-, Beta-, Gamma-, and Delta-Tocopherols*. Comité Européen de Normalisation (CEN), Brussels.

6 Comité Européen de Normalisation (2003) BS EN 14148. *Foodstuffs. Determination of Vitamin K1 by High Performance Liquid Chromatography*. Comité Européen de Normalisation (CEN), Brussels.

7 Luque-Garcia, J.L. and Luque de Castro, M.D. (2001) Extraction of fat-soluble vitamins. *Journal of Chromatography A*, **935**, 3–11.

8 Rizzolo, A. and Polesello, S. (1998) Chromatographic determination of vitamins in foods. *Journal of Chromatography A*, **624**, 103–152.

9 Blake, C.J. (2007) Status of methodology for the determination of fat-soluble vitamins in foods, dietary supplements and vitamin premixes. *Journal of AOAC International*, **90**, 897–910.

10 Rodriguez-Amaya, D.B. and Kimura, M. (2004) HarvestPlus Handbook for Carotenoid Analysis. HarvestPlus Technical Monograph 2. HarvestPlus, Washington, DC

11 Perales, S., Alegría, A., Barberá, R., and Farré, R. (2005) Review: determination of vitamin D in dairy products by high performance liquid chromatography. *Food Science and Technology International*, **11**, 451–462.

12 Byrdwell, W.C., DeVries, J., Exler, J., Harnly, J.M., Holden, J.M., Holick, M.F., Hollis, B.W., Horst, R.L., Lada, M., Lemar, L.E., Patterson, K.Y., Philips, K.M., Tarrago-Trani, M.T., and Wolf, W.R. (2008) Analyzing vitamin D in foods and supplements: methodologic challenges. *American Journal of Clinical Nutrition*, **88**, 554S–557S.

13 Xu, Z. (2008) Comparison of extraction methods for quantifying vitamin E from animal tissues. *Biosource Technology*, **99**, 8705–8709.

14 Shearer, M.J., Bach, A., and Kohlmeier, M. (1996) Chemistry, nutritional sources, tissue distribution and metabolism of vitamin K with special reference to bone health. *Journal of Nutrition*, **126** (Suppl.), 1181S–1186S.

15 Panel on Dietetic Products, Nutrition and Allergies (2008) Scientific Opinion of the Panel on Dietetic Products Nutrition and Allergies on vitamin K_2 added for nutritional purposes in foods for particular nutritional uses, food supplements and foods intended for the general population and vitamin K_2 as a source of vitamin K added for nutritional purposes to foodstuffs, in the context of Regulation (EC) No. 258/971. *The EFSA Journal*, **822**, 1–31.

16 Cook, K.K., Mitchell, G.V., Grundell, E., and Rader, J.I. (1999) HPLC analysis for *trans*-vitamin K_1 and dihydro-vitamin K_1 in margarines and margarine-like products using the C_{30} stationary phase. *Food Chemistry*, **67**, 79–88.

17 Jakob, E. and Elmadfa, I. (1996) Application of a simplified HPLC assay for the determination of phylloquinone (vitamin K_1) in animal and plant food items. *Food Chemistry*, **56**, 87–91.

18 Salo-Väänänen, P., Ollilainen, V., Mattila, P., Lehikoinen, K., Salmela-Mölsä, E., and Piironen, V. (2000) Simultaneous HPLC analysis of fat-soluble vitamins in selected animal products after small-scale extraction. *Food Chemistry*, **71**, 535–543.

19 Eitenmiller, R.R. and Landen, W.O. (1999) Multi-analyte methods for analysis of fat-soluble vitamins, in *Vitamin Analysis for the Health and Food Sciences*

(eds. R.R. Eitenmiller and W.O. Landen), CRC Press, Boca Raton, FL, pp. 185–220.

20 Heudi, O., Trisconi, M.-J., and Blake, C.-J. (2004) Simultaneous quantification of vitamins A, D_3 and E in fortified infant formulae by liquid-chromatography–mass spectrometry. *Journal of Chromatography A*, **1022** (1–2), 115–123.

21 Kamao, M., Tsugawa, N., Suhara, Y., Wada, A., Mori, T., Murata, K., Nishino, R., Ukita, T., Uenishi, K., Tanaka, K., and Okano, T. (2007) Quantification of fat-soluble vitamins in human breast by liquid-chromatography–tandem mass spectrometry. *Journal of Chromatography B*, **859**, 192–200.

22 van Breemen, R., Huang, C.-R., Yecheng, T., Sander, L.C., and Schilling, A.B. (1996) Liquid chromatography/mass spectrometry of carotenoids using atmospheric pressure chemical ionization. *Journal of Mass Spectrometry*, **31**, 975–981.

23 Sander, L.C., Sharpless, K.E., and Pursch, M. (2000) C_{30} stationary phases for the analysis of food by liquid chromatography. *Journal of Chromatography A*, **880**, 189–202.

24 Woollard, D.C., Indyk, H.E., Fong, B.Y., and Cook, K.K. (2002) Determination of vitamin K_1 isomers in foods by liquid chromatography with C_{30} bonded-phase column. *Journal of AOAC International*, **85**, 682–691.

25 Barker, S.A. (2007) Matrix solid phase dispersion (MSPD). *Journal of Biochemical and Biophysical Methods*, **70**, 151–162.

26 Bogialli, S. and Di Corcia, A. (2007) Matrix solid-phase dispersion as a valuable tool for extracting contaminants from foodstuffs. *Journal of Biochemical and Biophysical Methods*, **70**, 163–179.

27 Glaser, T., Lienau, A., Zeeb, D., Krucker, M., Dachtler, M., and Albert, K. (2003) Qualitative and quantitative determination of carotenoid stereoisomers in a variety of spinach samples by use of MSPD before HPLC–UV, HPLC–APCI-MS, and HPLC–NMR on-line coupling. *Chromatographia*, **57**, S19–S25.

Index

Page numbers in *italics* refer to figures or tables.